Hartmut Zohm

Magnetohydrodynamic Stability of Tokamaks

Related Titles

Wolf, R.

Atmospheric Pressure Plasma for Surface Modification

2013
ISBN: 978-1-118-01623-7
Also available in digital formats.

Harry, J.E.

Introduction to Plasma Technology
Science, Engineering and Applications

2011
ISBN: 978-3-527-32763-8
Also available in digital formats.

Parvulescu, V.I., Magureanu, M., Lukes, P. (eds.)

Plasma Chemistry and Catalysis in Gases and Liquids

2012
ISBN: 978-3-527-33006-5
Also available in digital formats.

Woods, L.C.

Physics of Plasmas

2004
ISBN: 978-3-527-40461-2
Also available in digital formats.

Woods, L.C.

Theory of Tokamak Transport
New Aspects for Nuclear Fusion Reactor Design

2006
ISBN: 978-3-527-40625-8
Also available in digital formats.

Stacey, W.M.

Nuclear Reactor Physics

2 edn.
2007
ISBN: 978-3-527-40679-1
Also available in digital formats.

Stacey, W.M.

Fusion
An Introduction to the Physics and Technology of Magnetic Confinement Fusion

2 edn.
2010
ISBN: 978-3-527-40967-9
Also available in digital formats.

Stacey, W.M.

Fusion Plasma Physics

2 edn.
2012
ISBN: 978-3-527-41134-4
Also available in digital formats.

Hartmut Zohm

Magnetohydrodynamic Stability of Tokamaks

WILEY-VCH
Verlag GmbH & Co. KGaA

Author

Hartmut Zohm
MPI für Plasmaphysik
Garching
Germany
zohm@ipp.mpg.de

Library of Congress Card No.: applied for

British Library Cataloguing-in-Publication Data
A catalogue record for this book is available from the British Library.

Bibliographic information published by the Deutsche Nationalbibliothek
The Deutsche Nationalbibliothek lists this publication in the Deutsche Nationalbibliografie; detailed bibliographic data are available on the Internet at <http://dnb.d-nb.de>.

Print ISBN: 978-3-527-41232-7
ePDF ISBN: 978-3-527-67736-8
ePub ISBN: 978-3-527-67734-4
Mobi ISBN: 978-3-527-67735-1
oBook ISBN: 978-3-527-67737-5

Cover Design Adam-Design, Weinheim, Germany
Typesetting Laserwords Private Limited, Chennai, India
Printing and Binding Markono Print Media Pte Ltd, Singapore

Printed on acid-free paper

Contents

Preface

This textbook is based on a University course on MHD stability of tokamak plasmas that I have been developing since 1996. MHD stability of tokamaks is an evolving field, and while a lot of specialist's knowledge exists, I found it very important for students to see that there is a solid theoretical foundation in common with other areas of plasma physics from which the understanding of tokamak MHD stability is developed. The book should hence serve to bridge the gap between a basic plasma physics course and forefront research in MHD stability of tokamaks. This means that there are some elements of review of the field's present status in it that will certainly develop over the coming years, but I found it important to point out where we have reached basic understanding and where there are still open ends at the time of writing the book.

At this point, I want to thank all the individuals who have contributed directly or indirectly to this book. First I would like to thank Karl Lackner for all the enlightening physics discussions and especially for always having time for me whenever I entered his office. Some of the most insightful physics arguments presented in this book actually originate from these discussions. Then, I want to thank the colleagues who have worked with me on MHD stability on ASDEX Upgrade for the last 20 years or so, namely Marc Maraschek, Anja Gude, Sibylle Günter and Valentin Igochine, trying to figure out where the experiment knows about the theory. It is also a pleasure to acknowledge the very helpful discussions with Per Helander on stellarator physics and Steve Sabbagh on RWM stability. In preparing the book, Hans-Peter Zehrfeld helped me through all the troubles of editing, formatting and organizing the text. He also contributed greatly to Chapter 2. A special thanks goes to Sina Fietz for her help with the figures and to Emanuele Poli for the thorough proofreading. Last but not least, a major part of the text was written during a stay at University of Wisconsin, Madison, and I want to thank Cary Forest and Chris Hegna for their hospitality and the very useful discussions about the topics of this book.

Hartmut Zohm
Garching, August 2014

1 The MHD Equations

1.1 Derivation of the MHD Equations

In this book, we will treat the description of equilibrium and stability properties of magnetically confined fusion plasmas in the framework of a fluid theory, the so-called Magnetohydrodynamic (MHD) theory. In this chapter, we are going to derive the MHD equations and discuss some of their basic properties and the limitations for application of MHD to the description of fusion plasmas. The derivation follows the treatment given in [1]. For a more in-depth discussion of the MHD equations, the reader is referred to [2]. Non-linear aspects of MHD are treated in [3]. A good overview of general tokamak physics can be found in [4].

1.1.1 Multispecies MHD Equations

As a magnetized plasma is a many-body system, its description cannot be done by solving individual equations of motion that would typically be a set of, say, 10^{20} equations[1] that are all coupled through the electromagnetic interaction. Hence, some kind of mean field theory is needed.

Starting point of our derivation is the kinetic equation known from statistical physics. It describes the many-body system in terms of a distribution function f_α in six-dimensional space d^3xd^3v, where

$$f_\alpha(\mathbf{x}, \mathbf{v}, t)\; d^3x\; d^3v \tag{1.1}$$

is the probability to find a particle of species α at $\mathbf{x}$ with velocity $\mathbf{v}$ at time t. Here, $\mathbf{x}$ and $\mathbf{v}$ are independent variables that, in the sense of classical mechanics, fully describe the system.

The basic assumption of kinetic theory is that fields and forces are macroscopic in the sense that they have already been averaged over a volume containing many particles (say, a Debye-sphere[2]) and the microscopic fields and forces at the exact

1) Here, we think of a typical fusion plasma of density 10^{20} particles per cubic metre.
2) The Debye length λ_D is the typical distance on which the electric field in a plasma is shielded so that its action is limited to a sphere of radius λ_D.

Magnetohydrodynamic Stability of Tokamaks, First Edition. Hartmut Zohm.

particle location can be expressed through a collision term giving rise to a change of f_α along the particle trajectories in six-dimensional space. We note that this has reduced the microscopic problem of the 10^{20} interactions to the proper choice of the collision term.

Evaluating the total change of f_α along the trajectories and keeping in mind that along these, $d\mathbf{x}/dt = \mathbf{v}$ and $d\mathbf{v}/dt = \mathbf{F}_\alpha/m_\alpha$, where $\mathbf{F}_\alpha$ is the force acting on the particle and m_α its mass, the kinetic equation can be expressed as

$$\frac{\partial f_\alpha}{\partial t} + \mathbf{v} \cdot \mathbf{\nabla} f_\alpha + \frac{q_\alpha}{m_\alpha}(\mathbf{E} + \mathbf{v} \times \mathbf{B}) \cdot \nabla_v f_\alpha = \left(\frac{\partial f_\alpha}{\partial t}\right)_{\text{coll}} \tag{1.2}$$

where we have assumed that the only relevant force is the Lorentz force and hence explicitly neglected gravity (which is a good approximation for magnetically confined fusion plasmas, but generally not true in Astrophysical applications). According to the above-mentioned description of mean field theory, the fields $\mathbf{E}$ and $\mathbf{B}$ will have to be calculated from Maxwells equations using the charge density and current resulting from appropriate averaging over the distribution function in velocity space as will be described in the following.

The kinetic equation is used to describe phenomena that arise from f_α not being a Maxwellian, which is the particle distribution in thermodynamic equilibrium to which the system will relax through the action of collisions. In fusion plasmas, this frequently occurs as the mean free path is often large compared to the system length as is for example the case for turbulence dynamics in a tokamak along field lines. Another important example is when the relevant timescales are short compared to the collision time, such as in RF (radio frequency) wave heating and current drive that can occur by Landau damping rather than collisional dissipation. Here, a description using the Vlasov or Fokker–Planck equation is needed.

However, in situations where f_α is close to Maxwellian, one can average the kinetic equation over velocity space to obtain hydrodynamic equations in configuration space. When doing so, one encounters so-called moments of f_α. The kth moment, which is related to the velocity average of $\mathbf{v}^k$, is given by

$$\int \mathbf{v}^k f_\alpha d^3v = n_\alpha \langle \mathbf{v}^k \rangle \tag{1.3}$$

These moments are related to the hydrodynamic quantities used to describe the plasma in configuration space. For the zeroth moment, we obtain

$$n_\alpha(\mathbf{x}, t) = \int f_\alpha(\mathbf{x}, \mathbf{v}, t) d^3v \tag{1.4}$$

which is the number density in real space. The first moment of f_α is related to the fluid velocity in the centre of mass frame by

$$\mathbf{u}_\alpha(\mathbf{x}, t) = \frac{1}{n_\alpha} \int \mathbf{v} f_\alpha(\mathbf{x}, \mathbf{v}, t) d^3v \tag{1.5}$$

For the second moment, it is of advantage to separate the particle velocity into the fluid velocity and the random thermal motion $\mathbf{w}$ according to

$$\mathbf{v} = \mathbf{u}_\alpha + \mathbf{w} \tag{1.6}$$

It is easy to show that $\langle \mathbf{w} \rangle = 0$, as expected for thermal motion, as

$$\langle \mathbf{w} \rangle = \langle \mathbf{v} \rangle - \langle \mathbf{u}_\alpha \rangle = \mathbf{u}_\alpha - \mathbf{u}_\alpha = 0 \tag{1.7}$$

However, the quadratic average is non-zero, representing the thermal energy via

$$\frac{1}{2} m_\alpha \int \mathbf{w}^2 f_\alpha d^3 v = \frac{3}{2} n_\alpha k_B T_\alpha = \frac{3}{2} p_\alpha \tag{1.8}$$

where k_B is the Boltzmann constant and we have used the definition of the thermal energy density and its relation to the pressure p_α for an ideal plasma. We note that this definition relies on the previous assumption that f_α is close to Maxwellian. More generally, the second moment is defined as a tensor of rank 2, the pressure tensor

$$\mathbf{P}_\alpha = m_\alpha \int \mathbf{w} \otimes \mathbf{w} f_\alpha d^3 v, \tag{1.9}$$

where $\otimes$ denotes the dyadic product. The non-diagonal terms of this tensor are related to viscosity, whereas from Eq. (1.8), it is clear that the trace of $\mathbf{P}_\alpha$ is equal to $3p_\alpha$, that is for an isotropic system, the diagonal elements of $\mathbf{P}_\alpha$ are just equal to the scalar pressure. Therefore, the pressure tensor is also often written as

$$\mathbf{P}_\alpha = p_\alpha 1 + \Pi_\alpha, \tag{1.10}$$

where 1 is the unit tensor and Π_α the anisotropic part of $\mathbf{P}_\alpha$.

We now integrate the kinetic equation (Eq. 1.2) over velocity space[3)] to obtain

$$\frac{\partial n_\alpha}{\partial t} + \nabla \cdot (n_\alpha \mathbf{u}_\alpha) = 0 \tag{1.11}$$

which is the equation of continuity for species α. Here, we have assumed that the velocity space average of the collision term is zero, meaning that the total number of particles is conserved for each species. Should this not be the case (e.g. by ionization or fusion), the right-hand side would consist of a source term $S(\mathbf{x}, t)$.

The next moment is obtained by multiplying the kinetic equation by $\mathbf{v}$ and integrating over velocity space. This yields the momentum balance

$$\begin{aligned} m_\alpha \frac{\partial (n_\alpha \mathbf{u}_\alpha)}{\partial t} &= -\nabla \cdot (m_\alpha n_\alpha \mathbf{u}_\alpha \otimes \mathbf{u}_\alpha + \mathbf{P}_\alpha) \\ &\quad + n_\alpha q_\alpha (\mathbf{E} + \mathbf{u}_\alpha \times \mathbf{B}) + \mathbf{R}_{\alpha\beta}, \end{aligned} \tag{1.12}$$

where the friction force $\mathbf{R}_{\alpha\beta}$ is the first moment of the collision term for collisions with species β. We note that only collision with unlike particles lead to a net friction force while collisions within one species, which are important for thermalization, do not transfer net momentum to that species. This form is also called the *conservative* form as, like the equation of continuity, it relates the temporal derivative of a quantity (in this case, the momentum) to the divergence of a flux.

3) When integrating over velocity space, it is useful to remember that t, $\mathbf{x}$ and $\mathbf{v}$ are independent so that the derivative with respect to t and x can be taken out of the integral. In addition, terms containing a v derivative are integrated partially and the surface term vanishes as $f_\alpha \to 0$ faster than any power of v for $v \to \infty$.

However, this equation can be rearranged using the continuity equation into a form in which the dyadic product of the velocity can be absorbed in the derivative on the left-hand side:

$$m_\alpha n_\alpha \left(\frac{\partial \mathbf{u}_\alpha}{\partial t} + \mathbf{u}_\alpha \cdot \nabla \mathbf{u}_\alpha \right) = -\nabla \cdot \mathbf{P}_\alpha + n_\alpha q_\alpha (\mathbf{E} + \mathbf{u}_\alpha \times \mathbf{B}) + \mathbf{R}_{\alpha\beta}, \tag{1.13}$$

which is usually called the *force balance*. Here, the operator

$$\frac{d}{dt} = \frac{\partial}{\partial t} + \mathbf{u}_\alpha \cdot \nabla \tag{1.14}$$

is called the *substantial* or *convective derivative* and measures the change along the trajectory of a fluid element in the laboratory frame. In ordinary hydrodynamics, Eq. (1.13) is called the *Euler equation* while equations in the co-moving frame are referred to as the *Lagrange description*.

The system of equations so far is not closed as a second moment appears in the first moment equation, just as the velocity as first-order moment occurs in the zeroth-order continuity equation. It is clear that this problem cannot be solved by adding the second moment of the kinetic equation as a third moment will appear. This is the closure problem of MHD, where at each step, an additional relation will be required to close the system. If we want to stop here, we obviously need a relation for the pressure, that is an equation of state. This could be the adiabatic equation

$$\frac{d}{dt} \left(\frac{p_\alpha}{\rho_\alpha{}^{\gamma_\alpha}} \right) = 0, \tag{1.15}$$

where γ_α is the adiabatic coefficient and we have assumed that we only deal with the scalar pressure in Eq. (1.13). Together with Maxwell's equations for the fields $\mathbf{E}$ and $\mathbf{B}$, we now have indeed a closed system. However, we will still simplify this system for a two-component plasma in Section 1.1.2.

1.1.2
One-Fluid Model of Magnetohydrodynamics

For the case of a two-component plasma consisting of one ion species and electrons, the system of two-fluid equations can be combined to give a set of one-fluid equations. Here, owing to the large mass difference between the two species, the mass and momentum are more or less contained in the ions, whereas the electrons guarantee quasineutrality and lead to an electrical current if their velocity is different from that of the ions. In the following, we will assume a hydrogen plasma, that is charge number $Z = 1$. Specifically, the one-fluid variables are the mass density

$$\rho = n_i m_i + n_e m_e \approx n m_i \tag{1.16}$$

where we have used charge neutrality ($n_e = n_i = n$), the centre of mass fluid velocity

$$\mathbf{v} = \frac{1}{\rho}(m_i n_i \mathbf{u}_i + n_e m_e \mathbf{u}_e) \approx \mathbf{u}_i \tag{1.17}$$

and the electrical current density

$$\mathbf{j} = en_i\mathbf{u}_i - en_e\mathbf{u}_e = en(\mathbf{u}_i - \mathbf{u}_e). \tag{1.18}$$

The one-fluid equations are obtained by adding or subtracting the continuity and force balance equations for the individual species and expressing them in the one-fluid variables, neglecting terms of the order m_e/m_i. In this process, addition will give a one-fluid equation for the velocity, whereas the subtraction will yield one for the current density.

Adding the continuity equations yields

$$\frac{\partial \rho}{\partial t} + \nabla \cdot (\rho \mathbf{v}) = 0 \tag{1.19}$$

that is a one-fluid continuity equation while subtracting them leads to

$$\frac{\partial \rho_{\mathrm{el}}}{\partial t} + \nabla \cdot \mathbf{j} = 0 \tag{1.20}$$

which is the continuity equation for the electrical current. As we assume the plasma to be quasi-neutral, the electrical charge density $\rho_{\mathrm{el}} = en_i - en_e$ vanishes and the equation just reads $\nabla \cdot \mathbf{j} = 0$.

Adding the force equations leads to

$$\rho \left(\frac{\partial \mathbf{v}}{\partial t} + \mathbf{v} \cdot \nabla \mathbf{v} \right) = -\nabla \cdot \mathbf{P} + \mathbf{j} \times \mathbf{B} \tag{1.21}$$

the Euler equation, where $\mathbf{P} = \mathbf{P_i} + \mathbf{P_e}$ as in an ideal plasma, the total pressure is the sum of the partial pressures of the individual species. As pointed out earlier, the fluid velocity is mainly the ion velocity. To determine the role that the electrons play, we can re-write the electron equation of motion in terms of the one-fluid velocity $\mathbf{v}$ to obtain

$$\mathbf{E} + \mathbf{v} \times \mathbf{B} = \frac{1}{\sigma}\mathbf{j} + \frac{1}{en_e}(\mathbf{j} \times \mathbf{B} - \nabla p_e) - \frac{m_e}{e}\frac{d\mathbf{u}_e}{dt} \tag{1.22}$$

which is Ohm's law for a plasma. One can see that here, not all two-fluid variables could be eliminated from the equation through $m_e \ll m_i$. However, we will argue in the following that the last two terms are usually small for our applications and can be neglected so that this problem will not appear in what follows.

Assuming that we will deal with the scalar pressure only, we can use the adiabaticity equation

$$\frac{d}{dt}\left(\frac{p}{\rho^\gamma} \right) = 0. \tag{1.23}$$

In ordinary hydrodynamics, neglecting the viscous part of the pressure tensor corresponds to infinite Reynolds number and hence the use of the Euler instead of the Navier Stokes equation that rules out a proper description of fluid turbulence. However, if we keep finite conductivity in Ohm's law, there is still dissipation in the system and the relevant dimensionless number becomes the magnetic Reynolds number (Chapter 8).

Finally, we use Maxwell's equations for **E** and **B**

$$\nabla \cdot \mathbf{B} = 0, \tag{1.24}$$

$$\nabla \times \mathbf{B} = \mu_0 \mathbf{j}, \tag{1.25}$$

$$\nabla \times \mathbf{E} = -\frac{\partial \mathbf{B}}{\partial t} \tag{1.26}$$

and we have a closed system of equations to describe a plasma as a single fluid. We note that we have neglected the polarization current in Ampere's law, eliminating phenomena with (phase) velocity close to that of the speed of light, such as electromagnetic waves that arise from this term. In addition, we do not need to solve explicitly an equation for $\nabla \cdot \mathbf{E}$ in the quasi-neutral plasma. Finally, counting the number of variables and equations reveals that one scalar equation seems obsolete; this is related to the fact that any solution that satisfies $\nabla \cdot \mathbf{B}$ in the beginning will always do so and hence this is rather a boundary condition than a separate equation.

1.1.3
Validity of the One-Fluid Model of Magnetohydrodynamics

The system of equations derived earlier relies on a number of assumptions that have been made during the derivation. Here, we briefly review them and point out the restrictions arising.

- By assuming that we can use a continuum description, we think of the plasma described as fluid elements that are infinitesimally small such that individual particles are not distinguished. This means that the typical 'extension' of the particle orbit, that is the Larmor radius r_L, is small compared to a typical system length L:

  $$r_{Li} = \frac{\sqrt{m_i k T_i}}{eB} \ll L. \tag{1.27}$$

  This is also known as the condition for a *magnetized plasma* and is usually very well fulfilled in the fusion plasmas under study here where typical ion Larmor radii are of the order of millimetres and the electron Larmor radius is even smaller by a factor $\sqrt{m_e/m_i}$, which is the reason why we have used the ion Larmor radius earlier. For the MHD instabilities treated in this book, it is important to remember that the validity of our results will break down for very small scales, and finite Larmor radius (FLR) effects set the limit to the applicability in the limit $L \to 0$.
- Defining a local temperature requires that f_α is close to a Maxwellian. This relies on considering timescales that are long compared to the collision time

  $$\tau_{coll} \sim T^{3/2}/n \ll \tau$$

  or, in terms of spatial scales, the mean free path λ_{mfp} being small compared to the system length:

  $$\lambda_{\mathrm{mfp}} \sim T^2/n \ll L.$$

These conditions are usually not well fulfilled in typical hot fusion plasmas, at least parallel to the magnetic field, as typical values for λ_{mfp} can easily reach the order of kilometres while L is usually of the order of metres such that the particles pass through the system many times before equilibrating. Hence, a kinetic description is often needed to describe the dynamics along field lines, as is the case for example in gyrokinetic description of turbulence. On the other hand, perpendicular to the field, the typical mean free path is the Larmor radius and the validity condition is fulfilled if Eq. (1.27) is fulfilled.

- In Ohm's law (Eq. (1.22)), the ratio of the 'Hall term' $(\mathbf{j} \times \mathbf{B} - \nabla p_e)/(en_e)$ to the term $\mathbf{v} \times \mathbf{B}$ can be estimated using the force balance $\mathbf{j} \times \mathbf{B} - \nabla p_e \approx \nabla p_i \approx p_i/L$ to be small if the typical velocity v fulfills the condition $v \gg r_{Li}/Lv_{\text{th},i}$. As we have already assumed $r_{Li}/L \ll 1$ (Eq. (1.27)), this condition is usually well fulfilled for the fast (v of the order of $v_{\text{th},i}$) MHD phenomena treated in our context. However, it breaks down for phenomena that are slow enough that the difference between ion and electron fluid matters, such as drift waves. Then, two-fluid theory will have to be used and diamagnetic effects will become important.
- Applying a similar argument to the term $(m_e/e)(d\mathbf{u}_e/dt)$ shows that it can be neglected whenever the typical length scale L is long compared to the electron Larmor radius $r_{L,e}$, which, as pointed out earlier, is fulfilled whenever Eq. (1.27) is fulfilled.
- Finally, while the remaining terms in Ohm's law are already true one-fluid terms, another important simplification can often be made considering that a hot plasma is a very good electrical conductor, meaning that the typical timescale for current diffusion

$$\tau_R = L^2 \mu_0 \sigma \sim L^2 T^{3/2} \gg \tau \tag{1.28}$$

can be of the order of seconds while typical MHD instabilities grow much faster. Hence, one can often also neglect the finite conductivity effects in Eq. (1.22), which formally corresponds to the limit $\sigma \to \infty$. The resulting Ohm's law

$$\mathbf{E} + \mathbf{v} \times \mathbf{B} = 0 \tag{1.29}$$

is the fundamental ingredient of what is called *ideal MHD*, and its consequences will be discussed in Section 1.1.2. We will come back to finite resistivity in Chapter 8.

We mention here that for toroidal confinement systems, dimensionless variables are often used for scaling arguments in a way comparable to the wind tunnel approach in ordinary hydrodynamics. It can be shown that, if the plasma can be assumed to be quasi-neutral[4], a consistent set of dimensionless variables is given by normalizing energy, length and time scales as follows: the kinetic plasma energy can be normalized by the magnetic field energy, a parameter known as β (Eq. (1.46)). For a typical length scale, we define the ratio between Larmor radius and system length as $\rho^* = r_L/L$. Finally, if we want to relate the time to the time between collisions, a convenient definition for a toroidal system is the so-called

4) This corresponds to the assumption of small Debye length, $\lambda_D/L \to 0$.

collisionality ν^*, which is the inverse ratio of the above-mentioned collision time to the time a particle takes to complete a characteristic orbit in a toroidal confinement system, the so-called banana transit time[5]. According to the discussion earlier, MHD corresponds to the limits $\rho^* \to 0$ and $\nu^* \to \infty$, with the caveat that in ideal MHD, the electrical conductivity is still high enough that the condition described by Eq. (1.28) holds. In this case, the typical time scale is not given by the collision time, but rather by the inertia of the plasma. This so-called Alfvén timescale is discussed in Section 1.2.

1.2 Consequences of the MHD Equations

The system of equations derived earlier describes a magnetized plasma as a one-component fluid. Before applying the equations to specific magnetic confinement schemes, we will point out some general consequences arising from the equations.

1.2.1 Magnetic Flux Conservation

An important consequence of the ideal Ohm's law is the fact that magnetic flux is conserved when moving with the plasma. To prove this statement, we consider a contour C moving through the plasma with velocity $\mathbf{u}_C$. The geometry is shown in Figure 1.1a. The change of magnetic flux $\Psi = \int \mathbf{B} \cdot d\mathbf{S}$ through this contour is given by

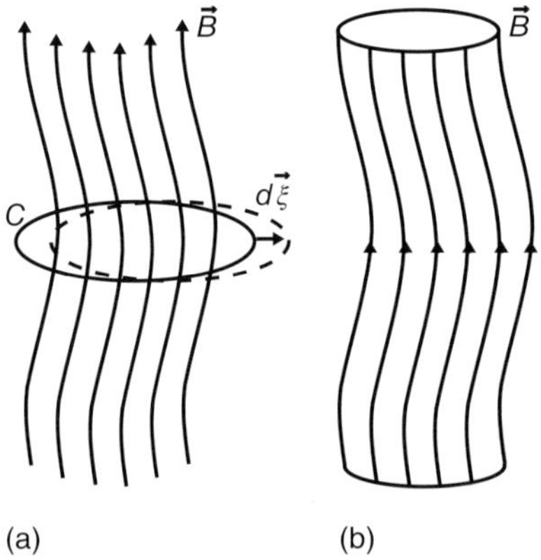

Figure 1.1 Geometry used for the derivation of flux conservation (a) and visualization of the concept of flux tubes (b).

5) In a toroidal system, there is a population of particles that bounce back and forth in the magnetic mirror created by the inhomogeneous B-field.

$$\frac{d\Psi}{dt} = \int \frac{\partial \mathbf{B}}{\partial t} \cdot d\mathbf{S} - \oint \mathbf{u}_C \times \mathbf{B} d\ell, \tag{1.30}$$

where the first term accounts for explicit change of $\mathbf{B}$ with time and the second term comes from the fact that the change of surface of C can be calculated by integrating the vector product of the curve's tangent $\mathbf{d}\ell$ and the displacement $d\xi$ (where $d\xi/dt = \mathbf{u}_C$) along the contour C.

Using Faraday's law (Eq. (1.26)) in combination with Ohm's law (Eq. (1.22)), in the first integral of Eq. (1.30) and applying Stokes' theorem, leads to

$$\frac{d\Psi}{dt} = \oint \left(\mathbf{v} - \frac{\mathbf{j}}{en_e} - \mathbf{u}_C\right) \times \mathbf{B} \cdot d\ell \tag{1.31}$$

where we have assumed the plasma to be an ideal conductor, $\sigma \rightarrow \infty$. In deriving Eq. (1.31), we have made use of the fact that the curl of a gradient vanishes, eliminating the $\nabla p_e/(n_e e)$ term which is true as long as $\nabla p_e/n_e = \nabla(p_e/n_e)$, that is for polytropic equations of state[6]. This means that the magnetic flux is constant if the contour is moved with the electron velocity[7]

$$\mathbf{u}_C = \mathbf{v} - \frac{\mathbf{j}}{en_e} = \mathbf{v}_e \tag{1.32}$$

Consequently, one can imagine the plasma consisting of 'flux tubes' that are small cylinders with their mantle defined by the magnetic field lines as depicted in Figure 1.1b. As the flux is constant in these cylinders, the flux tubes are convected with v_e. In some sense, magnetic field lines thus become real objects in an ideal plasma. Sometimes, the field lines are also said to be 'frozen into the plasma'[8]. An important consequence for the motion of flux tubes in ideal MHD is that they cannot intersect as this would change the flux within the individual tubes. Hence, the motion is constrained to not change the topology of the flux tubes. Such topological changes are only possible invoking additional terms in Ohms law outside of the ideal MHD model, such as finite resistivity or finite electron inertia. Resistive effects are treated in Chapter 8.

For the MHD instabilities treated in this book, it means that they are expected to move with v_e, which has to be determined from the force balance equation of the electrons and will in general consist of a combination of fluid velocity and diamagnetic velocity as we have

$$\mathbf{v}_{e,\perp} = \frac{\mathbf{E} \times \mathbf{B}}{B^2} + \frac{\nabla p_e \times \mathbf{B}}{en_e B^2} = \mathbf{v}_{E\times B} + \mathbf{v}_{e,\mathrm{dia}} \tag{1.33}$$

6) Note that this argument does in general not hold for the $\mathbf{j} \times \mathbf{B}$ term as it is only equal to ∇p in stationary force equilibrium while here, we are concerned with dynamic changes of the equilibrium.

7) We note that using the ideal Ohm's law (Eq. 1.29), the field lines are found to be frozen into the fluid velocity rather than the electron velocity, since then, no current, that is no difference between electron and ion velocity is considered.

8) This terminology is slightly misleading as the effect is due to the high electrical conductivity that comes from the plasma being quite hot!

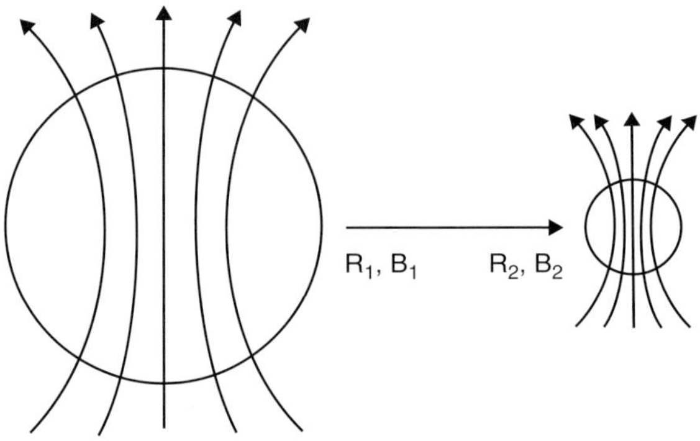

Figure 1.2 Schematic visualization of the consequence of flux conservation for the collapse of a neutron star: owing to the decrease of the equatorial surface, the magnetic field has to increase accordingly (remember that the density of field lines is representative for the field strength).

which follows from the vector product of Eq. (1.13) for the electrons with $\mathbf{B}$. Note that $\mathbf{v}_{E\times B}$ is the same for each species α and hence, the frame where $\mathbf{E} = 0$ is often referred to as the *rest frame of the plasma*.

An interesting illustration of the consequences of flux conservation is the collapse of a star of mass exceeding the critical mass for formation of a neutron star: during this process, the radius changes from $R_1 \approx 10^6$ to $R_2 \approx 10$ km while conserving magnetic flux, schematically shown in Figure 1.2. If we assume that we start with a dipole field of 10^{-5} T (roughly the Earth's magnetic field), we arrive at a final field of 10^5 T as flux conservation yields $B_2 = B_1(R_1/R_2)^2$. Such enormous magnetic field strengths can indeed be inferred from measuring the electromagnetic radiation coming from neutron stars, giving proof of the applicability of ideal MHD even under these extreme conditions.

1.2.2
MHD Equilibrium

A common application is the calculation of an equilibrium configuration using the MHD force balance (Eq. (1.21)). An equilibrium state is characterized by stationarity, that is $\partial/\partial t \to 0$. If we are interested in equilibria where the dynamic pressure coming from the term $(\mathbf{v}\cdot\nabla)\mathbf{v}$ can be neglected, the force balance reads

$$\nabla p = \mathbf{j}\times\mathbf{B} \tag{1.34}$$

stating that a pressure gradient can be sustained by currents possessing a component perpendicular to magnetic fields. Using Ampère's law for the current density, we can re-write the equilibrium force balance:

$$\nabla p = \frac{1}{\mu_0}(\nabla\times\mathbf{B})\times\mathbf{B} = -\nabla\frac{B^2}{2\mu_0} + \frac{1}{\mu_0}(\mathbf{B}\cdot\nabla)\mathbf{B} = -\nabla\frac{B^2}{2\mu_0} + \frac{B^2}{\mu_0}\boldsymbol{\kappa}, \tag{1.35}$$

where the curvature $\boldsymbol{\kappa}$ has been introduced according to

$$\boldsymbol{\kappa} = \frac{\mathbf{B}}{B} \cdot \nabla \left(\frac{\mathbf{B}}{B} \right), \quad \text{with} \quad |\boldsymbol{\kappa}| = \frac{1}{R_c}, \tag{1.36}$$

where R_c is the local radius of curvature of the field line. Hence, there are two contributions by which the magnetic field can exert a force on the plasma. The magnetic pressure $B^2/(2\mu_0)$ produces a restoring force when the field lines are compressed, whereas the field line tension $B^2/\mu_0\boldsymbol{\kappa}$ exerts a force in order to straighten out a field line once it is bent.

Using Eq. (1.35), we can also evaluate the condition under which the dynamic pressure can be neglected in the force balance. Obviously, the magnetic field mainly balances the kinetic pressure as long as

$$\nabla p \gg \rho \mathbf{v} \cdot \nabla \mathbf{v} \quad \rightarrow \quad \frac{p}{L} \gg \rho \frac{v^2}{L} \quad \rightarrow \quad \sqrt{\frac{p}{\rho}} \gg v \tag{1.37}$$

which states that the flow velocity should be much smaller than roughly the speed of sound.

1.2.3
Magnetohydrodynamic Waves

In this section, we address a particular point of the dynamics of ideal MHD. When deriving the set of MHD equations, we neglected the displacement current in Ampère's law, thus eliminating electromagnetic waves from the solutions. However, in the preceding section, we saw that MHD provides two kinds of restoring forces to displacement of a field line, magnetic pressure and field line tension. These give rise to MHD waves, the so-called Alfvén waves, which exist within the system of equations derived earlier. Figure 1.3 shows these two situations.

As a starting point, we linearize the force equation, assuming that all quantities can be written as a zeroth-order term that is constant in time and space and a small first-order perturbation that may vary in time and space. In addition, we set $\mathbf{v}_0 = 0$, assuming that flow does not play a role in the zeroth-order force balance.

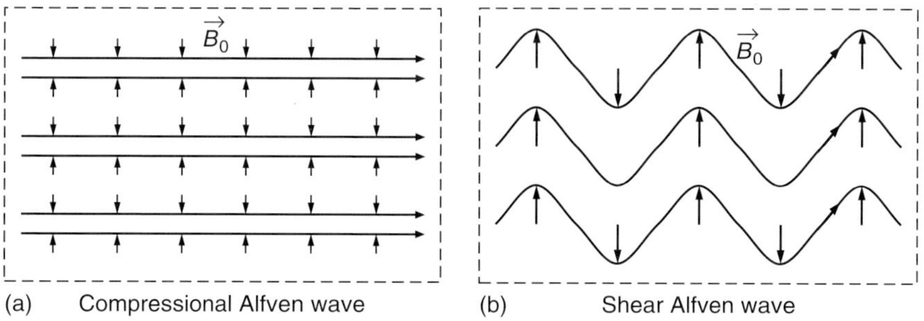

Figure 1.3 (a,b) Geometry for the derivation of compressional (a) and shear (b) Alfvén waves. The small arrows represent the perturbation of the system giving rise to an oscillation.

Linearizing (1.21), writing the right-hand side in the form (1.35) and differentiating with respect to time, we obtain

$$\rho_0 \frac{\partial^2 \mathbf{v}_1}{\partial t^2} = -\nabla \frac{\partial p_1}{\partial t} + \frac{1}{\mu_0}\left(\mathbf{B}_0 \cdot \nabla \frac{\partial \mathbf{B}_1}{\partial t} - \nabla\left(\mathbf{B}_0 \cdot \frac{\partial \mathbf{B}_1}{\partial t}\right)\right), \tag{1.38}$$

where the last two terms represent field line tension and pressure, respectively. In the following, we will hence distinguish two cases, namely pure compression (where the restoring force is given by magnetic pressure) and incompressible displacement (where the restoring force is due to field line tension). In reality, there will be waves that are a mixture of these two limiting cases.

1.2.3.1 Compressional Alfvén Waves

In this section, we assume that the plasma is compressed homogeneously in the direction perpendicular to the equilibrium magnetic field lines that are straight as we assumed $\mathbf{B}_0 =$ const.. This corresponds to Figure 1.3a and $\mathbf{B}_0 \cdot \nabla \to 0$ in Eq. (1.38). In ordinary hydrodynamics, this wave is the sound wave, propagating in the longitudinal direction due to the restoring force of kinetic pressure.

In order to re-write Eq. (1.38) in terms of $\mathbf{v}_1$, we express the first term on the right-hand side using the adiabatic law:

$$\frac{d}{dt}\left(\frac{p}{\rho^\gamma}\right) = 0 \;\rightarrow\; \frac{\partial}{\partial t}(\rho_0 p_1 - \gamma p_0 \rho_1) = 0 \tag{1.39}$$

Using the linearized continuity equation, we obtain

$$\frac{\partial p_1}{\partial t} = -\gamma p_0 \nabla \cdot \mathbf{v}_1 \tag{1.40}$$

stating that the change of pressure comes from the compression of a fluid element.

Next, we obtain an equation for $\mathbf{B}_1$ by combining Faraday's law and Ohm's law.

$$\frac{\partial \mathbf{B}_1}{\partial t} = -\nabla \times \mathbf{E}_1 = \nabla \times (\mathbf{v}_1 \times \mathbf{B}_0). \tag{1.41}$$

This can be rewritten using a vector theorem

$$\nabla \times (\mathbf{v}_1 \times \mathbf{B}_0) = (\mathbf{B}_0 \cdot \nabla)\mathbf{v}_1 - (\mathbf{v}_1 \cdot \nabla)\mathbf{B}_0 + \mathbf{v}_1(\nabla \cdot \mathbf{B}_1) - \mathbf{B}_0(\nabla \cdot \mathbf{v}_1). \tag{1.42}$$

In the case of pure compression, the first term on the right-hand side vanishes due to geometry (no change along the equilibrium field). The second term vanishes due to $\mathbf{B}_0 =$ const. As $\nabla \cdot \mathbf{B} = 0$ to each order, the third term generally vanishes. Hence, we arrive at

$$\frac{\partial \mathbf{B}_1}{\partial t} = -\mathbf{B}_0(\nabla \cdot \mathbf{v}_1). \tag{1.43}$$

Now, we can express (Eq. (1.38)) as follows:

$$\frac{\partial^2 \mathbf{v}_1}{\partial t^2} = \left(\gamma \frac{p_0}{\rho_0} + \frac{B_0^2}{\mu_0 \rho_0}\right) \Delta \mathbf{v}_1, \tag{1.44}$$

where we have used the fact that the perturbation does not introduce any vortices, that is $\nabla \times \mathbf{v}_1 = 0$, which leads to $\nabla(\nabla \cdot \mathbf{v}_1) = \Delta \mathbf{v}_1$. Equation (1.44) is a wave equation with phase velocity

$$v_{\mathrm{ph}} = \sqrt{\frac{\gamma p_0}{\rho_0} + \frac{B_0^2}{\mu_0 \rho_0}}. \tag{1.45}$$

There are two contributions, namely the contribution of kinetic pressure and that of magnetic pressure. Depending on their ratio, which in plasma physics is known as the *plasma beta*

$$\beta = \frac{2\mu_0 p}{B^2} \tag{1.46}$$

the wave will propagate at the speed of sound $c_s = \sqrt{\gamma p_0/\rho_0}$ for $\beta \gg 1$ or, at $\beta \ll 1$, with the Alfvén speed

$$v_A = \frac{B_0}{\sqrt{\mu_0 \rho_0}}. \tag{1.47}$$

More generally, the wave speed will be a combination of both and we call the wave a magneto-aoustic wave.

1.2.3.2 Shear Alfvén Waves

Now, we turn to the second case, depicted in Figure 1.3b, where we assume that the only restoring force is due to field line tension, that is the plasma is perturbed in an incompressible way, meaning $\nabla \cdot \mathbf{v} = 0$. In this case, the pressure perturbation vanishes (Eq. (1.40)) and in Eq. (1.42), the last term is zero while now, the first term on the right-hand side is non-zero as the plasma is perturbed along the field lines.

Hence, Eq. (1.38) becomes

$$\frac{\partial^2 \mathbf{v}_1}{\partial t^2} = \frac{B_0^2}{\mu_0 \rho_0} \nabla_{\parallel}{}^2 \mathbf{v}_1 \tag{1.48}$$

which again is a wave equation, this time for waves travelling along the equilibrium field lines, that is a transverse wave, analogous to the oscillation of for example a guitar string. The phase velocity is the Alfvén velocity v_A, this time without any contribution of the kinetic pressure as we have assumed the motion to be incompressible.

The Alfvén velocity is quite important for the dynamics of ideal MHD as it sets the 'natural' timescale limited by inertia. Estimating the Alfvén timescale as

$$\tau_A = L/v_A \tag{1.49}$$

and inserting typical parameters of magnetically confined fusion plasmas, it is of the order of $1-10\ \mu s$, that is quite fast, because of the small mass of the very low density plasma. Consequently, ideal MHD instabilities in tokamaks often grow so fast that they have to be slowed down by passive structures such as conducting wall elements in order to be accessible for magnetic feedback control. An example for this is the vertical displacement event (VDE) treated in Section 4.4 or the Resistive Wall Mode (RWM) treated in Section 7.4.

In ideal MHD, Alfvén waves will normally appear as a damping term for instabilites as their excitation is a sink of free energy. In a toroidal confinement system, the Alfvén spectrum usually is a continuum, that is there are no resonances at discrete frequencies that extend in radial direction and hence they lead to strong damping. However, owing to toroidicity, there also exists a kind of discrete Alfvén waves that can be excited under certain circumstances, for example by a population of fast particles. These can be quite important for future reactor grade fusion plasmas in which a large population of fast particles is expected, but their detailed treatment is beyond the scope of this book.

2
MHD Equilibria in Fusion Plasmas

In this chapter, we take a closer look at the equilibrium configurations of magnetically confined fusion plasmas. We start with linear devices as important insight can be gained from these cases that can be treated analytically. Then, we describe in more detail the tokamak equilibrium. This chapter closes with some remarks about the stellarator configuration.

2.1
Linear Configurations

Linear configurations do no longer play any significant role in magnetic confinement of fusion plasmas but can serve as a simple example of how the force balance is fulfilled. Hence, we discuss here the z-pinch and the screw pinch configuration, where the latter in its periodic form can serve as a model system for the tokamak in cases where toroidicity is not important, for example for the current driven modes treated in Chapter 4.

2.1.1
The z-Pinch

One of the simplest configurations of magnetic confinement is to run an axial current with current density j_z and magnitude I_p through a cylindrical plasma and rely on the $\mathbf{j} \times \mathbf{B}$ force due to the poloidal field B_θ generated by the plasma current. Owing to the cylindrical symmetry, this problem is best described in cylindrical coordinates as shown in Figure 2.1.

From Ampère's law, the poloidal field can be related to the current density by

$$\frac{1}{\mu_0 r}\frac{d}{dr}(rB_\theta) = j_z \tag{2.1}$$

and the force balance becomes

$$\frac{dp}{dr} = -j_z B_\theta = -\frac{1}{\mu_0 r}B_\theta\frac{d}{dr}(rB_\theta) = -\frac{B_\theta^2}{\mu_0 r} - \frac{d}{dr}\left(\frac{B_\theta^2}{2\mu_0}\right) \tag{2.2}$$

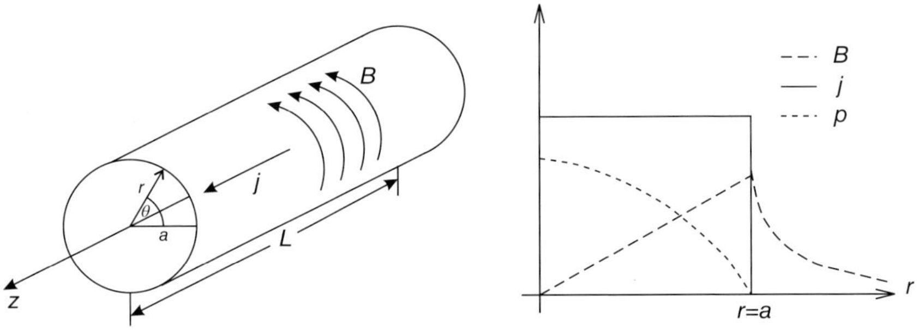

Figure 2.1 Schematic z-pinch configuration and co-ordinate system used for its analysis (a) and profiles $j(r), B_\theta(r)$ and $p(r)$ for a constant current density (b).

which now explicitly shows the contributions of field line tension (note that the radius of curvature is just r) and magnetic pressure associated with the poloidal field (Eq. 1.35).

If we specify the radial current distribution, we can calculate the poloidal field distribution. For example, for the simplest case j_z = const., we obtain

$$B_\theta(r) = \frac{\mu_0 I_p}{2\pi a^2} r \quad \text{for} \quad r \leq a \tag{2.3}$$

$$B_\theta(r) = \frac{\mu_0 I_p}{2\pi r} \quad \text{for} \quad r > a \tag{2.4}$$

$$p = \frac{\mu_0 I_p^2}{4\pi^2 a^2} \left(1 - \left(\frac{r}{a}\right)^2\right) \tag{2.5}$$

where we have used the boundary condition $p(r = a) = 0$ and a is the radius of the cylindrical plasma column. These profiles are shown in Figure 2.1b. We note that the pressure on axis is determined by the total current I_p and varies quadratically with it.

We can use the previously introduced quantity β as a measure of how efficient the magnetic field is used for confinement, as it will determine how much fusion power[1] can be generated for given magnetic field[2]. For this purpose, it is useful to define it as

$$\beta_\theta = \frac{2\mu_0 \langle p \rangle}{B_\theta^2(a)} \tag{2.6}$$

Evaluating Eq. (2.6) for the profiles given earlier yields $\beta_\theta = 1$, that is very efficient use of the magnetic field. In the z-pinch, this result is independent of the current profile. This can be seen by integrating by parts the expression for the

1) In the regime relevant for D–T fusion, P_{fus} roughly scales as p^2.
2) The coils needed to generate the magnetic field in a fusion device contribute a sizeable fraction of the cost, for example roughly 30% in ITER.

averaged pressure

$$\langle p\rangle = \frac{2}{a^2}\int_0^a r\,dr\,p(r) = -\frac{1}{a^2}\int_0^a dr\,r^2\frac{dp}{dr} + r^2p|_{r=0}^a \tag{2.7}$$

where the second term on the right-hand side is zero because of $r = 0$ and $p(a) = 0$ at the boundary. Next, we insert the relations

$$j_z(r) = \frac{1}{2\pi r}\frac{dI(r)}{dr} \text{ and } B_\theta(r) = \frac{\mu_0 I(r)}{2\pi r} \tag{2.8}$$

into $dp/dr = -j_z B_\theta$ in Eq. (2.7). Here, $I(r)$ is the total current flowing inside minor radius r. This leads to

$$\langle p\rangle = \frac{\mu_0}{(2\pi a)^2}\int_0^a dr\,I\frac{dI}{dr} = \frac{\mu_0}{(2\pi a)^2}\frac{1}{2}I_p^2 \tag{2.9}$$

which is equal to $B_\theta^2/(2\mu_0)$, giving $\beta_\theta = 1$.

This so-called Bennett relation shows that for given $\langle p\rangle$ and I_p, the z-pinch is in equilibrium at exactly one specific radius a. Hence, it will contract (pinch) to that radius when heated by the current and the configuration changes whenever the pressure changes. If we want to maintain the radius independent of the pressure, we have to add another degree of freedom in the form of an axial field B_z. This leads to the screw pinch discussed in Section 2.1.2.

However, there is another important reason for the addition of the axial field: consider a slightly inhomogeneous contraction of the z-pinch, that is a variation of a with z as shown in Figure 2.2a. In the narrower region, the poloidal field and hence the magnetic pressure will be stronger than in the neighbouring region as the total current still has the same value. Similarly, the field line tension will increase as the radius of curvature decreases. Hence, this initial perturbation

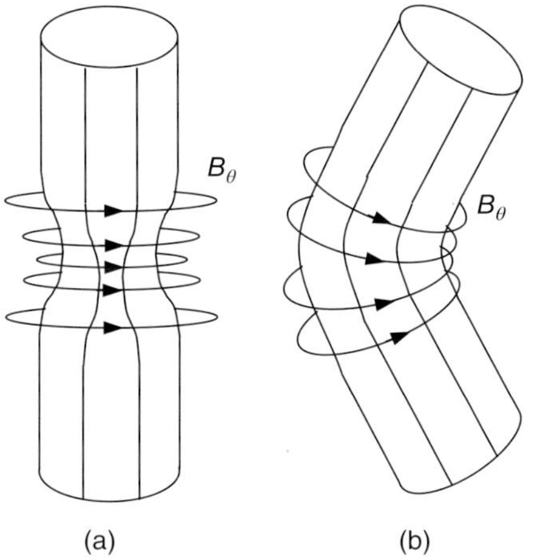

Figure 2.2 Sausage (a) and pinch (b) instability in a z-pinch. The initial perturbation leads to a force pointing in the direction of the initital perturbation, that is to an unstable situation.

will lead to a radial force that tends to increase the perturbation. The z-pinch is prone to this so-called sausage instability. A similar argument applies to the so-called kink instability shown in Figure 2.2b, where a kinking of the plasma column leads to higher magnetic pressure (indicated by the higher density of poloidal magnetic field lines in Figure 2.2) that reinforces the initial perturbation.

Obviously, this instability can be counteracted by adding an axial field as the perturbation will bend the axial field and thus lead to a restoring field line tension. Thus, in addition to providing flexibility to the configuration as described earlier, the axial field is also needed for stability. This is the reason why z-pinches are no longer used for experiments aiming at stationary magnetically confined fusion[3].

2.1.2 The Screw Pinch

In the screw pinch, we add an axial field to the z-pinch, leading to helical field lines as shown in Figure 2.3a. This field can be generated externally by coils surrounding the cylinder. Just like the field produced inside a solenoid of infinite length, it is constant in the cylinder and zero outside.

In some sense, the screw pinch can be viewed as a 'straight tokamak', and in fact, many of the stability arguments we will discuss later in the context of purely current driven modes (Chapter 4) can be derived using a screw pinch configuration that is assumed to be periodic in z with period $2\pi R_0$, mimicking the toroidal angle $z \to R_0\phi$. Here, R_0 is the major radius of the tokamak. We note that the description is inadequate whenever the toroidal curvature is important, as is for example the case for pressure-driven instabilities in a tokamak.

According to the discussion in the previous sections, the external field has to be large enough to suppress the sausage and kink instabilities. As will be derived in

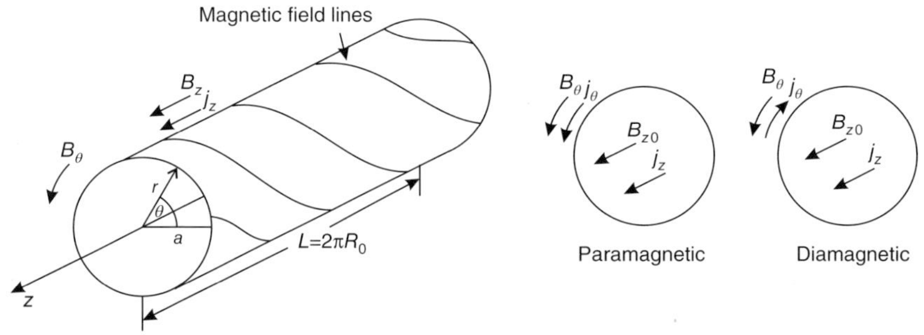

Figure 2.3 Schematic setup of a screw pinch (a) and currents contributing to the pressure balance (b) for a paramagnetic case ($\beta_p < 1$) and a diamagnetic case ($\beta_p > 1$).

3) z-Pinches still do play a role in inertial fusion applications.

Chapter 4, for a tokamak, this leads to a condition on the so-called safety factor q, with

$$q = \frac{\text{number of toroidal turns}}{\text{number of poloidal turns}} \tag{2.10}$$

following a field line around the torus. For example, a field line with $q = 1$ closes up on itself after going around the torus once toroidally, whereas a $q = 2$ field line has to go around twice toroidally to close up on itself poloidally.

Using the above-mentioned definition, in a screw pinch, q is just the ratio of toroidal to poloidal angle along the field line. As along a field line, we have $B_\theta / B_z = r\Delta\theta / R\Delta\phi$, we can derive

$$q = \frac{\Delta\phi}{\Delta\theta} = \frac{r}{R}\frac{B_z}{B_\theta} \tag{2.11}$$

In Chapter 4, it is shown that a rigorous limit for the kink instability in a tokamak is $q > 1$, the so-called Kruskal–Shafranov limit. In conventional tokamaks where $j(r)$ is peaked in the centre (following the conductivity profile that varies as $T_e^{3/2}$), the safety factor usually varies from around 1 on axis to values around 3–4 at the edge. Hence, for a tokamak of aspect ratio $R_0/a \approx 3$, we typically obtain $B_z \approx 10B_\theta$.

We mention here that another toroidal configuration exists that can be approximated by a periodic screw pinch, namely the so-called reversed field pinch (RFP). In this configuration, the poloidal and toroidal fields are comparable in magnitude, leading to $q \ll 1$ and stability is obtained by reversing the toroidal field close to the plasma periphery. This is briefly described in Section 4.3 in the context of external kink modes. Compared to a tokamak, the curvature of the poloidal field is much more important than that of the toroidal field and hence toroidicity is less important for the description of the RFP than for the tokamak. While quite interesting from the point of view of basic MHD studies, the fusion performance of the RFP is lagging behind that of the tokamak substantially and we will hence focus on the tokamak in this book. However, we mention the RFP several times when its physics is complementing the picture developed for the tokamak.

In the following chapters, we will see that the shape of the current profile plays an important role in determining the stability of the tokamak against current gradient driven modes. It is hence convenient to use a parametrization of the current profile for which $q(r)$ can be determined analytically in the periodic screw pinch. A convenient class of current profiles with this property is given by

$$j(r) = j_0\left(1 - \left(\frac{r}{a}\right)^2\right)^\mu \tag{2.12}$$

where μ parameterizes the current profile peakedness. This class of profiles can be integrated analytically to give the radial profile of $I(r)$, the current enclosed by the surface with radius r

$$I(r) = j_0\frac{\pi a^2}{\mu + 1}\left(1 - \left(1 - \left(\frac{r}{a}\right)^2\right)^{\mu+1}\right) \tag{2.13}$$

The corresponding profile of $B_\theta(r)$ is then given by

$$B_\theta(r) = \frac{\mu_0 I(r)}{2\pi r} = \frac{\mu_0 j_0}{2\pi r}\frac{\pi a^2}{\mu+1}\left(1-\left(1-\left(\frac{r}{a}\right)^2\right)^{\mu+1}\right) \tag{2.14}$$

and the q-profile can be calculated using Eq. (2.11). For $r/a = 1$, we obtain $q(a) = 2(\mu+1)B_z/(\mu_0 j_0 R_0)$, whereas in the limit $r/a \to 0$, we get $q(0) = 2B_z/(\mu_0 j_0 R_0)$ so that this class of q-profiles has the property

$$q(a) = (\mu+1)q(0) \tag{2.15}$$

Finally, we note that the magnetic shear $s = (r/q)(dq/dr)$ varies over radius according to the variation of the q-profile, but its edge value is always $s = 2$ since $s = 2 - rI'(r)/I(r)$ and $I'(r) = 0$ at the edge. Figure 2.4 shows typical profiles for a variation of μ.

The calculation of the MHD equilibrium in the screw pinch is done in a way analogous to the z-pinch, with the additional axial field entering according to

$$-\frac{1}{\mu_0}\frac{dB_z}{dr} = j_\theta \tag{2.16}$$

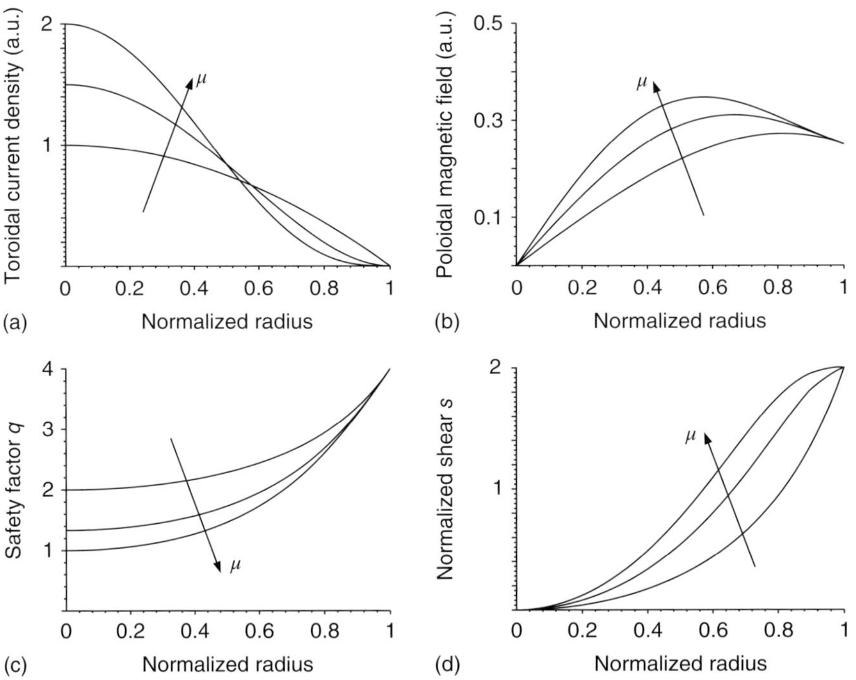

Figure 2.4 (a–d) Profiles of $j(r), B_\theta(r), q(r)$ and shear $s = (r/q)(dq/dr)$ for the class of current profiles $j(r) = j_0(1-(r/a)^2)^\mu$. The total current and hence $q(a)$ have been kept constant while μ has been chosen to be 1, 2 and 3.

Note that the constant part of the axial field introduced by the external coils, B_{z0} can be modified by an additional part B_{z1} coming from poloidal currents j_θ in the plasma. The force balance for the screw pinch reads

$$\frac{dp}{dr} = -\frac{d}{dr}\left(\frac{B_\theta^2 + B_z^2}{2\mu_0}\right) - \frac{B_\theta^2}{\mu_0 r} \tag{2.17}$$

In a screw pinch (tokamak) experiment, B_{z0} and j_z are imposed externally by the axial (toroidal) field coils and the applied axial (loop) voltage, whereas the current j_θ and hence B_{z1} is not directly controlled. According to Eq. (2.17), this current can contribute to the force balance by increasing the magnetic pressure (as the axial field is straight, it does not contribute to field line tension). We note that the vacuum axial field B_{z0} does not give rise to a force unless $j_\theta \neq 0$.

The result of this additional degree of freedom is that an equilibrium configuration can be found for fixed minor radius a for any pressure and hence β_θ-value[4]. If an initial plasma is formed filling a certain volume (e.g. a vacuum vessel in a tokamak discharge), a compression or expansion will, in ideal MHD, induce poloidal currents via the radial motion v_r across B_z that produce a j_θ directed such that $j_\theta \times B_z$ opposes that motion. This can be seen as a consequence of the incompressibility of the axial (toroidal) field that is generated by external coils. The sign of j_θ determines the sign of the resulting force. We can distinguish two cases:

- If $|dp/dr| < j_z B_\theta$, that is a case where a pure z-pinch would contract, $j_\theta \times B_z$ will be antiparallel to $j_z \times B_\theta$. The efficiency of the confinement with regard to B_θ is smaller than in the z-pinch, that is $\beta_\theta < 1$. As can be seen in Figure 2.3, the field B_{z1} generated by j_θ is parallel to the externally generated B_{z0} and the plasma is said to behave paramagnetic.
- If $|dp/dr| > j_z B_\theta$, the induced $j_\theta \times B_z$ is parallel to $j_z \times B_\theta$ and $\beta_\theta > 1$; the plasma behaves diamagnetic.

Another way to look at this is that in the paramagnetic case, the current in the external field coils is parallel to j_θ and hence attracts (expands) the plasma column, whereas in the diamagnetic case, it helps to compress it.

As will be discussed in the remainder of this book, in a tokamak, the total pressure is still limited to values of the order of $\beta_\theta = 1$ by the equilibrium limit and the occurrence of ideal and resistive MHD instabilities, and hence, the total β_{tot} is substantially smaller than 1. Using the numbers above, $B_z \approx 10B_\theta$ and $\beta_{\text{tot}} = 2\mu_0\langle p\rangle/B^2 \approx 2\mu_0\langle p\rangle/B_z^2 \approx 1/100\beta_\theta$. Hence, in tokamaks (and also in stellarators), typical β values are of the order of only several percent.

While the screw pinch offers more flexibility than the z-pinch, it still suffers from substantial end losses along the field lines that intersect the electrodes via which the axial plasma current is driven. This is the reason why magnetic confinement uses toroidal configurations that minimize these losses. In the following, we take a closer look at these configurations.

4) This statement does of course not consider the stability considerations discussed in the remaining chapters of this book.

2.2
Toroidal Configurations

In toroidal configurations, the topology of the plasma is that of a torus as shown in Figure 2.5. As

$$\mathbf{B} \cdot \nabla p = \mathbf{B} \cdot \mathbf{j} \times \mathbf{B} = 0 \tag{2.18}$$

and the same is true for $\mathbf{j}$, the vectors $\mathbf{j}$ and $\mathbf{B}$ lie in the surfaces $p = \text{const.}$. One can prove mathematically that these surfaces can form nested toroids in general geometry, indicating the existence of configurations such as the tokamak and the stellarator.

Topologically, two different sets of closed curves exist on the toroid surface, namely poloidal (short way around the torus) and toroidal (long way around the torus) curves. As $\nabla \cdot \mathbf{j} = \nabla \cdot \mathbf{B} = 0$, the value of the flux integrals

$$\Phi = \int_{S_{\text{tor}}} \mathbf{B} \cdot d\mathbf{S} \quad \text{(toroidal magnetic flux)} \tag{2.19}$$

$$\Psi = \int_{S_{\text{pol}}} \mathbf{B} \cdot d\mathbf{S} \quad \text{(poloidal magnetic flux)} \tag{2.20}$$

$$I_{\text{tor}} = \int_{S_{\text{tor}}} \mathbf{j} \cdot d\mathbf{S} = \frac{1}{\mu o} \oint_{C_{\text{pol}}} \mathbf{B} \cdot d\mathbf{s} \quad \text{(toroidal current)} \tag{2.21}$$

$$I_{\text{pol}} = \int_{S_{\text{pol}}} \mathbf{j} \cdot d\mathbf{S} = \frac{1}{\mu o} \oint_{C_{\text{tor}}} \mathbf{B} \cdot d\mathbf{s} \quad \text{(poloidal current)} \tag{2.22}$$

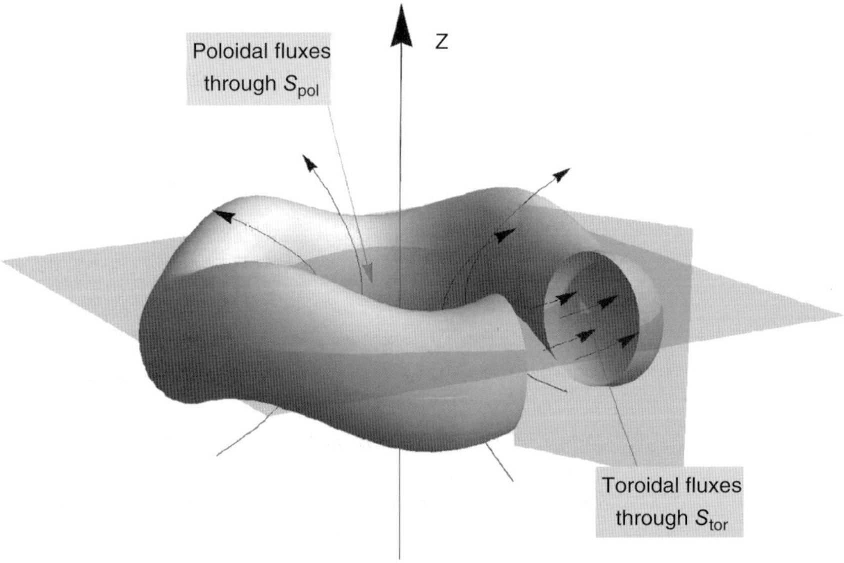

Figure 2.5 General toroidal geometry: two topologically different curves can be used to define flux integrals that can serve as flux surface labels. (Please find a color version of this figure on the color plates.)

for a given curve C is independent of the choice of integration surface S[5]. Furthermore, as $\mathbf{j}$ and $\mathbf{B}$ lie on the $p =$ const. surfaces, their flux integrals on the surface vanish, and hence, the flux surfaces can be labelled by either the toroidal or poloidal fluxes defined earlier. This flux surface label represents a generalized radial coordinate for arbitrarily shaped nested toroidal flux surfaces. In Section 2.2.1, we show that for axisymmetric geometry, we can derive a scalar partial differential equation (PDE) that describes the equilibrium in terms of the poloidal fluxes.

2.2.1 The Tokamak

The name tokamak is a Russian acronym for 'toroidal vessel with magnetic field coils'. In the tokamak, the toroidal field $\mathbf{B}_{\phi}$ is generated by external toroidal field coils, whereas the poloidal field $\mathbf{B}_{\mathrm{p}}$ is generated by a toroidal current I_{tor} induced by a change of flux in the central solenoid coil acting as primary winding of a transformer whose secondary winding is the plasma. As this current provides ohmic heating due to the finite electrical resistivity of the plasma, plasma generation is relatively easy in a tokamak and, owing to the good confinement properties, this line of magnetic confinement is the most advanced at present. On the other hand, the need to sustain the current by transformer action means that tokamak discharges are inherently pulsed unless the current can be driven by other means. This is a subject of present-day research. In addition to the toroidal field coils and the plasma current providing the basic components of the magnetic field, poloidal field coils must be used to provide force equilibrium and also to control the position of the plasma column in the vacuum vessel and the shape of the poloidal cross section.

The basic components of a tokamak are shown in Figure 2.6. In Section 2.2.1.1, we derive the equilibrium equation for the tokamak.

2.2.1.1 The Grad–Shafranov Equation

Figure 2.6 introduces the coordinates we will use for the calculation of general axisymmetric equilibria, which are cylindrical coordinates oriented such that the azimuthal angle ϕ represents the ignorable toroidal coordinate, whereas the poloidal plane is decomposed in R and Z.

Following the discussion in the previous section, we can freely chose the integration surface to calculate the poloidal fluxes and we use a circular surface oriented such that its surface normal points in Z-direction. Then, we can calculate the poloidal flux according to

$$\Psi(R,Z) = 2\pi \int_0^R B_Z(R',Z)R'dR' \tag{2.23}$$

5) To prove this, use Gauss' theorem for the sum of the integrals of two arbitrarily chosen surfaces bounded by the same curve C.

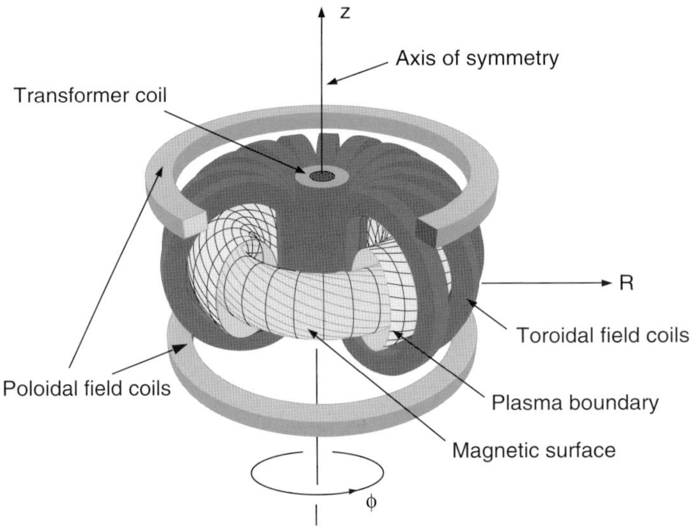

Figure 2.6 Schematic layout of a tokamak coil system as well as the co-ordinates used to derive the equilibrium equation (Eq. 2.33). (Please find a color version of this figure on the color plates.)

Differentiating with respect to the upper limit gives

$$B_Z(R,Z) = \frac{1}{2\pi R}\frac{\partial \Psi(R,Z)}{\partial R} \tag{2.24}$$

and using $\nabla \cdot \mathbf{B} = 0$ leads to

$$B_R(R,Z) = -\frac{1}{2\pi R}\frac{\partial \Psi(R,Z)}{\partial Z} \tag{2.25}$$

Hence, the poloidal magnetic fied can be written as

$$\mathbf{B}_p = B_R \nabla R + B_Z \nabla Z = \frac{1}{2\pi R}\left(-\frac{\partial \Psi}{\partial Z}\nabla R + \frac{\partial \Psi}{\partial R}\nabla Z\right) = \frac{1}{2\pi}\nabla \Psi \times \nabla \phi, \tag{2.26}$$

with $R\,\nabla\phi = \nabla Z \times \nabla R$. Using the integrated form of Ampère's law in Eq. (2.22), we can express B_ϕ by the poloidal current I_{pol}: Integrating along the intersection curve of a z = const-plane (Figure 2.5) with the Ψ = const. flux surface, we obtain

$$\mu_0 I_{\text{pol}} = 2\pi R B_\phi \rightarrow B_\phi = \frac{\mu_0 I_{\text{pol}}}{2\pi R} \tag{2.27}$$

and

$$\mathbf{B}_t = \frac{1}{2\pi}\mu_0 I_{\text{pol}} \nabla \phi \tag{2.28}$$

Note that for a vacuum toroidal field, the value of I_{pol} will be constant in between the legs of the toroidal field coil and zero outside, so that the externally generated B_ϕ is exactly proportional to $1/R$. The total magnetic field can now be represented by the poloidal fluxes Ψ of the magnetic field $\mathbf{B}$ and I_{pol} of the current density $\mathbf{j}$

through

$$\mathbf{B} = \mathbf{B_p} + \mathbf{B_t} = \frac{1}{2\pi}(\nabla\Psi \times \nabla\phi + \mu_0 I_{\text{pol}}\ \nabla\phi) \tag{2.29}$$

The corresponding current density $\mathbf{j}$ is given through Ampères law by

$$\mathbf{j} = \frac{1}{\mu_0}\nabla \times \mathbf{B} = \frac{1}{2\pi\mu_0}(\mu_0 \nabla I_{\text{pol}} \times \nabla\phi - \Delta^*\Psi\ \nabla\phi) \tag{2.30}$$

where

$$\Delta^*\Psi \equiv R^2\ \boldsymbol{\nabla} \cdot \frac{\nabla\Psi}{R^2} \tag{2.31}$$

Δ^* is the Stokes operator[6] acting on Ψ which reads in cylinder coordinates (R, ϕ, z)

$$\Delta^* = R\frac{\partial}{\partial R}\left(\frac{1}{R}\frac{\partial}{\partial R}\right) + \frac{\partial^2}{\partial Z^2} \tag{2.32}$$

Inserting current density $\mathbf{j}$ and magnetic field $\mathbf{B}$ in the basic force balance equation (Eq. (1.34)) leads (denoting derivatives with respect to Ψ by a prime) to

$$\Delta^*\Psi = -\mu_0 2\pi R j_\phi = -\mu_0(2\pi R)^2 p' - \mu_0^2 I'_{\text{pol}} I_{\text{pol}} \tag{2.33}$$

This equation is known as the *Grad–Shafranov equation.*

It is the analogue to the equilibrium equation in the screw pinch and contains, like the screw pinch, the contributions from both toroidal current (left-hand side) and diamagnetism (second term on the right-hand side). Again, for a pure vacuum B_ϕ, we have no contribution to the pressure balance as I_{pol} = const. in this case. It can be shown that the ratio of p' and I^2_{tor} is connected to the local β_{pol}. For making this connection more transparent, we introduce the volume V enclosed by the flux surface under consideration and, for any scalar quantity Q, a magnetic surface average by the following settings

$$V = \int_{V(x)\le V} d^3x \quad \text{and} \quad \langle Q\rangle_{MS} = \frac{d}{dV}\int_{V(x')<V} Q(x')d^3x' \tag{2.34}$$

From Eq. (2.33), we get for $j_t = j_\phi$

$$\frac{j_t}{2\pi R} = \frac{\mu_0 I_{\text{pol}}}{4\pi^2 R^2}\frac{dI_{\text{pol}}}{d\Psi} + \frac{dp}{d\Psi} \tag{2.35}$$

and after performing the flux surface averages

$$\left\langle \frac{j_t}{2\pi R}\right\rangle_{MS} = \frac{\mu_0 I_{\text{pol}}}{4\pi^2}\left\langle\frac{1}{R^2}\right\rangle_{MS}\frac{dI_{\text{pol}}}{d\Psi} + \frac{dp}{d\Psi} \tag{2.36}$$

From the definitions of toroidal flux Φ and toroidal current I_{tor}, the following relations hold:

$$\frac{d\Phi}{dV} = \frac{1}{2\pi}\langle\mathbf{B}\cdot\nabla\phi\rangle_{MS} = \frac{\mu_0 I_{\text{pol}}}{4\pi^2}\left\langle\frac{1}{R^2}\right\rangle_{MS} \tag{2.37}$$

$$\frac{dI_{\text{tor}}}{dV} = \frac{1}{2\pi}\langle\mathbf{j}\cdot\nabla\phi\rangle_{MS} = \left\langle\frac{j_t}{2\pi R}\right\rangle_{MS} \tag{2.38}$$

6) Δ^* was introduced by Stokes in hydrodynamics in connection with axisymmetric flows.

Inserting this into Eq. (2.36), we obtain the following equilibrium relation:

$$\frac{dI_{\text{tor}}}{dV} = \frac{d\Phi}{dV}\frac{dI_{\text{pol}}}{d\Psi} + \frac{dp}{d\Psi} \tag{2.39}$$

The safety factor is given by

$$q = -\frac{d\Phi}{d\Psi} \tag{2.40}$$

With the definition of the local β value according to

$$\beta_{\text{p(local)}} \equiv \frac{dp}{d\Psi}\frac{dV}{dI_{\text{tor}}} \tag{2.41}$$

we finally obtain

$$\beta_{\text{p(local)}} = 1 + q\frac{dI_{\text{pol}}}{dI_{\text{tor}}} \tag{2.42}$$

Hence, if $I_{\text{pol}} = \text{const.}$, it follows that $\beta_{\text{p(local)}} = 1$, equivalent to the relation derived for the screw pinch.

The Grad–Shafranov equation is a nonlinear elliptic PDE as the right-hand side can depend on Ψ in an arbitrary way through $p'(\Psi)$ and $I_{\text{pol}}(\Psi)$. While this makes it difficult to find analytical solutions, except for very special choices of the right-hand side[7], the structure allows a relatively straightforward iterative numerical solution. A practical method to solve the Grad–Shafranov equation is to prescribe the functions $p'(\Psi)$ and $I_{\text{pol}}(\Psi)$, for example as polynomials, as well as an initial guess of the flux surfaces $\Psi(R, Z)$. By doing so, the right-hand side of Eq. (2.33) becomes a known function of R and Z. The nth iteration step from a pre-scribed $\Psi_n(R, Z)$ to $\Psi_{n+1}(R, Z)$ consists then of solving the linear inhomogeneous PDE

$$\Delta^*\Psi_{n+1} = -\mu_0(2\pi R)^2 p'(\Psi_n) - \mu_0^2 I'_{\text{pol}}(\Psi_n) I_{\text{pol}}(\Psi_n) \tag{2.43}$$

This is a Poisson problem that can be solved for example by using a Green's function method[8]. It is important to note that in order to solve this equation on an (R, Z) grid, we have to specify boundary conditions. These can be for example the shape of the outermost flux surface, defining the boundary of the Poisson problem or the position of the minimum of Ψ, which corresponds to the magnetic axis of the plasma. The solution of the Poisson problem requires then to find a particular solution to the inhomogeneous problem, whereas the boundary conditions can be satisfied by adding solutions of the homogeneous problem $\Delta^*\Psi = 0$ as they do not change the right-hand side. Such solutions correspond to vacuum poloidal fields and can hence be generated by external current loops in toroidal direction, that is the poloidal field coils mentioned earlier. For example, by adding a pair of Helmholtz coils such as the blue coils shown in Figure 2.6, a vertical field B_Z is added that can control the radial position R_0 of the magnetic axis. In the following

7) A well-known representative of such analytically tractable solutions is the so-called Solov'ev equilibrium [5].
8) In praxi, the use of Green's second theorem provides an efficient method to reduce the calculation to a finite domain while still satisfying the boundary conditions at infinity [6].

sections, we take a closer look how the poloidal field coils provide position control and shaping of the poloidal cross section of the plasma.

2.2.1.2 Circular Cross Section

As outlined in Section 2.2.1.1, shape and position of the plasma column can be influenced by additional fields generated by the poloidal field coils. For a circular cross section, it is possible to derive an approximate analytical solution of the Grad–Shafranov equation that gives some insight into the way the force balance is established.

For this calculation, we will use the toroidal coordinate system r, θ, ϕ shown in Figure 2.7 that consists of nested tori with circular cross section centred at major radius R_0. The angles θ and ϕ are the poloidal and toroidal angles, respectively. We prescribe the last flux surface of the plasma to be a circle centred at R_0 with minor radius $r = a$. While the pressure is a flux function and hence constant on the flux surface, the surface of the inner half of the torus is smaller than that of the outer half, leading to a net force expanding the torus. Similarly, the magnetic force due to the plasma current in the torus tends to expand the ring and this combined 'hoop force' has to be balanced by the magnetic force. However, the poloidal magnetic field B_θ on this circle created by the plasma current alone will be lower on the torus outside than on the inside so that we expect that the solution of the Grad–Shafranov equation will exhibit a strengthening of B_θ on the outside and a weakening on the inside, which means that the centres of the circles inside the last surface will be shifted outward with respect to R_0. This effectively means that a vertical field has to be added by the external poloidal field coils.

According to this expectation, we make the following Ansatz for Ψ:

$$\Psi = \Psi_0 + \Psi_1 = \Psi_0(r) + \Psi_1(r)\cos\theta = \Psi_0(r) - \frac{d\Psi_0}{dr}\Delta(r)\cos\theta \tag{2.44}$$

that is Ψ consists of a component Ψ_0 that represents circles around R_0 and a shift $\Delta(r)$ that is assumed to be a first-order perturbation. Anticipating the outward

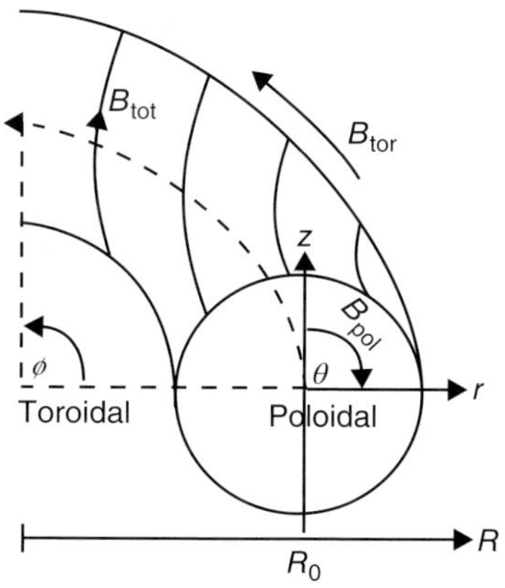

Figure 2.7 Definition of torus coordinates r, θ, ϕ around a point $R_0, z_0 = 0$ used in the derivation of equilibria with circular cross section.

shift of the inner flux surfaces, we have chosen the negative sign in our Ansatz such that Δ will be a positive quantity. According to our prescription of the last flux surface being a circle centreed at R_0, we know already that $\Delta(a) = 0$.

Now, the flux functions $p'(\Psi)$ and $I_{pol}(\Psi)$ are expanded in Ψ:

$$p'(\Psi) = p'(\Psi_0) + p''(\Psi_0)\Psi_1 \tag{2.45}$$

$$I_{pol}I'_{pol} = I_{pol}I'_{pol}(\Psi_0) + (I_{pol}I'_{pol})'\Psi_1 \tag{2.46}$$

The Grad–Shafranov equation in toroidal coordinates reads

$$\left(\frac{1}{r}\frac{\partial}{\partial r}r\frac{\partial}{\partial r} + \frac{1}{r^2}\frac{\partial^2}{\partial\theta^2}\right)\Psi - \frac{1}{R_0 + r\cos\theta}\left(\cos\theta\frac{\partial}{\partial r} - \sin\theta\frac{1}{r}\frac{\partial}{\partial\theta}\right)\Psi$$
$$= -\mu_0(2\pi(R_0 + r\cos\theta))^2 p'(\Psi) - \mu_0^2 I_{pol}(\Psi)I'_{pol}(\Psi) \tag{2.47}$$

Expanding this equation in inverse aspect ratio, $r/R_0 \ll 1$, the first term on the left-hand side is of zeroth order while the second is of first order. Similarly, on the right-hand side, $(R_0 + r\cos\theta)^2$ is expanded and we obtain a zeroth-order force balance:

$$\frac{1}{r}\frac{d}{dr}r\frac{d\Psi_0}{dr} = -\mu_0(2\pi R_0)^2 p'(\Psi) - \mu_0^2 I_{pol}(\Psi)I'_{pol}(\Psi) \tag{2.48}$$

and an equation for the first order

$$\left(\frac{1}{r}\frac{\partial}{\partial r}r\frac{\partial}{\partial r} + \frac{1}{r^2}\frac{\partial^2}{\partial\theta^2}\right)\Psi_1 - \frac{\cos\theta}{R_0}\frac{d\Psi_0}{dr}$$
$$= -\frac{d}{dr}(\mu_0(2\pi R_0)^2 p'(\Psi) - \mu_0^2 I_{pol}(\Psi)I'_{pol}(\Psi))\frac{dr}{d\Psi_0}\Psi_1 - 2(2\pi)^2\mu_0 r R_0 \frac{dp}{dr}\frac{dr}{d\Psi_0}\cos\theta \tag{2.49}$$

where we have used $d/d\Psi_0 = dr/d\Psi_0\, d/dr$. Now, we can replace Ψ_1 according to the Ansatz (Eq. 2.44) in the first-order equation. The factor $\cos\theta$ cancels everywhere, and after some algebra, where we also make use of the zeroth-order force balance, we arrive at

$$\frac{d}{dr}\left(rB_{\theta,0}^2\frac{d\Delta}{dr}\right) = \frac{r}{R_0}\left(2\mu_0 r\frac{dp}{dr} - B_{\theta,0}^2\right) \tag{2.50}$$

Here, we have used $B_{\theta,0} \approx 1/(2\pi R_0)d\Psi_0/dr$. Integration of this equation gives

$$\frac{d\Delta}{dr} = -\frac{r}{R_0}\left(\hat{\beta}_p(r) + \frac{1}{2}\hat{\ell}_i(r)\right) \tag{2.51}$$

where we have partially integrated the pressure term and defined a 'local' poloidal β

$$\hat{\beta}_p(r) = \frac{2\mu_0}{B_{\theta,0}^2(r)}\left(\frac{2}{r^2}\int_{r'=0}^{r} r'dr'p(r') - p(r)\right) = \frac{2\mu_0(\langle p\rangle - p(r))}{B_{\theta,0}^2(r)} \tag{2.52}$$

where $\langle\ \rangle$ represents the average over the poloidal cross section. For $r = a$, this definition coincides with that of β_p in the circular screw pinch (Eq. 2.6). Furthermore, the 'local' internal inductance is defined by

$$\hat{\ell}_i(r) = \frac{1}{B_{\theta,0}^2(r)}\frac{2}{r^2}\int_{r'=0}^{r} r'dr'B_{\theta,0}^2(r') = \frac{\langle B_{\theta,0}^2\rangle}{B_{\theta,0}^2(r)} \tag{2.53}$$

The latter is motivated by the definition of the internal inductance L_i through the relation between current and resulting field energy inside the plasma volume

$$\frac{1}{2}L_i I_p^2 = \int_{\text{plasma}} \frac{B_p^2}{2\mu_0} dV \tag{2.54}$$

such that the dimensionless internal inductance is

$$\ell_i = \frac{2L_i}{\mu_0 R_0} \tag{2.55}$$

which, for circular cross section, coincides with the definition (2.53), taken at $r = a$. We note here that a flat current profile circular cross-sectional tokamak has $\ell_i = 0.5$ while increasing peaking of the current profile leads to an increase in ℓ_i.

Equation (2.51) is an equation for the so-called Shafranov-shift that accounts for the above-mentioned need to strengthen the poloidal field on the outside of the torus in order to obtain force balance. To evaluate the Shafranov shift inside the plasma, the radial distributions of current and pressure must be known for integrating Eq. (2.51). The poloidal field B_θ can then be evaluated by using the first-order expansion of Ψ (Eq. 2.44)

$$\begin{aligned} B_\theta(r,\theta) &= \frac{1}{2\pi(R_0 + r\cos\theta)}\frac{d\Psi}{dr} \\ &\approx \frac{1}{2\pi R_0}\left(\left(1 - \frac{r}{R_0}\cos\theta\right)\frac{d\Psi_0}{dr} - \left(\frac{d\Psi_0}{dr}\frac{d\Delta}{dr} + \frac{d^2\Psi_0}{dr^2}\Delta\right)\cos\theta\right) \end{aligned} \tag{2.56}$$

For $r = a$, we know that $\Delta = 0$ so that the last term vanishes and we can evaluate the poloidal field on the last surface:

$$B_\theta(a,\theta) = \frac{\mu_0 I_p}{2\pi a}\left(1 + \frac{a}{R_0}\left(\beta_p + \frac{\ell_i}{2} - 1\right)\cos\theta\right) \tag{2.57}$$

where we have used the straight cylinder result for $d\Psi_0/dr$.

From this expression, it is clear that indeed, the expansion force due to pressure (β_p) and current (ℓ_i) acts to increase the poloidal field on the low-field side and weaken it on the high-field side, opposing the geometrical effect (represented by the term $-a/R_0 \cos\theta$). Such a field component can be represented by a vertical field that can be produced with a Helmholtz-like coil arrangement as indicated by the poloidal field coils in Figure 2.6. In order to calculate the necessary field strength, we have to obtain a solution for B_θ in the vacuum region and match it to the plasma solution. Following the procedure earlier, we assume that also for the vacuum field, we can write

$$\Psi_{\text{vac}} = \Psi_{\text{vac},0}(r) + \Psi_{\text{vac},1}(r)\cos\theta \tag{2.58}$$

The zeroth order can be found from solving Eq. (2.48) with the right-hand side set to zero, which gives

$$r\frac{d\Psi_{\text{vac},0}}{dr} = \text{const.} \rightarrow \Psi_{\text{vac},0} = -\mu_0 I_p R_0\left(\ln\frac{8R_0}{r} - 2\right) \tag{2.59}$$

where the constant has been chosen such that the flux is equal to $L_{ext}I_p$, where L_{ext} is the external inductance of the plasma column in large aspect ratio expansion.

Using this result, we obtain the first-order equation by inserting $\Psi_{vac,0}$ into Eq. (2.49) and setting the right-hand side to zero:

$$\left(\frac{1}{r}\frac{d}{dr}r\frac{d}{dr} - \frac{1}{r^2}\right)\Psi_{vac,1} = \mu_0 I_p \frac{1}{r} \tag{2.60}$$

which is an inhomogeneous differential equation for $\Psi_{vac,1}$. Using a polynomial Ansatz for the homogeneous part, $\Psi_{vac,1} \propto r^\alpha$ yields $\alpha^2 = 1$, that is

$$\Psi_{vac,1}^{hom} = \frac{c_1}{r} + c_2 r \tag{2.61}$$

The solution to the inhomogeneous differential equation can be obtained, for example by variation of constant, as $-(\mu_0 I_p/2)\, r(\ln \frac{8R_0}{r} - 1)$ so that the full solution reads

$$\begin{aligned}\Psi_{vac} &= -\mu_0 I_p R_0 \left(\ln \frac{8R_0}{r} - 2\right) \\ &\quad - \frac{\mu_0 I_p}{2}\left(r\left(\ln \frac{8R_0}{r} - 1\right) + \frac{c_1}{r} + c_2 r\right)\cos\theta \end{aligned} \tag{2.62}$$

Matching this to the plasma solution (2.57) gives the two conditions, $B_{\theta,plasma}(a,\theta) = B_{\theta,vac}(a,\theta)$ and $B_r(a,\theta) = 0$. The latter means that the circle at $r = a$ is a flux surface, that is $\Psi_{vac,1}(a) = 0$. Evauating the two conditions, the constants are found to be

$$c_1 = a^2\left(\beta_p + \frac{\ell_i}{2} - \frac{1}{2}\right) \tag{2.63}$$

$$c_2 = -\left(\beta_p + \frac{\ell_i}{2} - \frac{3}{2} + \ln \frac{8R_0}{a}\right) \tag{2.64}$$

Inserting these into Eq. (2.62) gives the flux function for $r \geq a$. In order to determine the vertical field necessary for equilibrium, we note that far away from the plasma, the term $c_2 r \cos\theta$ dominates the solution[9] and the resulting magnetic field

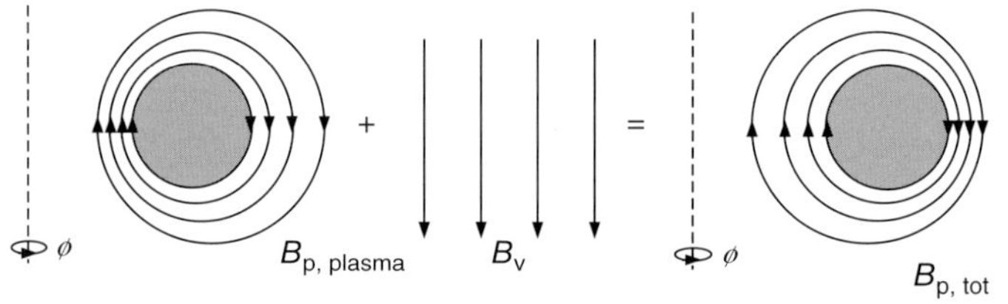

Figure 2.8 Addition of a vertical field B_v to the field $B_{p,plasma}$ generated by the plasma current alone provides force balance and leads to the Shafranov shift as can be seen from the resulting field $B_{p,tot}$.

9) The term $\ln 8R_0/r - 1$ is a large aspect ratio expansion of a function that goes to zero for $r \to \infty$.

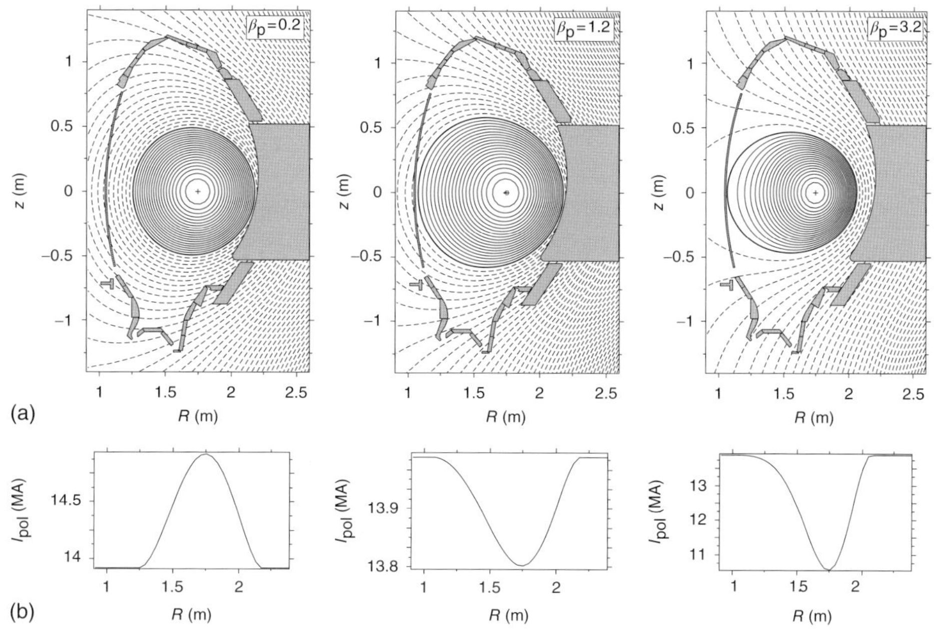

Figure 2.9 (a) Circular equilibria with increasing β_p, demonstrating the effect of increasing Shafranov shift. (b) The radial profiles of the poloidal current (note the suppressed zero that represents the current in the toroidal field coils) are shown. The change from paramagnetic to diamagnetic behaviour as β_p exceeds 1 can clearly be seen.

is a constant vertical field of magnitude

$$B_v = \frac{\mu_0 I_p}{4\pi R_0}\left(\beta_p + \frac{1}{2}\ell_i - \frac{3}{2} + \ln\left(\frac{8R_0}{a}\right)\right) \tag{2.65}$$

This is the vertical field needed for radial force balance in the large aspect ratio expansion with circular cross section. Figure 2.8 shows how the addition of the field produced by the plasma current and the vertical field results in the equilibrium configuration.

The addition of the vertical field naturally leads to a limit to β_p that can be maintained in force equilibrium with purely vertical field. This is shown in Figure 2.9 that shows a sequence of equilibria with circular cross section and increasing β_p. Here, the position of the plasma centre has been prescribed. From Figure 2.9, one can see that with increasing β_p, the magnetic field on the inner side is substantially weakened by the vertical field. For a purely vertical field, this leads to the appearance of a magnetic null point on the inner side, which limits the achievable β_p; a further increase of β_p will lead to rapid loss of plasma volume due to the inward moving separatrix. From Eq. (2.51), this so-called equilibrium limit can be estimated by setting $d\Delta/dr$ equal 1, that is the Shafranov shift approaches the minor radius a, and assuming $\beta_p \gg \ell_i/2$. This yields

$$\beta_{p,\max} \approx \frac{R_0}{a} \tag{2.66}$$

for the equilibrium limit. For usual values of the tokamak aspect ratio around 3 and $q(a) \approx 3$, this limit is usually not an issue as stability against ideal MHD modes will set a more strict limitation (see Section 7.5). However, for larger values of $q(a)$ or a/R_0, the equilibrium limit may play a role.

2.2.1.3 Arbitrary Cross Section

In modern tokamaks, additional poloidal field coils are used to shape the poloidal cross section in a way that it deviates from a circle. A prominent example is the generation of a so-called poloidal divertor configuration where a null point in the poloidal field is created by an additional current parallel to the plasma current in a poloidal field coil located above or below the vacuum vessel[10]. In Figure 2.10, such a configuration is shown on the right side in comparison to a circular equilibrium on the left side. For the circular equilibrium, the last closed flux surface is defined by contact of a circular flux surface with the wall. This part of the wall, usually called *limiter*, must be designed to accommodate the corresponding heat and particle fluxes and any wall material eroded there will end up in the plasma with a high probability. On the other hand, the last closed flux surface in the divertor configuration is defined by the separatrix, that is the flux surface on which the poloidal field null point, the so-called X-point, lies. Note that this surface is not in contact with the surrounding wall. The flux surfaces outside the separatrix, the so-called scrape-off-layer, are in contact with the wall, but the contact point can now be chosen such that it is at a safe distance to the plasma. Hence, divertor plasmas are less affected by impurity influx and exhaust of power and particles is usually done in such a configuration.

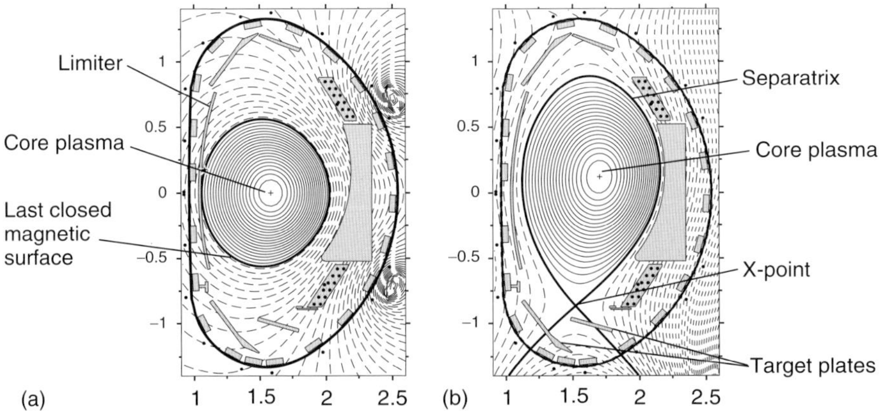

Figure 2.10 Limiter (a) and divertor (b) configuration produced by additional currents in poloidal field coils. The plasma cross section on the right is also elongated and triangular.

10) The resulting field is a quadrupole field; there are also configurations that apply a hexapole (so-called snowflake divertor)

In Figure 2.10, one can also see that the cross section of the plasma is elongated and triangular. These higher moments of plasma shape are introduced as they provide favourable stability properties, as will be pointed out for example in Section 5.2.1. It is clear that such shaped plasmas can no longer be described by a single radial coordinate, and hence, a generalized flux surface label is needed. A common definition is to normalize the poloidal flux function such that it is 0 in the centre and 1 at the separatrix and then use the square root of the flux as the flux itself varies roughly quadratically with increasing radius[11]:

$$\rho_p = \sqrt{\frac{\Psi - \Psi_{\text{axis}}}{\Psi_{\text{sep}} - \Psi_{\text{axis}}}} \tag{2.67}$$

where Ψ_{axis} and Ψ_{sep} are the values of Ψ at the magnetic axis and the separatrix, respectively. In addition, the definition of the safety factor given for large aspect ratio and circular flux surfaces (Eq. 2.11) can be generalized using the toroidal and poloidal fluxes to be given by Eq. (2.40). As the X-point of an equilibrium with a separatrix is a point where the poloidal field is zero, the field line approaches this point asymptotically and never crosses it (so-called stagnation point). Then, q diverges and hence is not a useful quantity close to the separatrix. For this reason, often the value of q at the flux surface containing 95% of the total poloidal flux inside the separatrix is used and called q_{95}.

The general definition of the poloidal β is

$$\beta_p = \frac{2\mu_0 \langle p \rangle_{Vol}}{\langle B_p^2 \rangle_{LCFS}} \tag{2.68}$$

For the internal inductance, there exist several generalizations of Eq. (2.54), with

$$\ell_i(3) = 2\langle B_p^2 \rangle V / (\mu_0^2 R_0 I_p^2) \tag{2.69}$$

being a quite common one.

Finally, the shape of the plasma is generally described by quantities related to the deviation from circular cross section. Commonly used definitions of the elongation κ and the triangularity δ of a flux surface are based on the positions $R_{\text{min}}, R_{\text{max}}, Z_{\text{min}}$ *and* Z_{max}, where R and z take on their minimum and maximum values on that flux surface:

$$\kappa = \frac{Z_{\text{max}} - Z_{\text{min}}}{R_{\text{max}} - R_{\text{min}}} \tag{2.70}$$

$$\delta_u = \frac{R_{\text{max}} + R_{\text{min}} - 2R(Z_{\text{max}})}{R_{\text{max}} - R_{\text{min}}} \tag{2.71}$$

$$\delta_l = \frac{R_{\text{max}} + R_{\text{min}} - 2R(Z_{\text{min}})}{R_{\text{max}} - R_{\text{min}}} \tag{2.72}$$

11) The linearity between square root of flux and radius is even better fulfilled for the toroidal flux, but the definition of the toroidal flux for cases with a separatrix is ambiguous outside the last closed flux surface and hence not practical.

where $R(Z_{\max}), R(Z_{\min})$ are the R-coordinates of the extrema of z and we have distinguished between upper and lower triangularity, considering the possibility of up-down asymmetric shapes.

We note here that the elongation of the plasma introduces a need for vertical position control as the equilibrium position between two coils on top and bottom that 'pull' on the plasma column is unstable to any small displacement in vertical direction. While active feedback is applied in present-day tokamaks to provide a stable equilibrium position, a loss of this control can lead to a so-called vertical displacement event (VDE) that in turn can lead to excessive heat loads and forces on the tokamak structure. This positional instability is discussed in detail in Section 4.4.

2.2.1.4 The Straight Field Line Angle

As discussed earlier, toroidicity and shaping have an important effect on the variation of B_p and B_t on the equilibrium flux surface. In a screw pinch with circular cross section, both are constant, and hence, the field line pitch angle is also constant along the field line. In a general toroidal equilibrium, there is a variation of the pitch along the field line. This is important for the description of MHD modes, which, as will become clear later in this book, often have constant phase along the field lines and can in general no longer be described by a single Fourier harmonic in poloidal angle once toroidicity is considered. In this section, we discuss this variation of pitch angle and provide a formalism how a general equilibrium can be mapped to a screw pinch like geometry such that the pitch angle is constant again. As this corresponds to 'straight' field lines, the corresponding angle θ^* is called the *straight field line angle.*

For the large aspect ratio, circular cross-sectional equilibrium treated earlier, B_t will vary in toroidal coordinates according to

$$B_\phi \approx \frac{B_{\phi,0}R_0}{R} = \frac{B_{\phi,0}}{1 + \frac{r}{R_0}\cos\theta} \tag{2.73}$$

and B_θ is given by Eq. (2.56), neglecting the term $d^2\Psi_0/dr^2\Delta(r)$, which is a good approximation at least in the edge, so that the equation of the field line (Eq. 2.11) becomes

$$\frac{d\phi}{d\theta} = \frac{rB_\phi}{RB_\theta} \approx \frac{rB_{\phi,0}}{R_0B_{\theta,0}}\frac{1}{(1+\frac{r}{R_0}\cos\theta)^2}\frac{1}{1+\frac{r}{R_0}(\hat{\beta}_p + \frac{\hat{\ell}_i}{2} - 1)\cos\theta} \tag{2.74}$$

Recalling that the constant part of the right-hand side was equal to q in the cylindrical case and expanding the denominator of Eq. 2.74 to first order in r/R_0, we can define the straight field line angle θ^* such that $d\phi/d\theta^* = \text{const.} = q$, which gives

$$d\theta^* \approx \frac{d\theta}{1+\frac{r}{R_0}\left(\hat{\beta}_p + \frac{\hat{\ell}_i}{2} + 1\right)\cos\theta} \approx d\theta\left(1 - \frac{r}{R_0}\left(\hat{\beta}_p + \frac{\hat{\ell}_i}{2} + 1\right)\cos\theta\right) \tag{2.75}$$

which can be integrated to yield the final expression

$$\theta^* = \theta - \frac{r}{R_0}\left(\hat{\beta}_p + \frac{\hat{\ell}_i}{2} + 1\right)\sin\theta \tag{2.76}$$

This so-called Merezhkin correction has to be considered when following the phase of MHD modes around the poloidal circumference of a toroidal plasma, for example when analysing mode structures with magnetic pick-up coils. Figure 2.11 shows, on the left-hand side, a so-called Poincaré plot of a poloidal cross section, where the intersection of the field line with the poloidal cross section has been plotted advancing the toroidal angle in equidistant steps. It is clear that the poloidal angle $\Delta\theta$ is not constant, varying according to Eq. (2.76).

The derivation of the straight field line angle can be generalized for shaped plasmas. Let us assume that the flux surfaces are parameterized in the poloidal plane R, Z through toroidal coordinates r and θ. Then, the condition for the straight field line angle reads

$$\frac{\mathbf{B}\cdot\nabla\theta^*}{\mathbf{B}\cdot\nabla\phi} = \text{const} \tag{2.77}$$

We are looking for a parameterization of θ^* in θ, so that we replace $\nabla\theta^* \rightarrow (d\theta^*/d\theta)\nabla\theta$. Using the representation of $\mathbf{B}$ given by Eq. (2.29), we arrive at

$$\frac{d\theta^*}{d\theta} = \text{const.}\frac{\mu_0 I_{\text{pol}}(\nabla\phi)^2}{\nabla\Psi\times\nabla\phi\cdot\nabla\theta} \tag{2.78}$$

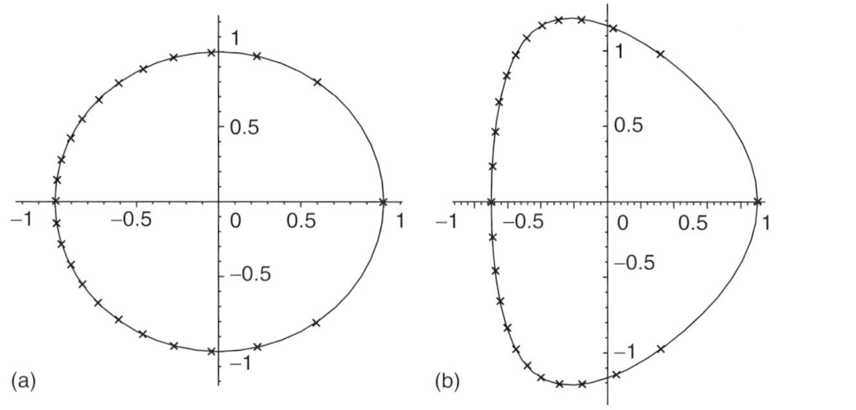

Figure 2.11 Poincaré plot of field lines in a tokamak configuration with circular cross section (a) and shaped flux surfaces (b). The crosses mark the poloidal position of a field line when advancing by equidistant steps in toroidal angle. As a result of toroidicity and shaping, the resulting steps in poloidal angle are not equidistant ($R_0/a = 3, \beta_p + \ell_i/2 + 1 = 2.5, S_2 = 0.2, S_3 = 0.1$ and the variation of S_2 and S_3 with radius has been assumed to be quadratic and cubic, respectively).

The expression $\nabla\Psi \times \nabla\phi \cdot \nabla\theta$ is the Jacobian of the transformation of torus coordinates Ψ, θ, ϕ to the cylindrical system R, ϕ, Z which, considering the symmetry along ϕ, can be written as

$$J = \frac{1}{R}\left(\frac{\partial\theta}{\partial R}\frac{\partial\Psi}{\partial Z} - \frac{\partial\Psi}{\partial R}\frac{\partial\theta}{\partial Z}\right) = \frac{1}{R}J_{R,Z} \tag{2.79}$$

and $J_{R,Z}$ is the Jacobian of the two-dimensional transformation from Ψ, θ to R, Z (remember that both coordinate systems share the ignorable coordinate ϕ). If we specify the flux surfaces in the R, Z-plane as a function of Ψ, θ, we need the inverse Jacobian $J_{\Psi,\theta}$ which is given by the inverse of the inverted function so that, considering the fact that I_{pol} is constant on a flux surface and $|\nabla\phi| = 1/R$, we finally arrive at

$$\theta^* = 2\pi\frac{\int_0^\theta d\theta' J_{\Psi,\theta'}(\theta')/R}{\int_0^{2\pi} d\theta' J_{\Psi,\theta'}(\theta')/R} \tag{2.80}$$

We have normalized the straight field line angle such that integrating θ^* around the poloidal circumference yields 2π, which also eliminates the constant.

As an example, we analyse an up-down symmetric parameterization according to

$$R(r,\theta) = R_0 + r\cos\theta + \Delta(r)\cos\theta - S_2(r)\cos\theta + S_3(r)\cos 2\theta \tag{2.81}$$

$$Z(r,\theta) = r\sin\theta + S_2(r)\sin\theta - S_3(r)\sin 2\theta \tag{2.82}$$

where the shape parameters S_2 and S_3 can be related to elongation and triangularity according to Eqs. (2.70) and (2.70):

$$\kappa = \frac{1 + \frac{S_2}{r}}{1 - \frac{S_2}{r}} \tag{2.83}$$

$$\delta = 4\frac{S_3}{r} \tag{2.84}$$

Here, r is a flux surface label that is related to Ψ through the shape of the current profile by $d\Psi/dr = B_\theta$. Hence, when evaluating the Jacobian in r, θ, there will be an additional factor $d\Psi/dr$. However, as this factor does not depend on θ (both Ψ and r are constant on flux surfaces), this factor will cancel in the calculation of the straight field line angle. This means that we have to evaluate the Jacobian

$$J_{r,\theta} = \frac{1}{r}\left(\frac{\partial R}{\partial r}\frac{\partial Z}{\partial\theta} - \frac{\partial r}{\partial\theta}\frac{\partial Z}{\partial r}\right) \tag{2.85}$$

for the parameterization given by Eqs (2.81) and (2.82). Treating R_0 as zeroth-order quantity, r as first-order quantity and Δ, S_2 and S_3 as second-order quantities, we can evaluate the Jacobian up to second order and integrate it according to Eq. (2.80) to obtain

$$\theta^* = \theta - \frac{r}{R_0}\left(\hat{\beta}_p + \frac{\hat{\ell}_i}{2} + 1\right)\sin\theta - \frac{r\kappa'}{(\kappa+1)^2}\sin 2\theta + \frac{1}{12}(r\delta' - \delta)\sin 3\theta \tag{2.86}$$

where the $\sin\theta$ correction reproduces the result derived above for circular cross section, and we see that elongation and trianguarity introduce corrections corresponding to $\sin 2\theta$ and $\sin 3\theta$, respectively. A corresponding flux surface and the Poincare plot of the field lines are shown in Figure 2.11b. We conclude that toroidicity and shaping introduce a variation of the field line pitch with poloidal angle. This correction is important for the coupling of MHD instabilities on different flux surfaces and we will come back to it in Section 7.2.

2.2.2
The Stellarator

In a stellarator (lat. stella = star), the poloidal field component is also created by external coils so that the plasma does not necessarily have to carry a net toroidal current. In this situation, a solution with a unidirectional poloidal field as in the tokamak is not possible since $\int \mathbf{B}\cdot d\mathbf{s} = \mu_0 I_p = 0$. However, if the poloidal field changes sign on the poloidal circumference, for example by varying as $\sin(\ell\theta)$, the condition can be fulfilled. Such a field can be generated using 2ℓ poloidal field coils with alternating current direction to approximately create the appropriate multipole moment. In order to generate helically winding field lines, it is necessary to wind these coils helically around the torus such that the external coil currents are approximately distributed according to

$$I(\theta) = I_0 \sin(\ell\theta - M\phi) \tag{2.87}$$

where M denotes the toroidal periodicity of the field. Clearly, the explicit ϕ-dependence shows that this configuration is no longer axisymmetric as is the case in the tokamak. Figure 2.12 shows the corresponding field lines resulting from such a current distribution in the $\theta - \phi$ plane for $\ell = 2, M = 4$.

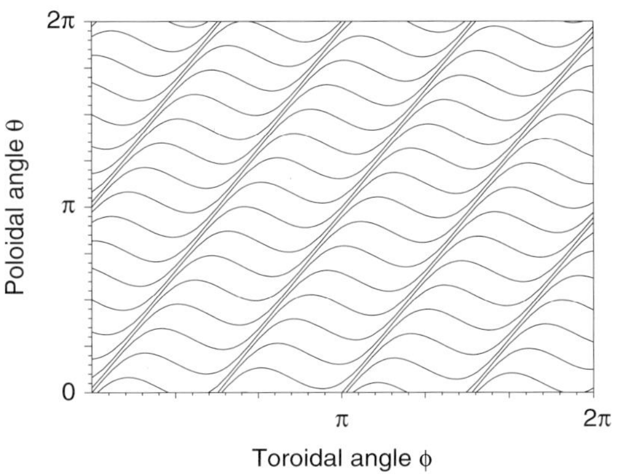

Figure 2.12 Map of field lines generated by helical windings according to Eq. (2.87) plotted in the ϕ, θ-plane ($\ell = 2, M = 4$).

The configuration of a 'classical' stellarator consists of an axisymmetric toroidal field and helical windings as discussed earlier. Depending on ℓ, the q-profiles[12] can vary appreciably. For example, the shear in an $\ell = 2$ stellarator is much smaller than that in a tokamak, whereas an $\ell = 3$ stellarator has a shear comparable to that of a tokamak, but the q-profile is reversed, that is q decreases from the centre to the edge. In Figure 2.13a, the coil configuration of a classical $\ell = 2$ stellarator is shown, consisting of a set of planar toroidal field coils and helical windings. Also shown in Figure 2.13, in the lower part, is a procedure by which this set of coils can be replaced by a set of modular coils that produces the same field but is not topologically interlinked, which is the case in the classical stellarator because of the helical winding. Such a set of modular coils offers significant advantages in terms of coil fabrication and maintenance, especially for large devices.

As can be seen from the example of the configuration of the W7-X stellarator shown in Figure 2.13b, the configurational space of stellarators offers a much wider variety than just the classical stellarator, and hence, the question of the optimum stellarator configuration for use as a future fusion reactor is not straightforward to answer. This is the goal of stellarator optimization, which optimizes the 3D plasma geometry according to a given set of criteria. These criteria can both be of physical nature, for example minimizing radial transport of heat and particles or maximizing the achievable β as well as of technological nature, such as minimizing stresses in coils and components or easing maintainability. The overall cost of the machine is of course another important criterion. For example, the W7-X stellarator shown in Figure 2.13b has been optimized with respect to neoclassical transport of both thermal and fast particles as well as to achieve high β.

While a stellarator configuration is no longer axisymmetric, symmetries do play an important role in its optimization. A problem for the classical stellarator is the inability to confine the fusion-born fast α-particles for a long enough time: while one can show that for tokamaks, this is always possible for high enough plasma current, the classical stellarator with its 3D field structure exhibits a class of particle orbits that, for high enough energy, that is low enough collisionality, will be lost from the device in very short time. For α-particles, this time would be much shorter than the slowing-down time, meaning that they are lost for the plasma self-heating. This problem is most pronounced for particles that are trapped in a magnetic mirror due to the variation of the magnetic field strength on the flux surface as they experience a large excursion from the surface due to drifts. As the ability of tokamaks to confine these trapped (the so-called banana) particles is linked to the axisymmetry, which leads to conservation of toroidal canonical angular momentum, it is natural to design a stellarator configuration such that there is an ignorable co-ordinate, which will lead to conservation of the corresponding angular momentum and hence to confined orbits. This idea leads to the concept of quasi-helical symmetry, which means that the magnetic field on a flux surface

12) In the stellarator, often the inverse of q, the rotational transform $\iota = 1/q$ is used to describe the field line geometry.

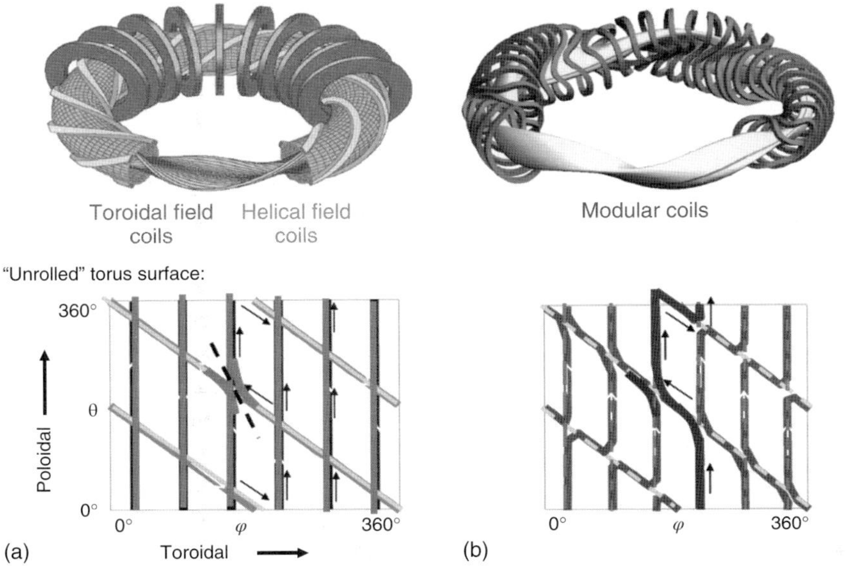

Figure 2.13 Coil configuration of a classical $\ell = 2$ stellarator (a) consisting of a set of planar toroidal field coils and a set of helical windings. In the lower panels, it is shown how this coil set can be replaced by a set of 3D-shaped modular coils that produce a stellarator field without helical windings (b, configuration of the W7-X experiment). (Please find a color version of this figure on the color plates.)

can be written as a sum of components with same helicity, that is

$$B = \sum_k B_{km,kn}(\Psi)e^{ik(m\theta - n\phi)} \tag{2.88}$$

that is can be written as $B = B(\Psi, m\theta - n\phi)$, which shows the analogy to the tokamak as it effectively becomes a function of two variables only (although the configuration is of course still 3D in space). As mentioned earlier, this property leads to the existence of an invariant of motion and hence to confined orbits even for collisionless particles. For $n = 0$, the configuration is called *quasi-axisymmetric*, and for $m = 0$, it is called *quasi-poloidally symmetric*.

A different optimization strategy is to design the field such that the drifting trapped particles drift in poloidally closed loops. Such so-called quasi-isodynamic configurations are a special case of the wider class of omnigenous magnetic fields, which are defined as configurations where all particles have zero bounce-averaged radial drift. A necessary, but not sufficient, requirement for a quasi-isodynamic field is that the lines of constant B must be poloidally closed. We mention here that owing to the 3D nature of stellarators, some of the constraints that were discussed for tokamaks are not necessarily valid here; for example the Shafranov shift can be minimized by minimizing the parallel current, relaxing the constraint of the equilibrium β-limit discussed for the tokamak in the context of Eq. (2.66) so that stellarators can have large aspect ratio without approaching this limit. An overview of stellarator physics is given in [7].

For all design principles mentioned earlier, it is important to note that while they serve as guidelines for optimizing a stellarator, they cannot be fulfilled in a mathematically exact way on all flux surfaces of a real design. Hence, an important aspect of stellarator optimization is to understand how important the effect of small deviations from the rigorous principles outlined earlier are. Moreover, while the optimization can be done numerically for neoclassical transport, MHD equilibrium and linear stability, it is not yet possible for turbulent transport, which is an inherently nonlinear problem, and hence, the question of the optimum stellarator configuration needs substantial experimental testing. Stellarator devices around the world follow different design strategies, and their experimental results will help to verify the design strategies outlined earlier.

Figure (2.14) shows four different experimental approaches. Figure(2.14)a shows the LHD device, which is similar to a classical stellarator except that the

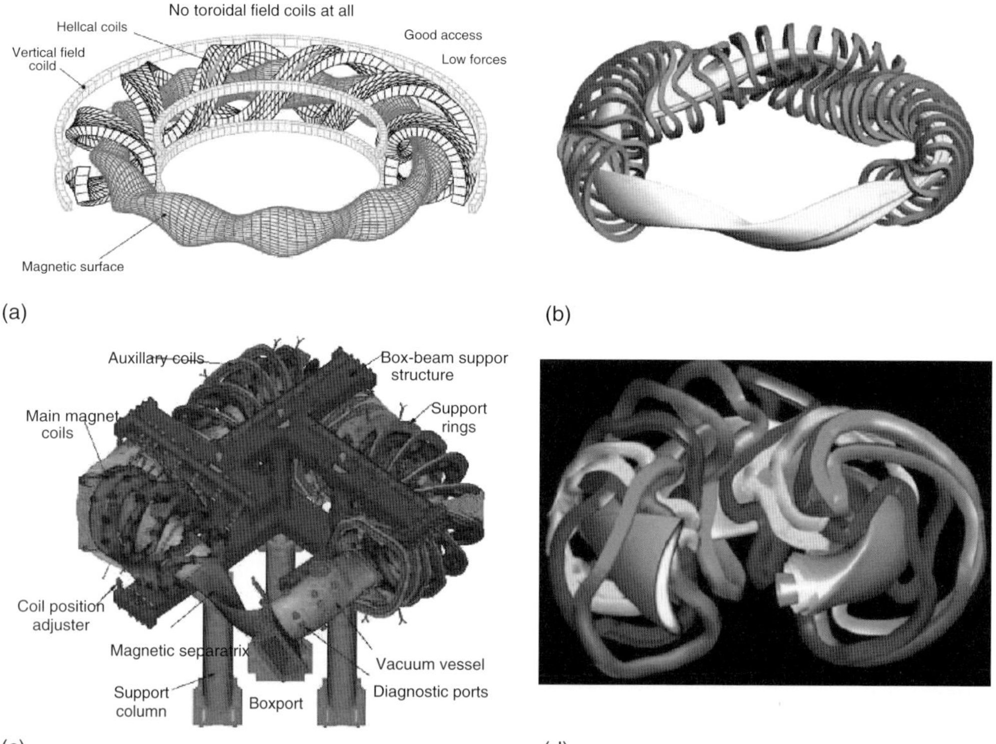

Figure 2.14 Different realizations of the stellarator configuration. (a) LHD (heliotron), (b) W7-X (quasi-isodynamic stellarator),(c) HSX (quasi-helically symmetric stellarator) and (d) NCSX (quasi-axisymmetric stellarator). (Please find a color version of this figure on the color plates.)

two helical windings carry current in the same direction (stellarator windings are pairs of opposing currents and therefore net current free). This concept is called a *heliotron* or a *torsatron*. Figure(2.14)b shows the W7-X device already introduced in Figure 2.13, which is an approximately quasi-isodynamic system with minimized parallel plasma currents (Pfirsch-Schlüter and bootstrap current). Figure(2.14)c shows HSX, which has a quasi-helical field, whereas Figure(2.14)d is a scheme of NCSX, a quasi-axisymmetric device that is supposed to rely also on the toroidal bootstrap current for the generation of its rotational transform[13]. It can be seen that LHD, HSX and W7-X all have substantially larger aspect ratio than typical tokamaks, making them relatively large devices, whereas NCSX has also been optimized for smaller, tokamak-like aspect ratio.

Owing to this large degree of freedom for optimization, there is not yet agreement on what the best suited geometry for a stellarator reactor is. The different approaches shown in Figure 2.14 should help to verify the physics basis for this decision. Similarly, the engineering challenges connected to the 3D nature are being addressed during construction of present-day devices and in conceptual studies for future reactor-grade machines. However, it is clear that if the stellarator turns out to demonstrate confinement properties comparable to that of tokamaks in terms of transport and stability, its intrinsic steady-state property due to the absence of the need to drive current as well as the absence of current driven instabilities such as disruptions (Chapter 10) makes it an attractive candidate for a fusion reactor based on magnetic confinement.

13) Construction of this device is presently stopped.

3
Linear Ideal MHD Stability Analysis

In Chapter 2, we have discussed equilibrium configurations that fulfilled the requirement of $\nabla p = \mathbf{j} \times \mathbf{B}$, that is no net force acting on any volume element. However, as discussed in the case of the z-pinch, this requirement does not guarantee stability, that is applying a perturbation to the system might lead to loss of equilibrium. A simple mechanical analogon of this situation is given by a particle in a potential $V(x)$, as shown in Figure 3.1. Equilibrium is defined by $-dV/dx = 0$ (remember that this is the force resulting from the potential), but a perturbation in the form of a displacement of the particle from its equilibrium position may lead either to a restoring force, which, in the absence of damping, leads to an oscillation around the equilibrium position, or to a force in the direction of the initial perturbation, leading to a growth of the perturbation and hence a loss of equilibrium.

One can see that the stability properties of the system are determined by the sign of the force, that is the second derivative of $V(x)$. In Figure 3.1a,b, we have assumed that the displacement leads to a potential change that is small with respect to typical equilibrium values and hence the potential can be approximated by a parabola, corresponding to a local Taylor expansion around the equilibrium point at which the first derivative vanishes. This corresponds to the concept of linear stability analysis as discussed in this chapter. However, if the perturbation becomes macroscopic, the particle will experience the detailed shape of the potential and the exponential growth may saturate, resulting in a new equilibrium position as indicated in Figure 3.1c. Such a situation can arise from the back reaction of the perturbation on the equilibrium itself, indicating that saturation is a nonlinear process. In Figure 3.1d, a situation is depicted where a linearly stable situation is nonlinearly unstable, that is for a perturbation with a large enough amplitude, the initially oscillating solution converts into an exponentially growing one.

Although very simple, these concepts can also be found in the linear MHD stability analysis that we present in the following sections.

Magnetohydrodynamic Stability of Tokamaks, First Edition. Hartmut Zohm.

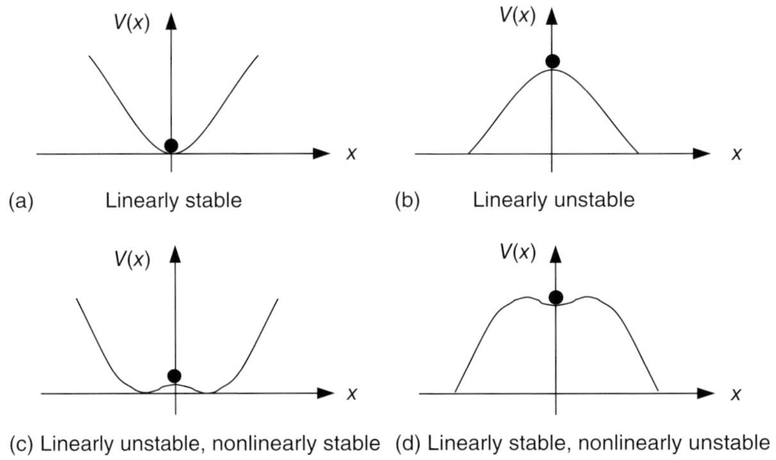

Figure 3.1 Mechanical system for demonstrating the different stability properties: in all situations, the particle is in equilibrium but its reaction to a small displacement will be very different.

3.1 Linear MHD Stability as an Initial Value Problem

In this section, we derive the equations governing linear stability of an MHD system. The steps we undertake are known from linear analysis of any system:

- Linearization of the equations, assuming the existence of a smallness parameter ϵ, allows to write all quantities involved in terms of a Taylor expansion around $f(\mathbf{x}_0)$:

$$f(\mathbf{x}) = f_0 + \epsilon f_1 + \epsilon^2 f_2 \ldots \tag{3.1}$$

  Equating the terms proportional to ϵ leads to a set of partial differential equations that is linear in the first-order terms and the coefficient functions are of zeroth order.
- A Fourier decomposition of the perturbed quantities in all variables that the zeroth-order quantities do not explicitly depend on (usually time and ignorable spatial coordinates that represent a symmetry of the system). This will replace the derivatives in these variables by multiplicative factors. As the equilibrium quantities by definition do not explicitly depend on time, this will always lead to at least a Fourier decomposition of the form

$$f_1(\mathbf{x}, t) = \sum_{n=-\infty}^{n=\infty} f_1^n(\mathbf{x}) e^{-i\omega_n t} \tag{3.2}$$

  For a linear system, the different harmonics denoted by n do not couple and each component can be treated individually. Hence, we will suppress the sum and the index n from now on.

- The remaining set of equations can be reduced, for example using Gauss' algorithm, to yield a linear partial differential equation that takes on the form of an Eigenvalue problem for ω[1] The Eigenvalues will determine the stability properties of the system, where a real ω corresponds to the stable case (oscillatory solution) and an imaginary ω to an unstable one (exponential growth).

Starting point of the analysis is the linearization of the ideal MHD equations derived in Chapter 1. In line with the mechanical analogon presented earlier, we introduce the displacement vector $\boldsymbol{\xi}(\mathbf{x})$ by

$$\mathbf{v}_1 = \frac{d\boldsymbol{\xi}}{dt} \tag{3.3}$$

representing a first-order perturbation of the system. For the most general case, we assume that the zeroth-order quantities depend on all spatial variables. In order to write the equations in ξ, we have to integrate them in time as in the original set of equations, $\mathbf{v}$ appears and not $\boldsymbol{\xi}$. Without the loss of generality, we can chose the starting conditions as $\boldsymbol{\xi}(\mathbf{x}, t = 0) = 0, \mathbf{B}_1(\mathbf{x}, t = 0) = 0, \rho_1(\mathbf{x}, t = 0) = 0$ and $p_1(\mathbf{x}, t = 0) = 0$ but assume that $\mathbf{v}_1(\mathbf{x}, t = 0) = 0$. This choice of boundary conditions corresponds to an initial value problem where the system passes through the equilibrium point at $t = 0$ with finite velocity $\mathbf{v}_1$ and has the advantage that when integrating the equations in time, as is done in the following, there is no additional term that describes the system at $t = 0$. As in Chapter 2, we will assume that there is no equilibrium flow, that is $\mathbf{v}_0 = 0$. The linearized and integrated equation of continuity becomes

$$\frac{\partial \rho_1}{\partial t} = -\nabla \cdot (\rho_0 \mathbf{v}_1) \;\rightarrow\; \rho_1 = -\nabla \cdot (\rho_0 \boldsymbol{\xi}) \tag{3.4}$$

that is a change in density is due to a compression of the volume element. Using this equation, the linearized adiabatic equation becomes

$$\frac{\partial p_1}{\partial t} = -p_0 \gamma \nabla \cdot \mathbf{v}_1 - \mathbf{v}_1 \cdot \nabla p_0 \;\rightarrow\; p_1 = -p_0 \gamma \nabla \cdot \boldsymbol{\xi} - \boldsymbol{\xi} \cdot \nabla p_0 \tag{3.5}$$

that is a change of pressure is due to either an adiabatic compression or moving the volume element into a region of different pressure.

Combining Faraday's law and Ohm's law yields

$$\frac{\partial \mathbf{B}_1}{\partial t} = \nabla \times (\mathbf{v}_1 \times \mathbf{B}_0) \;\rightarrow\; \mathbf{B}_1 = \nabla \times (\boldsymbol{\xi} \times \mathbf{B}_0) \tag{3.6}$$

According to this equation, displacing the plasma across the equilibrium magnetic field will induce an electric field that in turn gives rise to a change in magnetic field. We know from Chapter 1 that this will lead to the conservation of the magnetic flux in a flux tube.

The linearized equation of motion becomes

$$\rho_0 \frac{\partial \mathbf{v}_1}{\partial t} = \mathbf{j}_0 \times \mathbf{B}_1 + \mathbf{j}_1 \times \mathbf{B}_0 - \nabla p_1 \tag{3.7}$$

1) in the absence of dissipation, as is the case here, the Eigenvalue problem will at least be quadratic in ω.

Now, we can express p_1 through Eq. (3.5) by $\boldsymbol{\xi}$ and $\mathbf{j}_1$ using Ampère's law $\mu_0\mathbf{j}_1 = \nabla \times \mathbf{B}_1$ and arrive at the MHD force equation

$$\rho_0 \frac{\partial^2 \boldsymbol{\xi}}{\partial t^2} = \frac{1}{\mu_0}(\nabla \times \mathbf{B}_0) \times \mathbf{B}_1 + \frac{1}{\mu_0}(\nabla \times \mathbf{B}_1) \times \mathbf{B}_0 + \nabla(p_0\gamma\nabla \cdot \boldsymbol{\xi} + \boldsymbol{\xi} \cdot \nabla p_0) \quad (3.8)$$

We note that the perturbed magnetic field $\mathbf{B}_1$ still appears in the equation, but it can be related to $\boldsymbol{\xi}$ using Eq. (3.6). This means that Eq. (3.8) is a linear partial differential equation for $\boldsymbol{\xi}$, which, in symbolic form, can be written as

$$\rho_0 \frac{\partial^2 \boldsymbol{\xi}}{\partial t^2} = \mathbf{F}(\boldsymbol{\xi}) \quad (3.9)$$

where we have introduced the MHD force operator $\mathbf{F}(\boldsymbol{\xi})$. Applying the Fourier decomposition in time as outlined earlier leads to the generalized Eigenvalue problem

$$-\omega^2 \rho_0 \boldsymbol{\xi} = \mathbf{F}(\boldsymbol{\xi}) \quad (3.10)$$

where $\rho_0(\mathbf{x})$ plays the role of a weight function for the inner product of the Eigenfunctions ξ. An important property of the force operator is that it is self-adjoint , that is

$$\int \boldsymbol{\eta}^* \mathbf{F}(\boldsymbol{\xi}) dV = \int \boldsymbol{\xi}^* \mathbf{F}(\boldsymbol{\eta}) dV \quad (3.11)$$

where the asterisk denotes the complex conjugate. It is straightforward but tedious to prove this property term by term, and the interested reader is referred to the literature [1]. As an important consequence of this property, the Eigenvalues ω^2 are real and the Eigenfunctions represent a complete set of orthogonal functions, that is they have the property

$$\int \rho_0 \xi^*_{\mathbf{n}} \xi_{\mathbf{m}} dV = \delta_{mn} \quad (3.12)$$

As the Eigenvalues are real, the stability properties are given by

- $\omega^2 > 0$: then, ω is real and according to Eq. (3.2), the system is oscillating around the equilibrium point, that is it is stable.
- $\omega^2 < 0$: then, ω is purely imaginary and according to Eq. (3.2), both an exponentially growing and an exponentially decaying solution exist. Owing to the existence of the exponentially growing solution, the system is unstable.

Finding the solutions to the Eigenvalue problem will hence give us complete knowledge about the stability properties of the system. However, as one can imagine when looking at Eq. (3.8), there are only few cases where an analytical solution is possible, and in general, the Eigenvalue problem has to be solved numerically. While this is possible and commonly being done in linear MHD stability analysis of real systems, there exists a method that allows more physical insight into the stability properties of magnetically confined fusion plasmas, although usually at the expense of a loss of detailed information about the Eigenfunctions and Eigenvalues. This method, the so-called energy principle , is treated in Section 3.2.

3.2
The Energy Principle of Ideal MHD

In this section, we derive the energy principle, which is a variational formalism for determining MHD stability. It is based on Ritz' variational principle for self-adjoint systems that was originally developed for fluid mechanics and has also been applied for example in quantum mechanics to estimate Eigenvalues for systems that are no longer analytically tractable.

The basic idea of this formalism is to transform the partial differential equation representing the Eigenvalue problem into an integral expression by multiplying it by $\boldsymbol{\xi}^*$ and integrating over the volume:

$$\frac{\omega^2}{2}\int \rho_0|\xi|^2\, dV = -\frac{1}{2}\int \boldsymbol{\xi}^* \cdot \mathbf{F}(\boldsymbol{\xi})\, dV = \delta W(\boldsymbol{\xi}^*, \boldsymbol{\xi}) \tag{3.13}$$

The integral on the left-hand side, usually referred to as $K(\boldsymbol{\xi}^*, \boldsymbol{\xi})$, is the kinetic energy of the perturbed system. The right-hand side, δW, corresponds to the potential energy, that is the work done by displacing the system against the force $\mathbf{F}$. Formally, this can be seen as an equation for ω^2 by writing

$$\omega^2(\boldsymbol{\xi}, \boldsymbol{\xi}^*) = \frac{\delta W(\boldsymbol{\xi}, \boldsymbol{\xi}^*)}{K(\boldsymbol{\xi}, \boldsymbol{\xi}^*)} \tag{3.14}$$

but as the relation is valid for arbitrary $\boldsymbol{\xi}$, in general, ω^2 is no longer an Eigenvalue of Eq. (3.10). Such an Eigenvalue can of course be obtained by setting the variation $\delta\omega(\boldsymbol{\xi}, \boldsymbol{\xi}^*)^2 = 0$, where the corresponding Euler Lagrange equation is just the Eigenvalue equation (Eq. (3.10)). There is, however, a relation between the general $\omega^2(\boldsymbol{\xi}, \boldsymbol{\xi}^*)$ and the Eigenvalues that can be seen by decomposing $\boldsymbol{\xi}$ into the Eigenfunctions according to

$$\boldsymbol{\xi} = \sum_n a_n \boldsymbol{\xi}_n \tag{3.15}$$

and evaluating Eq. (3.14):

$$\omega^2 = -\frac{\sum_n \sum_m a_m^* a_n \int \boldsymbol{\xi}_m^* \cdot \mathbf{F}(\boldsymbol{\xi}_n) dV}{\sum_n \sum_m a_m^* a_n \int \rho_n \boldsymbol{\xi}_m^* \boldsymbol{\xi}_n dV} = \frac{\sum_n |a_n|^2 \omega_n^2}{\sum_n |a_n|^2} \tag{3.16}$$

where in the last step, we have used the Eigenvalue equation and the orthogonality relation of the Eigenfunctions. This means that ω^2 is a weighted sum of the Eigenvalues of the system. Hence, the stability criterion can be reformulated for ω^2 obtained using an arbitrary test function ξ:

- $\omega^2 < 0$: According to Eq. (3.16), at least one of the Eigenvalues ω_n^2 must be negative and hence the system is unstable.
- $\omega^2 > 0$: In this case, a general statement cannot be made, unless it is possible to prove that $\omega^2 > 0$ for all test functions. Then, the system is stable.

We note that as K is always positive, this means that the sign of ω^2 is determined by δW, that is

$$\delta W < 0 \text{ the system is unstable} \tag{3.17}$$

$$\delta W > 0 \text{ the system is stable} \tag{3.18}$$

very similar to the simple mechanical analogon presented in the introduction of this chapter. As can be seen from Eq. (3.14), the Euler Lagrange equation corresponding to $\delta(\delta W) = 0$ corresponds to the marginally stable Eigenvalue $\omega^2 = 0$. We will use this later when we discuss pressure-driven modes in Chapter 5.

Thus, the energy principle allows to determine the stability of the system, provided suitable test functions can be found. We have less knowledge of the system than in the case of solving explicitly the Eigenvalue problem (Eq. (3.8)) as we do not know the real Eigenvalues and Eigenfunctions. However, from Eq. (3.16), one can see that the fastest growing mode, corresponding to the most negative Eigenvalue, will make δW most negative, that is it represents an extremum for the variational form (3.14)[2)]. Thus, we know that there will always exist an unstable Eigenmode with at least the growth rate calculated for the particular test function that made δW negative. In the following, we will look at different forms of the energy principle that have proved to yield physical insight into the stability properties of ideal MHD in general and magnetic confinement devices in particular.

3.3 Forms of δW

It is straightforward to insert $\mathbf{F}(\boldsymbol{\xi})$ into the definition of δW (Eq. (3.13)), but the resulting expression is complex and does not allow for physics insight. However, the energy principle can be rewritten assuming that we deal with a plasma surrounded by a vacuum region that in turn is surrounded by a conducting wall. This situation, relevant to magnetically confined fusion plasmas, is usually called the *extended energy principle.* It introduces the following boundary conditions at the interfaces:

- at the wall, the radial component of $\boldsymbol{\xi}$ vanishes. If the wall is ideally conducting, the same is true for $B_\perp$[3)]. Outside the ideally conducting wall, all quantities vanish. This usually implies a shielding current flowing in the wall, leading to a jump of the parallel magnetic field across it.
- at the plasma–vacuum interface, the perpendicular component of $\mathbf{B}_1$ is continuous, whereas the parallel component may jump, implying a surface current. Moreover, the pressure balance must be fulfilled across the surface.

Applying these conditions and properly rearranging the terms, one finds that δW can be written as follows[4)]:

$$\delta W = \delta W_F + \delta W_S + \delta W_V, \tag{3.19}$$

where the index F stands for 'fluid', that is the bulk plasma, S for 'surface', that is the contribution from the plasma–vacuum interface and V denotes the 'vacuum' contribution. These individual contributions can be written as follows:

2) This is the original statement of Ritz' variational principle.
3) We will deal with a wall of finite electrical resistivity in Section 7.4.
4) The mathematically interested reader will find more details about the derivation of the different forms of δW in [1].

$$\delta W_F = \frac{1}{2}\int_{\text{Fluid}} \left(\frac{|\mathbf{B}_1|^2}{\mu_0} - \xi_\perp^* \cdot \mathbf{J}_0 \times \mathbf{B}_1 + \gamma p_0 |\nabla \cdot \boldsymbol{\xi}|^2 \right.$$
$$\left. + (\boldsymbol{\xi}_\perp \cdot \nabla p_0)(\nabla \cdot \boldsymbol{\xi}_\perp^*) \right) dV \tag{3.20}$$

which is the so-called standard form of δW_F [8]. For δW_V, we obtain

$$\delta W_V = \frac{1}{2}\int_{\text{Vacuum}} \frac{B_1^2}{2\mu_0}\, dV \tag{3.21}$$

that is the vacuum contribution is just the magnetic field energy related to the perturbation and thus always positive. The surface term is

$$\delta W_S = \frac{1}{2}\int_{\text{Surface}} |\mathbf{n} \cdot \boldsymbol{\xi}_\perp|^2\, \mathbf{n} \cdot \|\nabla \left(p_0 + \frac{B_0^2}{2\mu_0} \right)\|\, dS \tag{3.22}$$

where $\mathbf{n}$ is the normal vector on the surface and the double line represents the jump of the quantity across the plasma–vacuum interface. As described earlier, the total pressure is continuous across the interface, but its first derivative may jump, leading to a surface current. Conversely, if we assume that there are no surface currents, we can set $\delta W_S = 0$. Here and in what follows, the index $\perp$ refers to the direction perpendicular to the equilibrium magnetic field $\mathbf{B}_0$, and the index $||$ refers to the direction parallel to the equilibrium magnetic field.

These expressions for δW are the starting point for our analysis of linear ideal MHD stability. To analyse the stability of different configurations, they have to be expressed in the appropriate co-ordinates and the coefficient functions, which consist of zeroth-order quantities, have to be specified. They are the result of the zeroth-order force balance, that is they contain the full equilibrium information.

However, before proceeding to this step, we note that with some rearrangement, the volume integral over the plasma can be written as [9]

$$\delta W_F = \frac{1}{2}\int_{\text{Fluid}} \left(\frac{|B_{1\perp}|^2}{2\mu_0} + \frac{B_{0\perp}^2}{2\mu_0} |\nabla \cdot \boldsymbol{\xi}_\perp + 2\boldsymbol{\xi}_\perp \cdot \boldsymbol{\kappa}|^2 + \gamma p_0 |\nabla \cdot \boldsymbol{\xi}|^2 \right.$$
$$\left. -2(\boldsymbol{\xi}_\perp \cdot \nabla p_0)(\boldsymbol{\kappa} \cdot \boldsymbol{\xi}_\perp^*) - \frac{j_{0||}}{B_0}(\boldsymbol{\xi}_\perp^* \times \mathbf{B}_0) \cdot \mathbf{B}_1 \right) dV \tag{3.23}$$

where $\boldsymbol{\kappa}$, the curvature vector of the equilibrium magnetic field, has been defined in Eq. (1.36). The main step here is to separate the perpendicular current density that is related to the pressure gradient and the parallel current density that is force free[5] but represents a source of free energy.

This form of δW_F is called the *intuitive* form as it allows to characterize the different stabilizing and destabilizing contributions by their physics nature. It can be seen that all terms in the first line of Eq. (3.23) are positive and hence stabilizing. They correspond to the MHD waves derived in Chapter 1. The first term is the magnetic field energy associated with the perturbation and can be related to the

5) The term *force-free* relates to currents that flow along field lines and hence do not contribute to the force balance.

shear Alfvén wave. The second term is related to the compression of both equilibrium field and plasma; it is therefore due to the compressional Alfvén waves. The third term is related to (adiabatic) compression of the ideal plasma; it refers to the sound waves.. Obviously, these contributions are minimized by perturbations that fulfil $\nabla \cdot \boldsymbol{\xi} = 0$, that is the most unstable perturbations will usually be incompressible. We will make use of this fact later on when we minimize the energy functional for the tokamak and the screw pinch.

The second line of Eq. (3.23) contains terms that can be either positive or negative; they describe the principal sources of instability:

- Pressure-driven instabilities: the term $2(\boldsymbol{\xi}_\perp \cdot \nabla p_0)(\boldsymbol{\kappa} \cdot \boldsymbol{\xi}_\perp)$ describes the destabilizing contribution of a pressure gradient. It can be seen that its sign will depend on the relative orientation of $\boldsymbol{\kappa}$ and ∇p, such that if the vectors are parallel, the contribution will be negative and hence destabilizing, whereas for the antiparallel case, the term is positive and stabilizing. These situations have been dubbed 'bad' and 'good' curvatures and relate to the idea of the interchange instability. Consider the situation shown in Figure 3.2, where for simplicity we assume that there is a region filled with plasma surrounded by a region of magnetic field that confines it. If, in Figure 3.2a, we interchange plasma and magnetic field, the energy of the system is lowered as the plasma expands and the magnetic field lines will shorten, as shown in Figure 3.2b, leading to a lowering of the magnetic energy. If, however, the interchange will lead to a lengthening of field lines, as is the case for good curvature, the magnetic field energy will be increased and the interchange instability can be stabilized. This kind of instability is crucial for the plasma confinement in magnetic mirrors that, in their simplest form, correspond to the situation shown in Figure 3.2a. For toroidal systems, the inner part of the torus is a region of good curvature while the outer part has bad curvature. As we will see, this means that the overall stability of pressure-driven modes in a tokamak crucially depends on the integral of the curvature along the field line, summing up the contributions of both regions. This also means that toroidal effects are essential for the description of pressure- driven modes in toroidal systems.
- Current driven instabilities: the term $j_\parallel(\boldsymbol{\xi}_\perp \times \mathbf{B}_0) \cdot \mathbf{B}_1/B_0$ describes instabilities driven by the current density parallel to the equilibrium field. Prominent

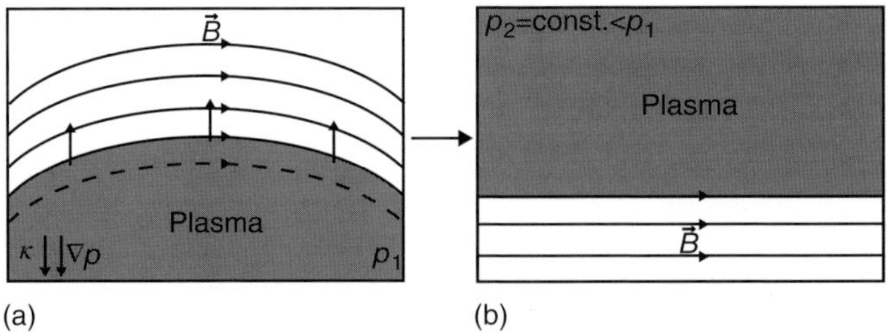

Figure 3.2 Sketch of the interchange instability: depending on the curvature relative to the plasma, the field lines are shortened (a) or lengthened (b) when interchanged with the plasma.

examples for these instabilities are the sausage and kink instabilities already discussed in Chapter 2 and shown in Figure 2.2. The fact that kink instabilities are essentially current driven can also be seen from the observation that they will occur in solid objects as well. For example, a straight copper wire carrying a current will start kinking and hence deform into a spiral above a critical current. A mechanical analogon to the kink instability occurs when wringing out a towel.

Although the distinction between pressure- and current-driven modes is very useful for a principal classification of MHD instabilities, we will see in the following chapters that instabilities in real systems often are a mixture of both.

3.4
The Ideal MHD Energy Principle for the Tokamak

For the analysis of MHD stability of the tokamak using the energy functional, it is necessary to express δW in the appropriate co-ordinates. For the general axisymmetric case, this is done using the flux functions introduced in Chapter 2, that is the poloidal flux function Ψ that serves as a flux surface label and the poloidal current $I_{\mathrm{pol}}(\Psi)$ and the pressure $p(\Psi)$. In equilibrium, these three quantities are connected by the Grad–Shafranov equation and hence will appear in the coefficient functions of the energy functional. The co-ordinates used are the (ignorable) toroidal angle ϕ, the direction perpendicular to the flux surfaces Ψ and the poloidal 'angle' co-ordinate χ with $\nabla\Psi \cdot \nabla\chi = 0$. The volume element in these coordinates is given by $d^3x = J\,d\Psi d\chi d\phi$, where J is the Jacobian. Owing to the axisymmetry of the tokamak, the displacement vector can be Fourier decomposed in toroidal angle according to

$$\underset{\sim}{\xi} = \underset{\sim}{\xi}(\Psi, \chi)e^{in\phi} \tag{3.24}$$

It can be shown [10] that, starting from the standard form (Eq. (3.20)), δW can be written as a function of the components of $\xi_\perp$, the part of the displacement vector perpendicular to the equilibrium field lines. The two independent components are expressed as scalar functions X and U in the direction of the flux surface normal vector and perpendicular to it according to

$$X = RB_p\xi_\Psi \tag{3.25}$$

$$U = \frac{\xi_\phi}{R} - \frac{\mu_0 I_{\mathrm{pol}}}{R^2 B_p}\xi_\chi \tag{3.26}$$

where B_p is the poloidal field. With these definitions, it is possible to write the volume part of the energy functional (Eq. (3.23)) for each toroidal mode number n as

$$\delta W_f = \pi \int J\,d\Psi d\chi \left(\frac{B^2}{R^2 B_p^2}\,|k_\parallel X|^2 + \frac{R^2}{J^2}\left|\frac{\partial U}{\partial \chi} - \mu_0 I_{\mathrm{pol}}\frac{\partial}{\partial \Psi}\left(\frac{JX}{R^2}\right)\right|^2 \right.$$
$$\left. + B_p^2\left|inU + \frac{\partial X}{\partial \Psi} + \frac{j_\phi}{RB_p^2}X\right|^2 - 2K|X|^2 \right) \tag{3.27}$$

where the stabilizing contribution of compression has already been eliminated by chosing ξ such that it is incompressible, that is $\nabla \cdot \boldsymbol{\xi} = 0$ This implies a particular choice of $\xi_{\|}$ but eliminates it from the expression and reduces the number of free scalar functions, relating to the components of ξ, from 3 to 2. Obviously, all terms in Eq. (3.27) are positive and hence stabilizing except for the last one that is related to the free energies in ∇p and $j_{\|}$ through the coefficient functions K and j_ϕ. They are related to the equilibrium quantities through

$$K = \frac{\mu_0 I_{\text{pol}} I'_{\text{pol}}}{R^2} \frac{\partial(\log R)}{\partial \Psi} - \frac{j_\phi}{R} \frac{\partial(\log (JB_p))}{\partial \Psi} \tag{3.28}$$

$$j_\phi = 2\pi R p' + \frac{\mu_0 I_{\text{pol}} I'_{\text{pol}}}{R} = -\frac{R}{J} \frac{\partial}{\partial \Psi} (JB_p^2) \tag{3.29}$$

where the last equation essentially is the Grad–Shafranov equation. The quantity $k_{\|}$ is the derivative along the field line

$$ik_{\|} = \frac{1}{JB}\left(\frac{\partial}{\partial \chi} + inv\right) \tag{3.30}$$

where

$$v(\Psi, \chi) = \frac{d\phi}{d\chi} = \frac{\mu_0 I_{\text{pol}} J}{R^2} \tag{3.31}$$

so that v is related to the safety factor q via

$$q(\Psi) = \frac{1}{2\pi} \oint v(\Psi, \chi) d\chi \tag{3.32}$$

This is the exact expression for the volume part of δW for a tokamak with arbitrary poloidal cross section . From the structure of the integral, it is clear that it will be difficult to draw any direct conclusion from it in an analytical way. On the other hand, as outlined earlier, numerical evaluation of δW is usually not very useful as it will not give information about the full spectrum of the force operator. However, in the following, we will use the expression (3.27) in two different limits to obtain physics insight. One is a local expansion in the vicinity of a flux surface, which is appropriate for localized modes at high n in the analysis of pressure-driven modes. This will be treated in Chapter 5. The other one is to use the periodic screw pinch as approximation for the tokamak, which allows to derive a form of δW that is quite suitable to discuss the basics of current-driven modes as will be done in Chapter 4. This derivation is outlined in the following.

The derivation of the energy principle for the general screw pinch can be found for example in [1]. It is derived using a Fourier decomposition in the poloidal angle with the periodicity constraint $f(\theta) = f(\theta + m2\pi)$ where the integer number m is called the *poloidal mode number*. In the general screw pinch, the wave vector k_z does not have such a constraint, but for the case of the periodic screw pinch resembling a toroidal system, an integer toroidal mode number n is introduced with $k_z = n/R_0$, with R_0 the torus' major radius such that $R_0\phi = z$. Thus, the Fourier decomposition in θ and z directions reads

$$\boldsymbol{\xi}(r, \theta, z) = \boldsymbol{\xi}(r) e^{i\left(m\theta - \frac{n}{R_0} z\right)} \tag{3.33}$$

and the mode numbers m and n are good quantum numbers[6] as the equilibrium quantities do not explicitly depend on θ and z. For the periodic screw pinch, the energy principle can be seen as a special case of Eq. 3.27 in which the flux surface label Ψ can be replaced by the minor radius r. The generalized poloidal coordinate χ becomes the cylindrical angle θ, the poloidal field B_p is constant on the flux surface, that is $B_\theta(r)$ and B now becomes the constant axial field B_z. The Jacobian is $J = r/B_\theta$ and a derivative with respect to Ψ is expressed as $d/d\Psi = 1/(RB_\theta)d/dr$. The pressure balance (Eq. (3.29)) then becomes the relation derived for the screw pinch (Eq. (2.17)). Using these transformations, $\boldsymbol{\xi}$ is decomposed into

$$\boldsymbol{\xi} = \xi_r \hat{\mathbf{e}}_r + \eta \hat{\mathbf{e}}_\eta + \xi_\| \hat{\mathbf{e}}_\| \tag{3.34}$$

where $\hat{\mathbf{e}}_\eta$ is perpendicular to both $\hat{\mathbf{e}}_r$ and $\hat{\mathbf{e}}_\|$, that is it represents the direction perpendicular to the equilibrium field lines on the flux surface.

In a first step, the incompressibility condition $\nabla \cdot \boldsymbol{\xi}$ is used to eliminate $\xi_\|$ from the expression[7]. Then, as η lies in the flux surface, it only appears algebraically in the expression for δW because of the Fourier decomposition (Eq. (3.33)). Thus, δW can be minimized explicitly with respect to η and by inserting this condition, the final δW only depends on ξ_r, which for simplicity is denoted as ξ in the following.

The final result for the periodic screw pinch can be written as

$$\delta W = \frac{2\pi^2 R_0}{\mu_0} \int_0^a (f\xi'^2 + g\xi^2) dr + \left. \left| \frac{2\pi^2 B_z^2}{\mu_0 R_0} \xi^2 \left(\frac{n^2 - \frac{m^2}{q^2}}{\frac{n^2}{R_0^2} + \frac{m^2}{r^2}} + \frac{r^2}{m} \Lambda \left(\frac{m}{q} - n \right)^2 \right) \right| \right|_{r=a} \tag{3.35}$$

where a is the plasma minor radius, q the safety factor defined for the circular cross-sectional cylinder in Eq. (2.10) and the functions f and g given by

$$f = r \frac{B_z^2}{R_0^2} \frac{\left(\frac{m}{q} - n \right)^2}{\frac{n^2}{R_0^2} + \frac{m^2}{r^2}} \tag{3.36}$$

$$g = \frac{2\mu_0 p'}{1 + \left(\frac{mR_0}{nr} \right)^2} + r \frac{B_z^2}{R_0^2} \left(\frac{m}{q} - n \right)^2 \left(1 - \frac{1}{\left(\frac{nr}{R_0} \right)^2 + m^2} \right) + \frac{2n^2 B_z^2}{r R_0^4} \frac{n^2 - \frac{m^2}{q^2}}{\left(\frac{n^2}{R_0^2} + \frac{m^2}{r^2} \right)^2} \tag{3.37}$$

In Eq. (3.35), the first term outside of the integral represents a contribution from the fluid term δW_F that comes from integration by parts in the derivation, whereas the second term outside the integral represents the vacuum contribution. In this contribution, the term Λ represents the stabilizing effect of a perfectly conducting

6) As in quantum mechanics, a 'good' quantum number is defined such that there is no coupling between components of different quantum number.
7) In Eq. (3.27), this has already been done by omitting the term $\gamma p \nabla \cdot \boldsymbol{\xi}$.

wall at $r = r_{\text{wall}}$ and can be written as

$$\Lambda = -\frac{mR_0K_a}{naK'_a}\frac{1 - K'_{r_{\text{wall}}}I_a/(I'_{r_{\text{wall}}}K_a)}{1 - K'_{r_{\text{wall}}}I'_a/(I'_{r_{\text{wall}}}K'_a)} \tag{3.38}$$

where $K_r = K_m(nr/R_0)$ and $I_r = I_m(nr/R_0)$ are modified Bessel functions and the prime denotes the radial derivative.

It can be seen from Eq. (3.37) that for $dp/dr < 0$, which is the usual case in toroidal confinement, the pressure contribution is always destabilizing. This is due to the fact that in the screw pinch, the curvature is only due to the poloidal field and always destabilizing as it is 'badly' curved. As outlined earlier, in a tokamak, the toroidal field is also curved and exhibits both good and bad curvature regions, so that it significantly contributes to stability. We will come back to this when we treat pressure-driven modes in a tokamak in Chapter 5 where the full toroidal geometry has to be considered using Eq. (3.27) instead of Eq. (3.35).

In both Eqs (3.36) and (3.37), terms containing $(m/q - n)$ appear. These vanish at surfaces where $q = m/n$, pointing out the special role of these surfaces for MHD instabilities in toroidal systems: while in general, perturbations lead to a bending of field lines, which implies a stabilizing contribution, those perturbations that have constant phase along field lines on rational surfaces with integer m/n avoid this and are hence usually the most unstable ones. In ideal MHD, these perturbations can be seen as standing Alfvén waves that do not propagate with respect to the plasma. The surfaces with integer m/n are for that reason often called *resonant surfaces*. Figure 3.3 shows an example for such a situation, assuming an $m = 9, n = 3$ mode on the $q = 3$ surface of a tokamak.

We note that the expression given earlier is exact and hence applies to any cylindrical configuration periodic in z. In particular, it describes current-driven modes in the tokamak as well as the RFP, whereas, as pointed out earlier, for pressure-driven modes in the tokamak, toroidicity has to be included in the analysis. In Chapter 4, we will apply an expansion in aspect ratio typical for tokamaks to analyse the stability of the tokamak against current-driven instabilities.

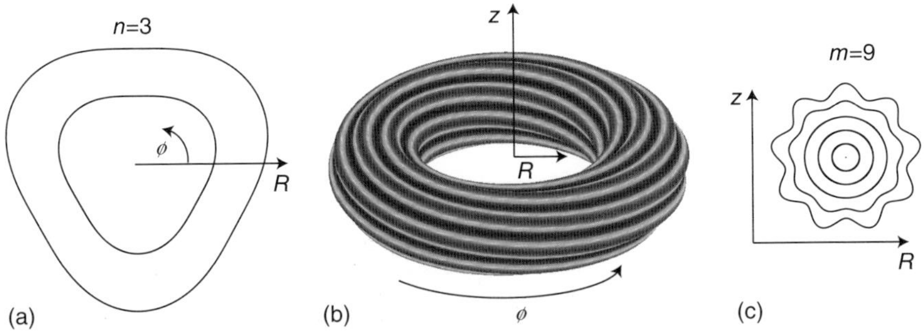

Figure 3.3 Schematic representation of an $m = 9, n = 3$ mode following the field lines on the $q = 3$ surface. (a) Horizontal cut through the 3D plot in (b) and (c) the poloidal cross section (in both plots, the amplitude of the perturbation is exaggerated compared to the middle one). (Please find a color version of this figure on the color plates.)

4
Current Driven Ideal MHD Modes in a Tokamak

In this chapter, we examine the stability of tokamaks against current driven MHD instabilities using the periodic screw pinch energy principle derived in Chapter 3. As outlined there, this is a good approximation as the curvature of the toroidal field is not too important for these modes. Experimentally, current driven modes lead to a number of restrictions to the tokamak operational space, which we will outline using experimental examples where applicable. While we first examine modes that are resonant on surfaces with $q = m/n$, the last section deals with another type of ideal instability related to the plasma current, the positional vertical instability occurring when the poloidal cross section is elongated by the external poloidal field coils.

4.1
Expression for δW in Tokamak Ordering

For the analysis of current driven modes in a tokamak, we adopt the so-called tokamak ordering, that is we assume a small inverse aspect ratio. In order to examine the physics of purely current driven modes, we neglect the pressure term in Eq. (3.37), corresponding to the assumption of negligible β. More specifically, this means Taylor expansion of δW as given by Eq. (3.35) in (r/R_0) assuming

$$\delta W = \mathcal{O}\left(\frac{r}{R_0}\right)^2, \tag{4.1}$$

$$\beta = \mathcal{O}\left(\frac{r}{R_0}\right)^2 \tag{4.2}$$

We note here that in the expansion, we also assume $m/n \geq 1$, which is a good assumption for a tokamak with modes having roughly constant phase along the field lines, but will not apply in an RFP. Applying this procedure to Eq. (3.36) yields

$$f(r) \approx \frac{r^3}{R_0^2} B_z^2 \left(\frac{1}{q} - \frac{n}{m}\right)^2, \tag{4.3}$$

$$g(r) \approx \frac{r}{R_0^2} B_z^2 (m^2 - 1) \left(\frac{1}{q} - \frac{n}{m}\right)^2 \tag{4.4}$$

In the second line, we have neglected $((nr)/(R_0 m))^2 2\mu_0 p'$ against terms of order $(r/R_0)^2 B_z^2$, consistent with Eq. (4.2), which makes the pressure term a fourth-order contribution to δW. We will come back to this when we treat finite pressure effects where this term has to be kept as it is not negligible for localized modes in the vicinity of a rational surface where $(1/q - n/m) \to 0$.

Applying the same procedure to the terms outside the integral in Eq. (3.35) leads to

$$\delta W_B = \frac{2\pi^2 B_z^2}{\mu_0 R_0} \xi_a^2 a^2 \left(\frac{n^2}{m^2} - \frac{1}{q_a^2} + \Lambda m \left(\frac{1}{q_a} - \frac{n}{m} \right)^2 \right) \tag{4.5}$$

where $\xi_a = \xi(a)$ and $q_a = q(a)$ and we have introduced the notation δW_B for this term as it represents the contribution from finite ξ on the plasma boundary. Combining the three terms leads to the following expression:

$$\begin{aligned} \delta W = {} & \frac{2\pi^2 B_z^2}{\mu_0 R_0} \int_0^a [(r\xi')^2 + (m^2 - 1)\xi^2] \left(\frac{n}{m} - \frac{1}{q} \right)^2 r\, dr \\ & + \frac{2\pi^2 B_z^2}{\mu_0 R_0} \xi_a^2 a^2 \left(\frac{n^2}{m^2} - \frac{1}{q_a^2} + \Lambda m \left(\frac{1}{q_a} - \frac{n}{m} \right)^2 \right) \end{aligned} \tag{4.6}$$

It can be seen that the integral term cannot become negative, and hence to this order, stability will depend on the boundary term. It is therefore practical to distinguish the case of external kinks, that is $\xi_a \neq 0$, and internal kinks $\xi_a = 0$. In the latter case, the plasma boundary is not perturbed by the mode and $\delta W_B = 0$. As we will see below, stability is then determined by the next order terms, that is of order $(r/R_0)^4$.

4.2
External Kinks in a Tokamak with $\beta = 0$

Here, we assume $\xi_a > 0$, that is the perturbation affects the plasma boundary. We first analyse the case without conducting wall, which can be considered as the most pessimistic assumption. However, we will see that tokamak plasmas are usually quite stable against external kinks in the case $\beta \to 0$ even without conducting wall, and the effect of wall stabilization is mainly important when the effect of finite β is included[1]. Hence, it will be treated in detail in Chapter 7 where the combined effect of pressure and current drive is discussed.

4.2.1
Modes with $m = 1$

It can be seen that the integral term in Eq. (4.6) can be zero if we chose a constant displacement $\xi = \xi_0$ and $m = 1$. In this case, we only have to discuss the sign

1) This is different in the RFP where external kinks also play an important role at $\beta \to 0$.

of δW_B. In the limit of a far away wall, its stabilizing effect vanishes ($\Lambda = 1$) and inserting $m = 1$ yields for this term

$$\delta W = \frac{4\pi^2 B_z^2}{\mu_0 R_0} n \left(n - \frac{1}{q_a} \right) a^2 \xi_a^2 \tag{4.7}$$

with the constant displacement $\xi_a = \xi_0$. For the tokamak, it means that

$$q_a > \frac{1}{n} \tag{4.8}$$

is the condition for stability, limiting the maximum current that can be driven in the plasma for given magnetic field B_z. This is most restrictive for $n = 1$, and as we have assumed $m = 1$, the so-called Kruskal–Shafranov limit takes on the form

$$q_a > 1 \quad \rightarrow \quad I_p < \frac{2\pi a^2}{\mu_0 R_0} B_z \tag{4.9}$$

where we have used the cylindrical definition of q (Eq. (2.11)). For a typical tokamak with a circular cross section, aspect ratio of 3, minor radius of $a = 0.5$ m and a field of, say, $B_z = 3$T, this limits the maximum plasma current to values of the order of 1 MA[2]. We note here that by shaping the plasma cross section as shown in Chapter 2, the safety factor is increased for same values of B_z and I_p because of a longer poloidal circumference, leading to a higher allowable value of I_p than in the circular case.

In present-day tokamak operation, this condition is usually not limiting the maximum achievable current as the occurrence of the (2,1) external kink mode (see the following discussion) is usually preventing operation at q_a values below 2 already. However, recent experiments using active feedback control of this mode by internal coils that counteract the mode growth have allowed to demonstrate tokamak operation at $q_a < 2$, coming closer to the Kruskal–Shafranov limit. Figure 4.1 shows time traces of a tokamak discharge with circular cross section that disrupts at $q_a = 2$ without control because of the occurrence of a (2,1) external kink mode[3] but can be run stably below $q_a = 2$ when active magnetic feedback control is used [11].

Finally, we note that a close fitting ideally conducting wall will effectively lead to $\xi_a \rightarrow 0$, which also implies $\delta W_B \rightarrow 0$. In this case, the Kruskal-Shafranov limit is removed and internal modes will set the limit. While for tokamaks, this will restrict operation to $q > 1$ everywhere, a new window of operation at much lower q_a occurs if one allows for a reversal of the q-profile, characteristic of the RFP. This is further discussed in Section 4.3. The effect of a conducting wall, including the case of finite conductivity, is treated in more detail in Chapter 7.4.

2) As the thermal confinement quality is found to increase roughly linearly with plasma current in tokamaks, there is tendency to maximize I_p in tokamak scenarios.
3) Owing to the close fitting shell used in the experiment, the mode occurs as a resistive wall mode rather than an ideal kink (Section 7.4).

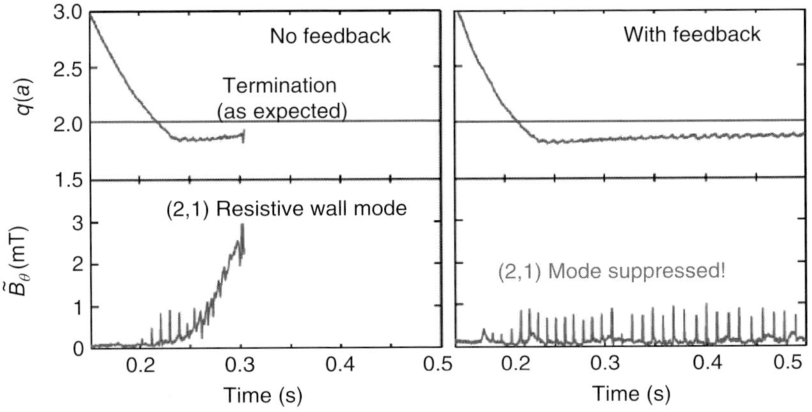

Figure 4.1 Tokamak discharge with circular cross-section run in the RFX reversed field pinch (in tokamak mode). As the current is increased, a disruption occurs at $q_a = 2$ due to the occurrence of an external (2,1) kink (a), whereas with active magnetic feedback control, the mode can be avoided (b). (Please find a color version of this figure on the color plates.)

4.2.2
Modes with $m \geq 2$

For poloidal mode numbers $m \geq 2$, the integral term in Eq. (4.6) will always make a positive contribution and stability is determined by a balance between the potentially destabilizing part of δW_B and the stabilizing terms.

In Figure 4.2, we plot δW_B versus q_a for a number of modes with $n = 1$ and different m, assuming the wall to be at infinity, that is $\Lambda = 1$. It can be seen that windows of negative W_B exist for each mode below $q_a = m/n$ and that the lowest

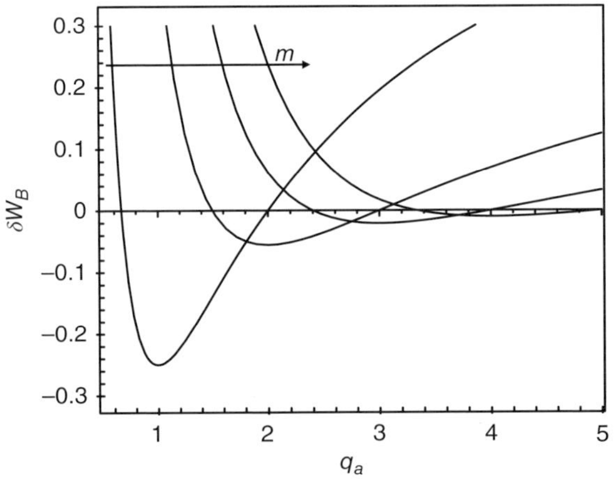

Figure 4.2 Contribution of δW_B to δW for $n = 1$ as a function of q_a for $m = 2, 3, 4, 5$. The potential windows of instability overlap.

m is most negative, indicating that it is potentially the most unstable mode. This can be understood as for

$$q_a < \frac{m}{n} \tag{4.10}$$

the term δW_B can be destabilizing, that is the external kink mode can be unstable if the resonant surface is located outside the plasma. Turning this statement around, we find that in the limit of negligible β, the external kink will always be stable if the resonant surface is located inside the plasma. We will come back to the case with finite β in Chapter 7.

In Figure 4.2, it can also be seen that the instability windows overlap, so it is essential to evaluate the integral term in Eq. (4.6) to determine if there are stable windows due to the stabilizing contribution of the fluid term δW_F. From the structure of the integral, it can be seen that it is minimized by a test function that has small amplitude in the bulk and increases towards the edge, where the term $(n/m - 1/q(r))^2$ decreases as the resonant surface is approached. However, an analytical treatment is in general not possible. Hence, the ideal kink analysis has been done using the class of current profiles introduced in Chapter 2 by Eq. (2.12) and integrating the Eigenvalue equation numerically [12]. The analysis shows that the peakedness of the current profile has a decisive influence on the stability of the external kink.

In Figure 4.3, the regions of stability are plotted as function of q_a and μ. While for peaked current profiles, $\mu > 2.5$, the entire q_a range is predicted to be stable down to the Kruskal–Shafranov limit a broader current distribution leads to instability windows, with the lowest m modes the most unstable ones, and for $\mu < 1$, no stable window is found in the entire range. We note that this is an idealized picture

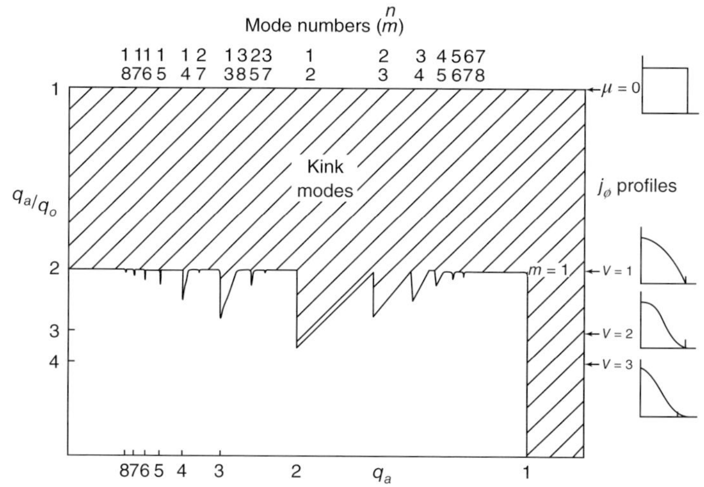

Figure 4.3 Stability diagram for external kinks at vanishing β as function of current peaking μ and edge safety factor q_a. *Source:* Wesson 1978 [12]. Reproduced with permission by the IAEA.

and, as shown earlier, the $q_a = 2$ limit is usually a hard one in experiments, no matter how peaked the current profiles are, unless active feedback control is applied.

It turns out that for tokamak operation with a 'conventional' q-profile, that is one where the current distribution is peaked in the centre[4], external kink modes are not an issue as long as $q_a > 2$, unless β is so high that it cannot be neglected any longer. Purely current driven external kinks have however been found to occur in fast current ramps, provided the ramp rate is much larger than the inverse of the resistive current redistribution rate, as these lead to a broad current profile. An example is shown in Figure 4.4 where a fast current ramp provokes ideal external kinks of mode number ($m = q_a, n = 1$) as q_a passes through the corresponding integer values.

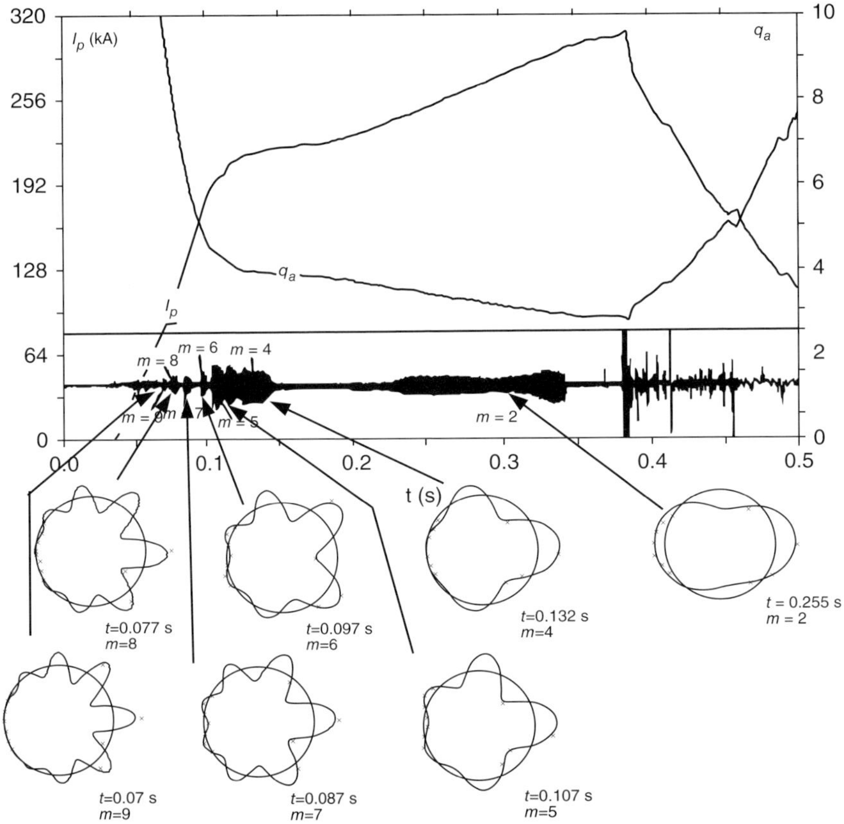

Figure 4.4 Occurrence of ideal external kinks in the current ramp up of a tokamak discharge in the ASDEX tokamak. The (2,1) mode occurring at reduced ramp rate is a resistive mode that terminates the discharge disruptively. This instability is treated in Chapter 10.

4) This is the case if the current profile is dominated by the ohmic contribution as the electrical conductivity is the highest in the centre (Chapter 7).

In summary, the external kink does not provide too strong restrictions to the conventional tokamak operational space in the case of negligible β and operation at $q_a > 2$. However, in the case of broad current profiles and finite β, this is different as we will see in Chapter 7.

Another case that is significantly different from the one treated earlier is the case of current profiles with finite edge current density $j(a) \neq 0$, which is excluded by the analysis using the profile class $j = j_0((1 - r/a)^2)^\mu$. In the cylindrical low β analysis, this will always lead to instability for large m, that is for a very localized test function, leading to the so-called peeling mode [13]. However, the detailed stability of the peeling mode depends crucially on the contribution of finite β and toroidicity and is hence not treated here in detail. Rather, we come back to the peeling mode in more detail when we treat the ELM instability in Chapter 6.

4.3
Internal Kink Modes

In this section, we treat internal kink modes, that is modes with $\xi(a) = 0$. For this case, only the integral term in Eq. (4.6) has to be considered. Similar to the derivation of the Kruskal–Shafranov limit, we set $m = 1$ and see that for a constant displacement, $\xi = \xi_0$, the integral will be zero. However, a constant displacement is in contradiction to the assumption of zero displacement at the plasma edge. This motivates the use of a test function that is constant inside $q = 1$ and zero outside and the drop occurs at $q = 1$[5], where the finite $d\xi/dr$ is compensated by $(n/m - 1/q)^2$ being zero.

This test function is shown in Figure 4.5. Using this function, $\delta W = 0$ to order $(r/R_0)^2$, as was assumed in deriving Eq. (4.6).

This means that the stability of the internal $(1, 1)$ kink in a tokamak is determined by the next order, which is $(r/R_0)^4$. According to the ordering (Eq. 4.2), this means including the pressure term and going to next order in expanding g, whereas the contribution due to f is still strictly zero because of the structure of the test function, $\xi' = 0$.

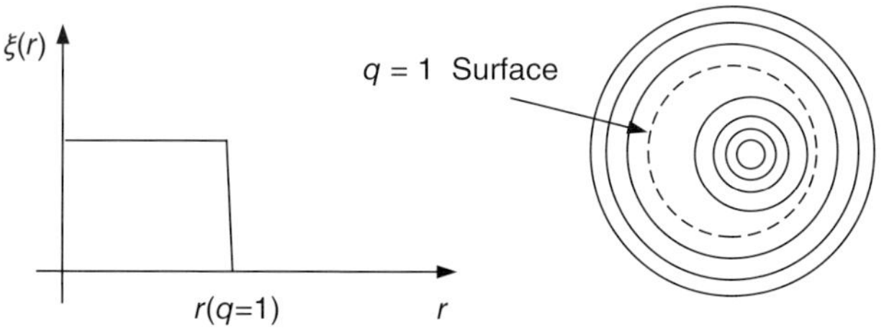

Figure 4.5 Test function used to minimize δW for internal kink analysis.

5) Again, we use $n = 1$ as it is the most restrictive condition in a tokamak where q decreases during the current ramp up, starting from ∞.

After expanding, the contribution to fourth order becomes

$$\delta W_{4,\text{cyl}} = \frac{2\pi^2 B_z^2}{\mu_0 R_0} \xi_0^2 n^2 \int_0^{r_1} r\,dr \left(r\beta' + \frac{r^2}{R_0^2} \left(\left(n - \frac{1}{q} \right) \left(3n + \frac{1}{q} \right) \right) \right) \tag{4.11}$$

where the integration is carried out from the centre to the $q = 1$ surface as $\xi = 0$ outside $q = 1$.

It is clear that the effect of finite β is usually destabilizing as $\beta' < 0$ for pressure profiles that decrease monotonically from the centre to the edge. The second term gives a negative contribution if $q(r) < 1$, and hence for monotonically increasing q-profiles, a necessary stability condition is

$$q(0) > 1 \tag{4.12}$$

Indeed, if $q(0)$ falls below 1 in tokamak discharges, an $(m = 1, n = 1)$ mode is often observed to occur, with topology similar to that shown in Figure 4.5, that is perturbing the plasma only inside $q = 1$. We note that this perturbation has the same topology as the plasma torus, explaining why the $(m = 1, n = 1)$ mode is a special case in toroidal systems. The condition given by Eq. (4.12) can be seen as a limit to the central current density as expansion of $q(r)$ near the axis and taking the limit $r \to 0$ yields

$$q(0) = \frac{2B_z}{\mu_0 R_0 j_0} \quad \to \quad j(0) < \frac{2B_z}{\mu_0 R_0} \tag{4.13}$$

as limit for the central current density $j(0)$. The occurrence of the $(m = 1, n = 1)$ mode at $q < 1$ is tightly connected to the sawtooth instability, which is treated in Chapter 11.

We also note here that the structure of Eq. (4.11) allows another region of stability to internal modes if we assume that the q-profile can reverse over the radius, as is the case for the RFP. Then, the two terms in brackets will give stabilizing contributions in different windows of $q(r)$ and overall, a window of stability at $q < 1$ can be found by carefully optimizing the profiles[6] [14].

So far, we have discussed the stability of kink modes only in the cylindrical case. It turns out that toroidal effects, especially due to the pressure, are entering to order $(r/R_0)^4$ in the tokamak so that they have to be considered in the analysis of the internal kink. Specifically, toroidal mode coupling of the $(1, 1)$ mode to the $(2, 1)$ mode includes a correction that can be treated analytically for circular cross section and a special choice of the q-profile. An involved calculation assuming a parabolic q-profile with $(q(0) - 1) \ll 1$ yields [15]

$$\delta W_4 = \left(1 - \frac{1}{n^2} \right) \delta W_{4,\text{cyl}} + \delta W_{4,\text{tor}} \tag{4.14}$$

with

$$\delta W_{4,\text{tor}} = \frac{3n^2 r_1^4}{R_0^2} \frac{2\pi^2 B_z^2}{\mu_0 R_0} \xi_0^2 (1 - q_0) \left(\frac{13}{144} - \hat{\beta}_p^2 \right), \tag{4.15}$$

6) Note that for an RFP, the expansion in inverse aspect ration will be slightly different from the tokamak as n/m is not small.

where $\hat{\beta}_p$ is the poloidal β inside the $q = 1$ surface. For $n = 1$, the cylindrical contribution vanishes and stability is determined entirely by the toroidal contribution. As a result, there is a critical $\hat{\beta}_p$ below which the (1, 1) mode should be stable, but it is low under typical conditions, so that the limit (Eq. (4.13)) is still approximately valid.

We note, however, that the fact that the internal kink stability is determined by higher order terms means that subtle effects can play a role in determining the precise stability boundary. These include finite resistivity, two fluid effects, finite Larmor radius stabilization and the effect of a fast (i.e. $v \gg v_{\text{thermal}}$) particle population that can be both stabilizing and destabilizing to the internal kink. These effects will be treated in conjunction with the analysis of the sawtooth instability in Chapter 11.

In summary, the ideal kink modes limit the maximum value of I_p by the Kruskal–Shafranov limit and the maximum current density on axis by the internal kink. As mentioned earlier, the energy confinement time in tokamaks increases with I_p, so this also sets a limit to confinement. Furthermore, the peakedness of the current profile, which is limited by the limit to the central current density, determines the available magnetic shear, which is beneficial for stabilizing localized pressure-driven modes as will be shown in Chapter 5.

4.4 $n = 0$ Modes: The Vertical Displacement Event (VDE)

In this section, we consider another type of ideal MHD instability that does not rely on helical deformation of the plasma but rather is a positional instability of the whole plasma column. In Chapter 2, it was shown that the radial position of a plasma is defined by adding a vertical field from external poloidal field coils that compensates the hoop force. In a purely vertical field, the plasma does not experience a vertical force and the vertical position is not well defined. This can be cured by adding a small radial field such as shown in Figure 4.6a. Here, a small displacement in vertical direction will lead to a restoring force and the equilibrium position z_0 is defined by $B_R(z_0) = 0$. Note that in this case, we have assumed that B_R is so small that it does not noticeably change the shape of plasma cross section. Obviously, the vertical stability depends on the curvature of the additional field created by the poloidal field coils. It is convenient to express this by the field index n

$$n = -\frac{R}{B_z}\frac{\partial B_z}{\partial R} \tag{4.16}$$

For the stable situation shown in Figure 4.6, the radial field is positive (outward) for $z > z_0$ and negative (inward) for $z < 0$ so that $\partial B_r/\partial z > 0$ and as for a vacuum field $\partial B_r/\partial z = \partial B_z/\partial R$, the field index is positive (recall that, for positive plasma current, B_v according to Eq. (2.65) is in the negative z-direction.). Hence, the stability criterion against the positional vertical instability is

$$n > 0 \tag{4.17}$$

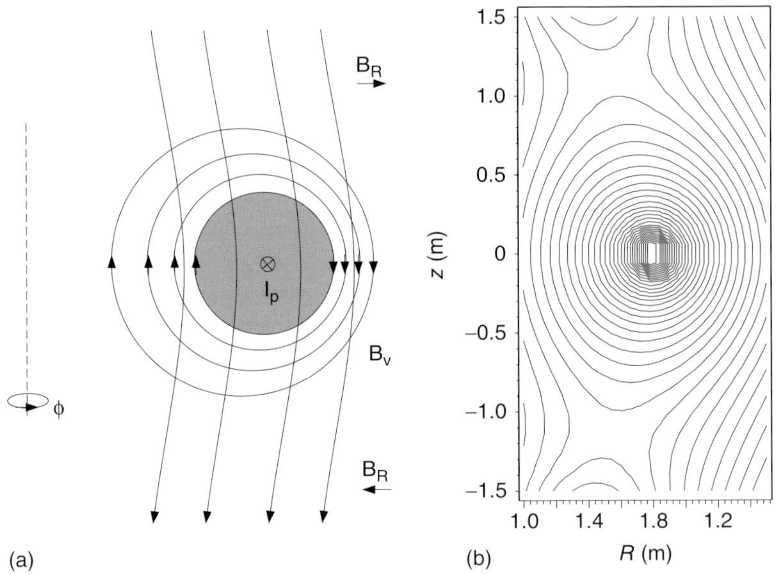

Figure 4.6 (a) Addition of a small radial field component to the vertical field provides a restoring force against vertical displacement. (b) With additional coils above and below the plasma, the cross section can be elongated, but vertical stability is lost (see text).

This becomes important when elongating the plasma cross section in order to exploit the favorable effects of shaping on stability (Section 7.5) and transport since for given a and B_t, q_a increases with elongation κ, that is a higher I_p can be run without violating the Kruskal–Shafranov criterion. To see this qualitatively, we can consider the plasma as a loop carrying current I_p. We assume that in equilibrium, the loop is located at $R = R_0, z = 0$ and B_R has to vanish so that no net vertical force remains.

In order to elongate the plasma, we need a quadrupole field that can for example be created by external coils above and below the plasma[7]. Figure 4.6b shows a flux pattern in which the plasma current has been approximated by a current filament at $R_0 = 1.8m, z_0 = 0m$, and additional poloidal field coils are located at $R_{PF} = 1.4m, z_{PF} = \pm 2$ m. All coils carry equal current. The effect of elongation is clearly visible. While the radial field added by the two coils cancels for $z = z_0$, it is directed such that for a small displacement in z-direction, the resulting vertical force will point in the direction of the initial perturbation, that is the equilibrium is unstable in the vertical direction[8]. For poloidal field coil currents parallel to I_p, the resulting $B_R(z)$ is such that the field index n is now negative, indicating instability. This is the basic mechanism leading to the so-called vertical displacement

7) We note that the quadrupole moment can be enhanced by running antiparallel currents in PF coils located to the left and right of the plasma current, effectively 'squeezing' the plasma column.
8) Another way to see this is that the force between plasma and PF coil will increase with decreasing distance and as elongation means that the coils 'pull' on the plasma with equal force at $z = 0$, any small perturbation will lead to a motion towards the closer PF coil.

events (VDEs) in tokamaks with elongated cross section, that is a loss of control of the vertical position. The detrimental consequences of VDEs are discussed in detail in Chapter 10 in the context of disruptions.

To analyse for a general external field the stability against a small vertical displacement, we expand the radial field around (R_0, z_0) and calculate the destabilizing $\mathbf{j} \times \mathbf{B}$ force in vertical direction assuming that the plasma current is a filament at R_0, z_0:

$$F_{\text{destab}} = 2\pi R_0 I_p \frac{\partial B_R}{\partial z}|_{(R_0,z_0)}(z - z_0) = -2\pi I_p B_z n(z - z_0) \tag{4.18}$$

where we have used the definition of the field index given earlier.

In general, the field B_z will be proportional to the poloidal field generated by the plasma current, with the precise value determined by the force balance through the Grad–Shafranov equation. Similar to the case with circular cross section treated in Section 2.2.1.2, it will depend on inverse aspect ratio a/R_0, the parameter $\beta_p + \ell_i/2$ and the detailed plasma shape. In the approximation that the plasma is a current carrying wire at R_0, z_0, we assume these details to be described by a coefficient α_S[9] such that we can define, in analogy to Eq. (2.65),

$$B_z = \alpha_S \frac{\mu_0 I_p}{4\pi R_0} \tag{4.19}$$

From the equation of motion, we obtain the linear growth rate using the exponential growth ansatz $z - z_0 = \bar{z}\exp(\gamma t)$

$$m_p \frac{d^2 z}{dt^2} = F_{\text{destab}} = -n\alpha_S \frac{\mu_0 I_p^2}{2R_0}(z - z_0) \;\rightarrow\; \gamma^2 = -\frac{v_{\text{A,pol}}^2}{R_0^2}\alpha_S n \tag{4.20}$$

where the poloidal Alfvén velocity has been defined as $\mu_0 I_p/(2\pi a)/\sqrt{\mu_0 \rho}$. Equation (4.20) indicates instability for $n < 0$, consistent with Eq. (4.17). The growth rate is hence of the order of the poloidal Alfvén time $\tau_{\text{A,pol}} = R_0/v_{\text{A,pol}}$, which is of the order of μs for typical tokamaks and thus inaccessible for feedback control by coils.

In order to change this situation, it is common to provide passive conducting elements that slow down the motion of the plasma. The stabilizing effect of a passive element, such as the vacuum vessel, comes from the current induced in it because of the change in magnetic field when the plasma moves vertically. By Lenz' rule, this current is directed opposite to I_p and its magnitude can be derived using the mutual inductance M_{cp} between plasma current and external conductor so that the flux balance for the conductor of self-inductance L_c reads

$$\Psi_c = M_{\text{cp}} I_p + L_c I_c \tag{4.21}$$

and the circuit equation when the plasma vertical position changes at fixed I_p reads

$$\frac{d\Psi_c}{dt} = I_p \frac{\partial M_{\text{cp}}}{\partial z}\frac{dz}{dt} + L_c \frac{dI_c}{dt} = -R_c I_c \tag{4.22}$$

9) This implies that B_z is roughly constant, but $\partial B_z/\partial R$ is not and determines n.

where we have introduced the electrical resistance R_c of the passive conducting structure. Inserting the exponential growth assumption, we get an equation for I_c:

$$I_c = -I_p \frac{\partial M_{\rm cp}}{\partial z} \frac{z}{L_c} \frac{\gamma \tau_R}{\gamma \tau_R + 1} \tag{4.23}$$

where the resistive timescale of the conductor is given by $\tau_R = L_c/R_c$. The radial field generated by I_c at the position of the plasma can be expressed through the flux $\Psi_p = M_{\rm pc} I_c + L_p I_p$

$$B_R = -\frac{1}{2\pi R_0} \frac{\partial \Psi_p}{\partial z} = -\frac{1}{2\pi R_0} \frac{\partial M_{\rm pc}}{\partial z} I_c = \alpha_c \frac{\gamma \tau_R}{\gamma \tau_R + 1} \frac{z}{R_0} \frac{\mu_0 I_p}{4\pi R_0} \tag{4.24}$$

where the current I_c has been expressed by I_p using Eq. (4.23) and we have defined the coefficient α_c in the same way as α_S above by

$$\alpha_c = \frac{2R_0}{\mu_0 L_c} \left(\frac{\partial M_{\rm cp}}{\partial z} \right)^2 \tag{4.25}$$

using the fact that due to reciprocity, $M_{\rm cp} = M_{\rm pc}$. With this definition, the stabilizing force due to the current I_c becomes

$$F_{\rm stab} = -\alpha_c \frac{\gamma \tau_R}{\gamma \tau_R + 1} \frac{\mu_0 I_p^2}{2} \frac{z}{R_0} \tag{4.26}$$

and adding this to the force balance (Eq. 4.20) from above using the exponential ansatz, we obtain the dispersion relation

$$\gamma^2 \tau_{\rm A,pol}^2 + \alpha_S n + \alpha_c \frac{\gamma \tau_R}{\gamma \tau_R + 1} = 0 \tag{4.27}$$

For $\tau_R = 0$, we recover the result without passive elements. For $\tau_R \gamma \to \infty$, corresponding to an ideally conducting wall, we see that a window of stability exists for negative n if the second term is smaller than the third, that is for sufficiently strong coupling between plasma and wall, described by the mutual inductance. Finally, in the regime where the induced currents slow down the motion to the resistive timescale of the wall, we can formally set $\tau_A \to 0$ and see that the growth rate becomes

$$\gamma \tau_R = -\frac{\alpha_S n}{\alpha_c + \alpha_S n} \tag{4.28}$$

which, in the unstable regime ($n < 0$), is of the order of the resistive timescale for strong coupling to the passive conductor, that is $|\alpha_S n| \leq \alpha_c$, and exhibits a singularity as the coupling gets weaker. This singularity comes from formally setting $\tau_{\rm A,pol} = 0$; if we use the full dispersion relation (Eq. (4.27)), this point corresponds to the case where wall stabilization is lost and the growth rate becomes ideal. Figure 4.7 shows the full dispersion relation for different values of τ_R/τ_A together with the approximation (Eq. (4.28)) for one particular case.

For $\tau_R/\tau_A = 1$, the system is always unstable on the Alfvén timescale, whereas for increasing τ_R/τ_A, the growth rate is reduced in the region $|\alpha_S n| < \alpha_c$ and

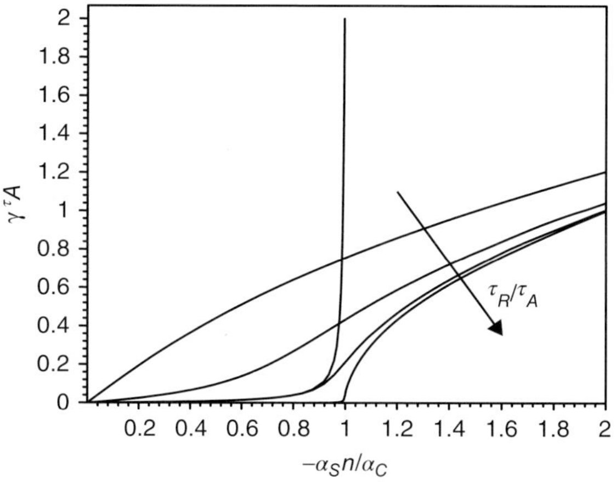

Figure 4.7 Growth rate of the VDE instability in the presence of a conducting wall for different values of τ_R/τ_A (1,10,100,1000) as function of $-\alpha_S n/\alpha_c$. For $-\alpha_S n \to \alpha_c$, the growth rate changes from resistive to ideal. Also shown is the singular case neglecting the plasma inertia for $\tau_R/\tau_A = 100$, which in this regime deviates from the full solution only close to the transition from the resistive to the Alfvénic regime.

approaches zero for large τ_R/τ_A, indicating a growth rate of the order of the inverse of τ_R. For $|\alpha_S n| > \alpha_c$, the system is always unstable on the Alfvén timescale.

We note here that the work done by displacing the plasma is $\delta W = F\delta z$ and hence the growth rate can also be expressed as $\gamma\tau_R = \delta W_{\text{no wall}}/\delta W_{\text{with wall}}$. We will come back to this point when we treat the resistive wall mode in Chapter 7.4.

For practical application, this means that there must always be active feedback control to avoid a VDE for elongated plasmas, but the precise calculation of the requirements has to include the real geometry that brings in two important aspects:

- In a finite aspect ratio tokamak, the flux surfaces are 'naturally' elongated, that is for a strictly vertical field, the cross section is not exactly circular but will exhibit a finite elongation $\kappa_{\text{nat}} > 1$. Hence, there is a window of stability at $1 \leq \kappa \leq \kappa_{\text{nat}}$ in which no feedback control is required. From calculations using a strictly vertical field in toroidal geometry, a fit as function of the aspect ratio $A = R/a$ yields $\kappa_{\text{nat}} - 1 \approx 0.5/(A - 1)$, which shows that the effect is small at usual A (e.g. $\kappa_{\text{nat}} = 1.25$ at $A = 3$) but can be quite important for low aspect ratio 'spherical' tokamaks.
- In the treatment earlier, we have assumed that the plasma moves rigidly without changing shape. In reality, the fast vertical motion cannot compress the toroidal field and is hence not rigid. This effect has to be considered when calculating the growth rate to determine the stability margin for a given feedback system.

While present-day tokamaks are designed to avoid VDEs during routine operation, they often occur in the context of the disruptive instability and are hence discussed in more detail in Chapter 10.

5
Pressure Driven Modes in a Tokamak

In this chapter, we treat pressure driven modes. In Chapter 4, we had explicitly neglected the p' term, arguing that from a mathematical point of view, neglecting β in expanding Eq. (3.36) means that $\beta = \mathcal{O}\left(\frac{r}{R_0}\right)^2$, which is in general well fulfilled. There is however an important exception to this argument in the vicinity of resonant surfaces, as there, according to Eqs (4.3) and (4.4), f and g, which represent the stabilizing contribution of field line bending, vanish, and the pressure term will dominate in the region where $1/q - n/m \approx 0$. Clearly, the width of that region will depend on the radial variation of the q-profile and hence we will restrict our analysis of pressure driven modes to radially localized instabilities. For typical q-profiles, the region over which the pressure dominates is indeed small, verifying the assumption of a local analysis[1]. The assumption $q = m/n$ implies that the modes have constant phase along the field lines. While this is strictly true for the localized interchange modes treated in Section 5.1, it does no longer hold for the ballooning modes treated later in this chapter.

5.1
Localized Interchange Modes in the Screw Pinch

In a first step, we treat interchange modes, that is pressure driven modes that become unstable in regions of unfavourable curvature as shown in Figure 3.2. We start with the analysis of the screw pinch, where the poloidal field is curved unfavourably while the axial field is not curved. Hence, we expect a destabilizing contribution in the region where the field line bending term vanishes, that is close to the rational surface as pointed out earlier. Consequently, we restrict our analysis to a typical region $x = r - r_s$ where r_s is the radius of the resonant surface. Then, q is expanded according to

$$q(x) = q(r_s) + q'(r_s)x + \cdots \quad \text{with} \qquad q(r_s) = \frac{m}{n} \tag{5.1}$$

1) This assumption and its validity is discussed in more detail when global pressure driven modes are treated in Chapter 7.

Magnetohydrodynamic Stability of Tokamaks, First Edition. Hartmut Zohm.

and the field line bending term becomes

$$\frac{1}{q} - \frac{n}{m} \approx \frac{1}{q_s} \frac{1}{1 + \frac{q'}{q_s} x} - \frac{n}{m} \approx -\frac{q'}{q_s^2} x \tag{5.2}$$

With this relation, we can evaluate δW in the vicinity of the resonant surface. As we are assuming a localized perturbation, the surface and vacuum terms in Eq. (3.35) do not play a role and only δW_F has to be evaluated. If we furthermore use the tokamak ordering $nr/(mR_0) = B_\theta / B_z \ll 1$, we obtain

$$f(x) \approx \frac{r_{\text{res}}^3}{q_s^2} \frac{B_z^2}{R_0^2} \left(\frac{q'}{q_s} \right)^2 x^2 \tag{5.3}$$

$$g(x) \approx \frac{r_{\text{res}}^2}{R_0^2} \frac{2\mu_0 p'}{q_s^2} + \frac{r_s B_z^2}{R_0^2} (m^2 - 1) \frac{q'^2}{q_s^4} x^2 \tag{5.4}$$

Inserting these into δW, we note that since $f(x)$ is multiplied by $(d\xi/dx)^2$, it is effectively of zeroth order in x and hence the second term in $g(x)$ can be neglected so that δW becomes

$$\delta W = \frac{2\pi^2 B_z^2}{\mu_0 R_0} \frac{r_s^2}{q_s^2} \int dx \left(s_s^2 \left(x \frac{d\xi}{dx} \right)^2 + \beta' r_s \xi^2 \right) \tag{5.5}$$

which shows the balance between stabilizing field line bending that increases with magnetic shear $s = rq'/q$, effectively decreasing the region in which the pressure term dominates, and the destabilizing normalized pressure gradient β', which is negative for pressure profiles which decrease with radius. Different from the analysis of current-driven modes, it is not obvious how to chose a test function that makes δW negative. Hence, we make use of the fact that a functional $\delta W = F(\xi, \xi', r)$ is minimized by the solution of the corresponding Euler–Lagrange equation

$$\frac{\partial F}{\partial \xi} - \frac{d}{dx} \frac{\partial F}{\partial \xi'} = 0 \tag{5.6}$$

which, using Eq. (5.5), leads to

$$D_S \xi + \frac{d}{dx} \left(x^2 \frac{d\xi}{dx} \right) = 0 \tag{5.7}$$

where

$$D_S = -r_s \beta' / s_s^2 \tag{5.8}$$

which, as pointed out earlier, is usually a positive quantity. This is an ordinary linear differential equation that can be solved with a polynomial ansatz $\xi(x) = c_1 x^{\ell_1} + c_2 x^{\ell_2}$ so that

$$D_S + \ell + \ell^2 = 0 \rightarrow \ell_{1,2} = -\frac{1}{2} \pm \frac{1}{2} \sqrt{1 - 4D_S} \tag{5.9}$$

The solution $\xi(r)$ minimizes δW close to the resonant surface. Hence, inserting ξ into δW should give an indication of the stability of the system against localized interchange modes. For $D_S > 1/4$, the exponent is complex, $\ell_{1,2} = -1/2(1 \pm$

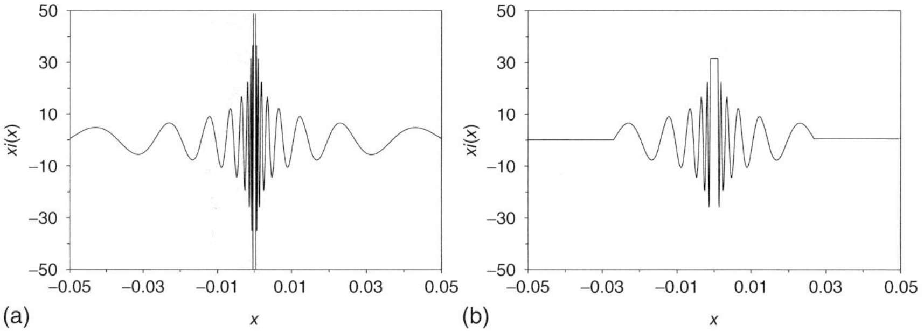

Figure 5.1 (a) Test function minimizing δW for localized interchange modes and (b) test function constructed from it that leads to a negative δW without singularity at $x = 0$.

$i\sqrt{4D_S - 1}$) and ξ becomes

$$\xi(x) = \frac{1}{\sqrt{|x|}} \left(c_1 e^{i\frac{1}{2}\sqrt{4D_S - 1}\ln x} + c_2 e^{-i\frac{1}{2}\sqrt{4D_S-1}\ln x} \right) \tag{5.10}$$

By chosing the constant properly, the solution can be expressed in pure sine or cosine terms, and with $c_1 = c_2 = c/2$, it becomes

$$\xi(x) = c\frac{1}{\sqrt{|x|}} \cos(\frac{1}{2}\sqrt{D_S - 1/4}\ \ln |x|) \tag{5.11}$$

This function is plotted on the left-hand side of Figure 5.1. It is evident that it has a singularity from the $1/\sqrt{x}$ term at $x = 0$ and hence cannot be used as test function.

A proper test function can however be constructed by realizing that as ξ minimizes δW, it has the property

$$\frac{d}{dx}(x^2\xi\xi') = x^2\xi'^2 + \xi(2x\xi' + x^2\xi'') = x^2\xi'^2 - D_S\xi^2 \tag{5.12}$$

where in the last step we have made use of the fact that ξ is a solution of the Euler Lagrange equation (Eq. (5.7)). Hence, the contribution to δW from integration over an interval x_1 to x_2 can be written as

$$\delta_{12} = \int_{x_1}^{x_2} (x^2\xi'^2 - D_S\xi^2)dx = x^2\xi\xi'|_{x_1}^{x_2} \tag{5.13}$$

Thus, while the function itself is not a suitable test function, such a function can be constructed using it on an interval where on one end, $\xi = 0$, and on the other end, $\xi' = 0$. Owing to the oscillatory nature of the function, this is always possible[2)]. A suitable test function is shown in Figure 5.1b, where in the vicinity of $x = 0$, it has been replaced by a positive constant ξ_0 matching at a point $\pm x_2$ where $\xi' = 0$ and

2) This is related to the more general theorem by Newcomb that states that oscillatory solutions indicate an unstable situation.

it has been set to 0 outside a point x_1 where $\xi = 0$. Using this test function, the only contribution to δW will be

$$\delta W = -\int_{-x_2}^{x_2} D_S \xi_0^2 dx < 0 \text{ as } D_S > 0 \tag{5.14}$$

It follows that in the case treated here, that is $D_S > 1/4$, the system is unstable to localized interchange modes. For the case $D_S < 1/4$, we see from Eq. (5.9) that ℓ is real and negative, that is the solution is singular at $x = 0$, but this time it is not of oscillatory nature and hence its contribution to δW cannot be made to vanish ($\xi' \neq 0$ and $\xi \neq 0$) in the vicinity of the rational surface. Re-expressing D_S by the physical quantities, this yields the so-called Suydam criterion, that is the plasma is unstable if

$$-\frac{8\mu_0 p' r_s}{B_z^2} > \left(r_s \frac{q'}{q_s}\right)^2 \tag{5.15}$$

which highlights the stabilizing nature of the magnetic shear.

5.2
Localized Pressure Driven Modes in the Tokamak

The Suydam criterion describes the stability against localized interchange modes in the screw pinch. However, it cannot be applied to the tokamak as there, also the curvature of the toroidal field has to be considered. Comparing the contributions to the force balance (1.35), we find

$$\frac{B_\phi^2}{\mu_0 R} \approx \left(\frac{qR}{r}\right)^2 \frac{B_\theta^2}{\mu_0 R} = q^2 \frac{R}{r} \frac{B_\theta^2}{\mu_0 r} \tag{5.16}$$

that is the toroidal curvature can actually dominate the effect of the poloidal curvature ($q^2 R/r \gg 1$), but it is stabilizing on the high field side and destabilizing on the low field side of the torus and hence evaluation of its effect requires proper toroidal averaging. Thus, the evaluation of localized interchange stability in the tokamak has to start from the energy functional in toroidal geometry (Eq. (3.27)). While the procedure is similar to that outlined earlier, the algebra is quite involved [10] and here, we only sketch the important steps.

To start with, we assume again that owing to the stabilizing effect of field line bending away from a rational surface, pressure driven modes are localized around the resonant surface. Formally, this can be accounted for by examining the limit $n \to \infty$, which, in Eq. (3.27), means that inU must be compensated by $\partial X/\partial \Psi$, implying that the variation of X perpendicular to the flux surface becomes so strong for $n \to \infty$ that $1/n\partial X/\partial \Psi$ remains finite. Assuming furthermore that U can be expanded in terms of $1/n$, an iterative prescription can be found that relates the minimizing U to $\partial U/\partial \chi$ using Eq. (3.30) and a minimization to first order in $1/n$ yields an algebraic condition expressing U in terms of the coefficient functions and the perpendicular variable X and its derivative $\partial X/\partial \Psi$. Inserting this leads to an expression for δW that is valid for localized modes and only contains X as free function to be used for the minimization:

$$\delta W = \pi \int J d\Psi d\chi (\frac{B^2}{R^2 B_p^2}|k_{\|}X|^2 + R^2 B_p^2 |\frac{1}{n}\frac{\partial(k_{\|}X)}{\partial\Psi}|^2$$

$$-2p' \left(\frac{\kappa_n}{RB_p}|X|^2 - i\frac{\mu_0 I_{\rm pol}\kappa_g}{B^2}\frac{X}{n}\frac{\partial X^*}{\partial\Psi} \right)) \tag{5.17}$$

Here, we have neglected a term proportional to the gradient of the parallel current density as it represents the kink drive in which we are not interested here[3] and introduced κ_n and κ_g, which are the components of the curvature vector defined by (1.36) when decomposed into a component normal to the flux surface, called the *normal curvature*

$$\kappa_n = \frac{RB_p}{B^2}\frac{\partial}{\partial\Psi}\left(\mu_0 p + \frac{B^2}{2}\right) \tag{5.18}$$

and the component lying in the flux surface in the 'poloidal' direction, the so-called geodesic curvature

$$\kappa_g = -\frac{1}{JB_p B^2}\frac{\partial}{\partial\chi}\left(\frac{B^2}{2}\right) \tag{5.19}$$

In Eq. (5.17), the terms in the first line are positive and involve the parallel derivative. They can hence be identified as the stabilizing contribution of field line bending. The second line is the contribution of the pressure drive, where it can be seen that the sign of the curvature will decide if it is stabilizing or destabilizing. As for the Suydam criterion, we hence have to balance field line bending versus pressure drive, but this time considering the variation of the curvature along the field line[4]. Similar to the screw pinch analysis, the field line bending vanishes if the modes are aligned with the field line (i.e. $k_{\|} = 0$ in Eq. (5.17)), but due to the more complex dependence of the pressure drive, we will find that modes that have a small but finite $k_{\|}$ can be the most unstable ones under some circumstances as they allow the amplitude to be localized in the unfavourable curvature region. These so-called ballooning modes will be treated in the following, but before, we discuss the equivalent of the Suydam criterion for the tokamak, which leads to the so-called Mercier criterion.

5.2.1 Interchange Modes in a Tokamak

Starting from the energy principle for localized pressure driven modes (Eq. (5.17)), the criterion for stability against localized interchange modes can be derived by choosing a test function that is localized around a rational surface and has

3) We will come back to this term when we treat localized current driven modes, that is peeling modes, in Chapter 6.
4) Note that for the screw pinch, $\kappa_g = 0$ and $\kappa_n = (B_p^2/B^2)(1/r)$, and with $k_{\|}X = 0$ but $\partial(k_{\|}X)/\partial\Psi \neq 0$, we recover the form of δW that led to the Suydam criterion.

constant phase along the field lines there. The difference to the screw pinch is that the poloidal mode number is not a good quantum number any more and hence the phase variation along the poloidal angle χ has to be chosen according to

$$X(\Psi, \chi, \phi) = \overline{X}(\Psi, \chi) e^{in(\phi - \int_0^{\chi} \nu(\Psi_0, \chi') d\chi')} \tag{5.20}$$

where Ψ_0 denotes the resonant surface. Expanding the function $\overline{X}(\Psi, \chi)$ around the rational surface in the radial variable $x = n^2(\Psi - \Psi_0)$, that is assuming that the radial localization is quite pronounced (of order n^{-2}), it is allowed to assume that the phase variation is the same in the vicinity of the resonant surface[5]. In this ordering, $\overline{X}_0$ is a function of Ψ only while $\overline{X}_1$ depends on both Ψ and χ. Minimizing δW with respect to $\overline{X}_1$, one can express it in terms of $\overline{X}_0$ only, and the resulting form of δW is quite similar to that obtained for the screw pinch:

$$\delta W = \frac{\pi^3}{n^2} \left(\frac{dq}{d\Psi_0} \right)^2 \left(\oint \frac{\nu B^2}{B_p^2} d\chi \right)^{-1} \mu_0 I_{\text{pol}} \int dx \left[x^2 \left(\frac{d\overline{X}_0}{dx} \right)^2 - D_M \overline{X}_0^2 \right] \tag{5.21}$$

where $d/d\Psi_0$ is the derivative with respect to the flux surface label Ψ taken at the resonant surface $\Psi = \Psi_0$. Formally, this can be minimized in the same way as described for the screw pinch, that is by deriving an Euler Lagrange equation, and the stability criterion is again related to the transition from monotonic to oscillatory solutions, appearing at $D_M < 1/4$. However, in the toroidal case, the stability index D_M appearing in Eq. (5.21) is given as an average over the flux surface $\Psi = \Psi_0$ according to

$$D_M = \frac{dp/d\Psi}{\mu_0 I_{\text{pol}} (2\pi dq/d\Psi_0)^2} \left[\left(\frac{dp}{d\Psi_0} \oint \frac{J}{B_p^2} d\chi - \oint \frac{\partial J}{\partial \Psi_0} d\chi \right) \oint \frac{\nu B^2}{B_p^2} d\chi \right.$$
$$\left. + \left(2\pi \mu_0 I_{\text{pol}} \frac{dq}{d\Psi_0} - \mu_0 I_{\text{pol}} \frac{dp}{d\Psi} \oint \frac{\nu}{B_p^2} d\chi \right) \oint \frac{\nu}{B_p^2} d\chi \right] \tag{5.22}$$

The criterion $D_M < 1/4$ was, in this form, first derived by Mercier [16] and is hence called the *Mercier criterion*, with D_M the Mercier index. It is a local criterion that must be evaluated separately on each flux surface (the distance to the neighbouring flux surfaces enters through the magnetic shear). A more intuitive form of the Mercier criterion can be obtained by rewriting it in terms of the normal curvature κ_n given by Eq. (5.18)

$$D_M = \frac{\mu_0 dp/d\Psi_0}{(dq/d\Psi_0)^2} \left\langle \frac{R^2 B_p^2}{J B^2} \right\rangle^{-2} \left(2 \left\langle \frac{R B_p \kappa_n}{B^2} \right\rangle + \left\langle \frac{\Lambda}{B^4} \right\rangle - \left\langle \frac{1}{B^2} \right\rangle \left\langle \frac{\Lambda}{B^2} \right\rangle \right) \tag{5.23}$$

where we have made use of the Grad–Shafranov equation in the form (3.29), used the definition

$$\Lambda = \mu_0 I_{\text{pol}} \left(\mu_0^2 I_{\text{pol}} \frac{dp}{d\Psi_0} - \frac{R^2 B_p^2}{J} \frac{\partial \nu}{\partial \Psi_0} \right) \tag{5.24}$$

5) This will not be true any more for the ballooning mode, leading to a different form of δW and to the need to introduce the 'ballooning angle'.

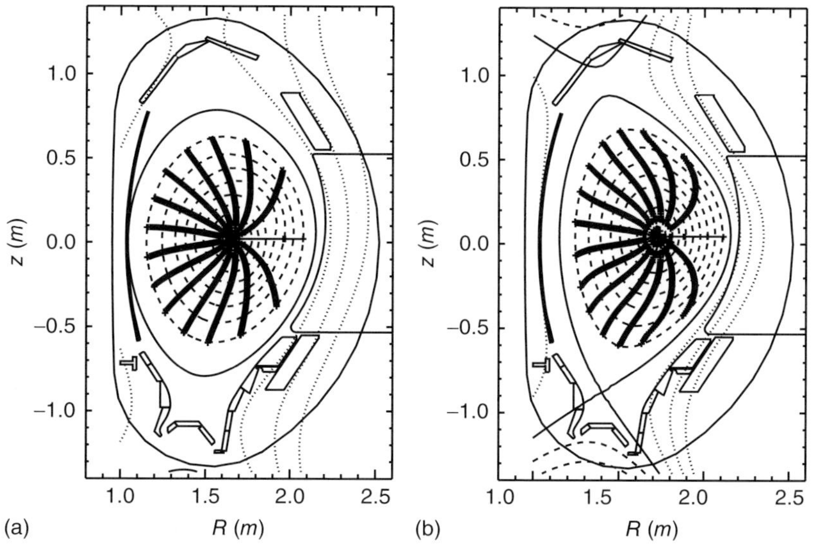

Figure 5.2 (a and b) Shaping of the poloidal cross section by adding triangularity (b) increases the fraction that the field lines spend in the region of favourable curvature (high field side), indicated by the density of puncture points in the Poincare plot. Consequently, the configuration in (b) is expected to have superior interchange stability.

and the appropriate average over the flux surface is defined by

$$\langle f \rangle = \oint \frac{fB^2}{R^2B_p^2} J\,d\chi \left(\oint \frac{B^2}{R^2B_p^2} J\,d\chi \right)^{-1} \tag{5.25}$$

One can see that for the screw pinch, where $B_p = B_\theta(r), R = R_0, B = B_z$ and $q = rB_z/(R_0B_z)$ the terms $\langle \Lambda/B^4 \rangle - \langle 1/B^2 \rangle \langle \Lambda/B^2 \rangle$ cancel and we obtain again the Suydam criterion.

While the Mercier criterion can be tested numerically in a straightforward way for any given tokamak equilibrium, it is possible to obtain a surprisingly simple analytical expression for a torus with cylindrical flux surfaces. Inserting the relations derived in Section 2.2.1.2, the plasma is unstable to localized interchange modes if

$$-\frac{8\mu_0 p'}{r_{\rm res}B_z^2}(1 - q^2) > \left(\frac{q'}{q}\right)^2 \tag{5.26}$$

It can be seen that the toroidal curvature contributes an extra stabilizing term proportional to $p'q^2$ and for $q > 1$, this dominates such that all surfaces will be Mercier stable. While the exact q-value for which the stabilization by the torus effect dominates varies with shape, the general tendency is found for arbitrary cross section. This can already be expected from Eq. (5.16), which states that the toroidal curvature will dominate the poloidal curvature by a factor q^2R/r.

In practical cases, the Mercier criterion usually restricts q_0, but often, the q_0 limit for global MHD modes, such as the (1,1) internal kink, dominates the stability properties and Mercier stability is not too restrictive. However, the general finding that there can be a net stabilizing effect by averaging the curvature over the flux surface is found to be important for a number of micro- and macroinstabilities. For example, the stability properties of plasmas with positive triangularity are found to be improved as the effect of triangularity is to increase the fraction a field line spends in the region of good curvature when spiralling around the torus as can be seen from the Poincare plot in Figure 5.2. We note here that the average of the curvature along the field line is related to the 'magnetic well', which measures the sign of the radial derivative of the average curvature and is a general figure of merit for interchange stability. The Mercier criterion and the magnetic well are also important concepts in stellarator optimization, although their mathematical form is more complex there than in the axisymmetric case.

It is important to note that the Mercier criterion is necessary but not sufficient for stability against pressure driven modes as it was derived assuming a special test function. In fact, while the Mercier criterion indicates that localized pressure driven modes which are aligned with the field lines should not be a big concern for tokamaks, there is a class of instabilities that is more restrictive by localizing its amplitude in the region of bad curvature at the expense of introducing some field line bending. These so-called ballooning modes are treated in Section 5.2.2.

5.2.2
Ballooning Modes

In Section 5.2.1, we derived the Mercier criterion for localized interchange modes under the assumption that the mode under consideration has constant phase along the field line. The result was that for large enough q, the stabilizing contribution from the toroidal curvature on the high field side of the tokamak can dominate the destabilizing contributions from the low field side and from the poloidal curvature, leading to unconditional stability against localized interchange modes. However, as was already indicated in Section 5.2.1, this does not necessarily mean that the plasma is stable to any kind of localized pressure driven mode. More specifically, if we give up the condition that the mode should be exactly aligned with the field lines, we introduce on the one hand a stabilizing contribution from field line bending, on the other hand, localized modes on adjacent resonant surfaces can now interfere in a way that the amplitude cancels in the region of good curvature but becomes large in the region of bad curvature. The concentration of the amplitude on the low field side, the so-called ballooning effect, is shown in Figure 5.3, where the left side shows a single mode of high poloidal mode number ($m = 25$ in this case), while the right side shows the addition of modes in the range $m = 20-30$ that have been assumed to be coupled such that the add up on the midplane low field side to produce the same amplitude there but virtually cancel on the high field side.

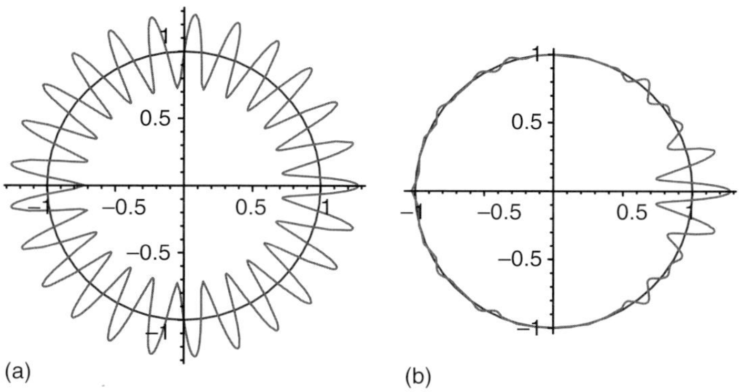

Figure 5.3 Illustration of the effect of outward balloning by superposition of modes with different poloidal mode number m: (a) a single mode with $m = 25$ is compared with (b) a superposition of 11 modes of equal amplitude with m ranging from 20 to 30, arranged in phase to interfere constructively on the midplane outside.

Clearly, the assumption of the coupling of several modes with different m but same n[6] is in contradiction to the localization on a single resonant surface, and we can estimate the radial distance x involved for the coupling of modes with poloidal mode number $\Delta m = m_1 - m_2$ and same n to be

$$\frac{\Delta m}{n} = \Delta q \approx q'x \quad \rightarrow \quad x \approx \frac{\Delta m}{q'n} \tag{5.27}$$

that is in the limit $n \rightarrow \infty$, it will still go to zero. This motivates to start the analysis of ballooning stability from the energy functional derived for $n \rightarrow \infty$ (Eq. (5.17)). As we still assume axisymmetry, n is a good quantum number and we again assume a Fourier decomposition in toroidal angle as given by Eq. (5.20). However, different from the interchange case, we now assume that the radial variation is described by a local variable $x \propto n(\Psi - \Psi_0)$, that is a less strict radial localization than for the interchange mode where we assumed an n^2 relation between x and Ψ. As a result, one finds that the energy functional for this case cannot easily be reduced to a form in which the χ integration is only done on the resonant surface as the radial extent is large enough that the deviation from $k_{\|} \neq 0$ has to be considered. Hence, the Euler–Lagrange equation becomes a partial differential equation involving both χ and x and is not straightforward to solve.

In order to reduce the resulting Euler–Lagrange equation to an ordinary differential equation, one can make use of the fact that, according to the previous discussion, the mode varies rapidly perpendicular to the field lines but slowly along it. This justifies an Eikonal ansatz where the rapid variation of the phase S is put into a phase factor exp iS and the coefficient function contains the slow variation. In our case, this means that we have to adapt the phase from the form given by Eq. (5.20), which assumed no radial variation of the field line pitch, to a

6) The toroidal mode number is still a 'good' quantum number since toroidal symmetry is maintained.

form which accounts for it using the relation (5.27) to expand it around the resonant surface defined by $\Psi = \Psi_0$

$$S(x,\chi,\phi) = n\phi - n\int_0^{\chi}\left(\nu(\chi',\Psi_0) + \frac{\partial\nu(\chi',\Psi_0)}{\partial\Psi_0}(\Psi-\Psi_0)\right)d\chi'$$
$$= n\phi - n\int_0^{\chi}\nu(\chi',\Psi_0)d\chi' - \int_0^{\chi}\frac{\partial\nu(\chi',\Psi_0)}{\partial\Psi_0}d\chi' x \qquad (5.28)$$

Using this ansatz in the energy functional, expanded in $1/n$ as discussed earlier, the Euler–Lagrange equation becomes

$$\frac{1}{J}\frac{d}{d\chi}\left[\frac{1}{JR^2B_p^2}\left(1+\left(\frac{R^2B_p^2}{B}\int_0^{\chi}\frac{\partial\nu}{\partial\Psi_0}d\chi'\right)^2\right)\frac{\partial\hat{X}}{\partial\chi}\right]$$
$$+\frac{2\mu_0}{RB_p}\frac{dp}{d\Psi_0}\left(\kappa_n - \frac{\mu_0 I_{\mathrm{pol}}RB_p^2}{B^2}\kappa_g\int_0^{\chi}\frac{\partial\nu}{\partial\Psi_0}d\chi'\right)\hat{X} = 0 \qquad (5.29)$$

However, there is a conceptual difficulty with Eq. (5.29) as the solution $\overline{X}(\chi,\Psi_0)$ we look for must be periodic in the 'poloidal' co-ordinate χ, which can in general not be fulfilled over the finite x-range assumed in the ansatz for S. To guarantee this periodicity, we have written the Euler–Lagrange equation for a function $\hat{X}$ that is defined to extend over the whole range $-\infty < \chi < \infty$ and our periodic solution $\overline{X}$ is then constructed by the relation

$$\overline{X}(\chi,\Psi_0) = \sum_{\ell=-\infty}^{\infty}\hat{X}(\chi+\ell 2\pi,\Psi_0) \qquad (5.30)$$

which assumes that $\overline{X}$ vanishes at $\pm\infty$ such that the sum converges. This procedure is known as the *ballooning transform* and it is easily shown that the functions $\hat{X}$ and $\overline{X}$ have the same Eigenvalues for a differential operator that is periodic in χ. This means that we have derived an ordinary differential equation which can be solved on each flux surface in the domain $-\infty < \chi < \infty$ and then stability can be tested by constructing a test function in analogy to the procedure used for derivation of the Suydam and Mercier criteria, that is we look for an oscillatory solution which then can be used to prove instability. Conversely, if the solution is not oscillatory, the plasma is stable against ballooning modes[7].

While the solution of Eq. (5.29) can only be obtained numerically except for special cases, it is at least possible to give the explicit form of the ballooning equation for the case of a large aspect ratio tokamak with circular flux surfaces where χ becomes θ and Ψ is related to r. If we neglect the Shafranov shift, but keep the variation of the shear due to pressure and toroidicity, this yields the following form

$$\frac{d}{d\theta}\left((1+(s\theta-\alpha\sin\theta)^2)\frac{d\hat{X}}{d\theta}\right) + \alpha((s\theta-\alpha\sin\theta)\sin\theta+\cos\theta)\hat{X} = 0 \qquad (5.31)$$

7) In practice, one rather solves numerically the full Eigenvalue problem with $\omega \neq 0$ to determine the stability.

where s is the previously defined normalized magnetic shear and α is the normalized pressure gradient

$$\alpha = -\frac{2\mu o R_0}{B^2} q^2 \frac{dp}{dr} \tag{5.32}$$

where a comparison with Eq. (5.29) shows that the pressure does not only enter as driving term but also through the variation of the shear on a flux surface. Numerical solutions of this equation are shown in Figure 5.4 for $s = 0.1$ and $s = 1$ for different values of the normalized pressure gradient α. For low-pressure gradient, the function $\hat{X}$ does not cross zero and hence the system is stable. For the highest pressure gradient tested, the function has a zero crossing, indicating instability to ballooning modes. It is clear that with this method, a marginally stable point can be found where the solution approaches zero for $\chi \to \infty$. For the particular s values chosen here, this is at $\alpha = 0.335$ and $\alpha = 0.63$ which lie on the stability boundary shown in Figure 5.5.

From this example, it is clear that one can generate stability diagrams in the $s - \alpha$ plane by testing stability for many points and interpolating the stability boundary. Such an $s - \alpha$-diagram then characterizes, for each flux surface, the stability against infinite n ballooning modes. For the case of a large aspect ratio tokamak with circular flux surfaces, it is shown in Figure 5.5.

In the $s - \alpha$ diagram shown in Figure 5.5, one notices that for increasing pressure gradient α, the allowable value before instability sets in increases with shear s, as could be expected based on the discussion of the localized interchange criterion. For $s > 0.5$, the stability boundary can be roughly described by $\alpha = 0.6s$. However, there is also another region of stability at larger α. This somewhat surprising result is due to the stabilizing effect of the variation of the magnetic shear on the flux surface that creates a 'magnetic well' much in the sense of the effect of shaping that was discussed for the interchange instability, yielding a larger contribution from the high field side when averaging the curvature along a field line. It relates to the

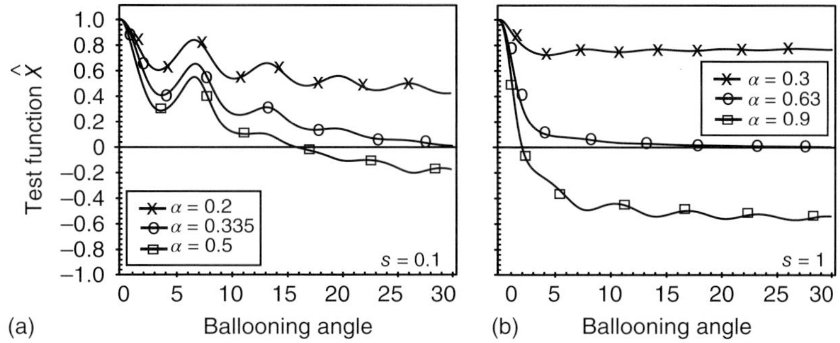

Figure 5.4 Solutions $\hat{X}$ of the ballooning mode equation for large aspect ratio tokamak with circular flux surfaces. (a and b) The three solutions have the same shear, (a) $s = 0.1$ and (b) $s = 1$, but different pressure gradient. The transition from stable (no zero crossing) to unstable (zero crossing) solutions occurs at $\alpha = 0.335$ and $\alpha = 0.63$, respectively.

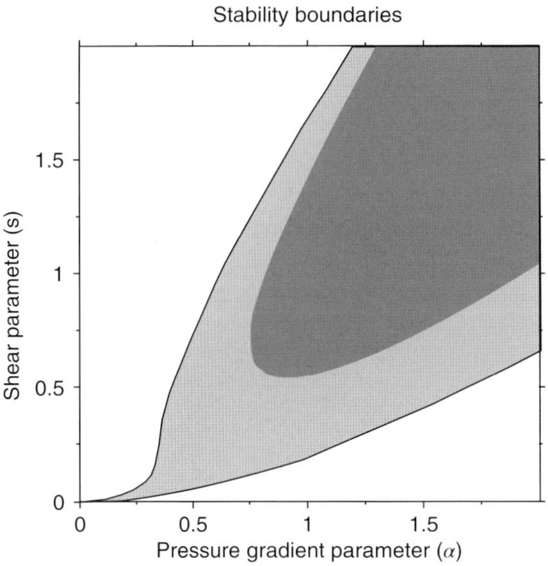

Figure 5.5 Ballooning stability diagram in a plane of normalized pressure α versus normalized shear s for a large aspect ratio toroidal equilibrium with circular flux surfaces. For concentric circles, the first and second stability regions are not connected (light shaded instability region) while fully including the Shafranov shift opens up a path to the region of second stability (dark shaded region). *Source:* Courtesy of H.-P. Zehrfeld, IPP. (Please find a color version of this figure on the color plates.)

above-mentioned fact that the pressure enters in the expression of the local shear in Eq. (5.31). From Figure 5.5, it can be seen that for concentric circles, which have constant poloidal field but varying shear on a flux surface, there is no connection to this region of 'second stability' against ballooning modes, but fully including the Shafranov shift opens up a connection between the two stability regions in the region of low s. This connection region can be further expanded for equilibria with shaped cross section. The second stability region is important for the analysis of edge stability in tokamaks in the context of edge localized modes (ELMs, see Chapter 6) , where the local shear can be low and the pressure gradient is high.

So far, we have outlined the procedure to test ballooning mode stability on individual flux surfaces at high n, but we have not attempted to determine the Eigenfunction as we used a test function approach in the limit $n \rightarrow \infty$. However, the formalism outlined earlier can be extended to yield the full Eigenvalue equation for finite radial wave number, corresponding to large, but finite n. It can be shown analytically that a typical Eigenfunction for one value of n[8] will consist of several localized modes with width $1/n$ and a broader envelope.

8) As outlined earlier, linear stability analysis can be done separately for each n as due to axisymmetry, different toroidal mode numbers cannot couple.

For the approximations made earlier, typical for the core plasma of tokamaks, the envelope is a Gaussian of width $n^{-1/2}$, actually requiring an expansion of the Eigenvalue equation in $n^{-1/2}$ rather than n^{-1} if the mode structure should be obtained. For ballooning modes in the edge, which will be treated in Chapter 6 when we discuss ELMs, the envelope is actually of width $n^{-1/3}$, requiring yet a different expansion. An example for such a linear Eigenfunction is shown in Figure 5.6 for an edge ballooning mode. It can clearly be seen that the shape of each individual Eigenfunction is similar, but their amplitude is scaled according to the envelope.

In summary, purely pressure driven modes are usually localized around the resonant surface of interest and their stability is determined by a balance of pressure drive and stabilization by field line bending. Owing to the strong contribution from toroidal curvature, it is essential to include toroidicity into the analysis. While the Mercier stability criterion is usually not a big concern for tokamaks, ballooning mode stability will restrict the achievable pressure gradient to values that are comparable with those observed experimentally. However, the analysis of the ideal MHD limit to the achievable β has to involve current gradient driven instabilities as well since their stability properties can also be changed at finite β. These effects are discussed in the next two chapters both for edge stability (ELMs) and global β-limits.

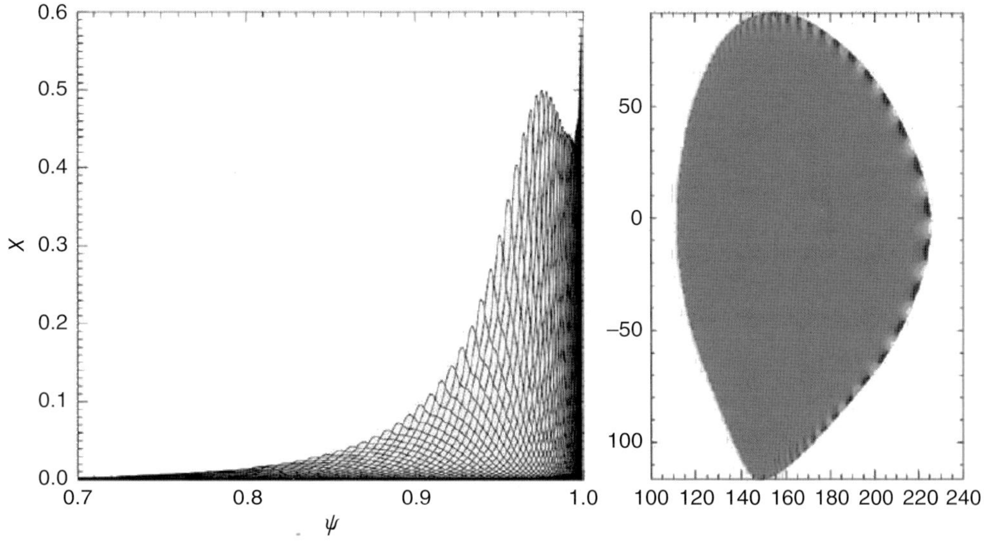

Figure 5.6 Finite *n* ballooning Eigenfunction in the edge of a tokamak plasma (JET-like parameters). For given *n*, there is a spectrum of poloidal wave numbers for which localized Eigenfunctions are calculated. These are combined under an envelope of larger width than the individual modes to form the full Eigenfunction. The ballooning effect is evident from the spatial representation on the right side. *Source:* Courtesy of S. Saarelma, CCFE. (Please find a color version of this figure on the color plates.)

6
Combined Pressure and Current Driven Modes: Edge Localized Modes

In the preceding chapters, we have distinguished between current and pressure driven modes. In practice, performance limiting instabilities are often due to a combination of both drive mechanisms, as was already indicated in the analysis of external kink modes in Chapter 4. In the following, we will analyse important limitations to tokamak operation arising from such ideal MHD instabilities In this chapter, we treat the limitations to edge pressure and current due to the occurrence of edge localized modes (ELMs). In Chapter 7, we show how global limits to β arise from a combination of ideal MHD instabilities.

ELMs are an instability occurring in the edge of H-mode discharges This tokamak operation regime is characterized by the occurrence of a narrow region of steep gradients of density and temperature just inside the last closed flux surface. In this region, the so- called edge transport barrier, turbulent transport of heat and particles is strongly reduced compared to the rest of the plasma [17]. Although the precise mechanism of the formation of the H-mode edge transport barrier is not yet known, it is clear that the suppression of turbulence there is connected to the occurrence of a highly sheared rotationthat seems to be responsible for the turbulence suppression, by tearing apart or thinning out the turbulent eddies.

While the detailed structure of the edge transport barrier is quite complex, it can be approximated as a so-called pedestal in the pressure profile described by a (maximum) pressure gradient ∇p_{max} and width Δ_{ped}. This is indicated in Figure 6.1 where experimentally determined profiles of the edge electron and ion temperature as well as the electron density are shown together with a sketch of the pressure pedestal.

As indicated in Figure 6.1b, the pedestal gives the boundary condition for the core profiles As core temperature profiles are found to be quite 'stiff', that is can roughly be described by $\nabla T/T =$ const. because of the onset of turbulent micro-instabilities above this threshold, the achievable pedestal top temperature is very important for the performance of a tokamak discharge While the pedestal properties are certainly to some extent described by the transport properties, their correct description will also involve understanding their MHD stability properties because of the presence of the ELM instabilities that periodically remove part of the pedestal energy on the timescale of $\tau_{crash} \leq$ ms, which is then restored on a longer timescale. In the following, we will first give a phenomenological description of ELMs and then describe the models that have

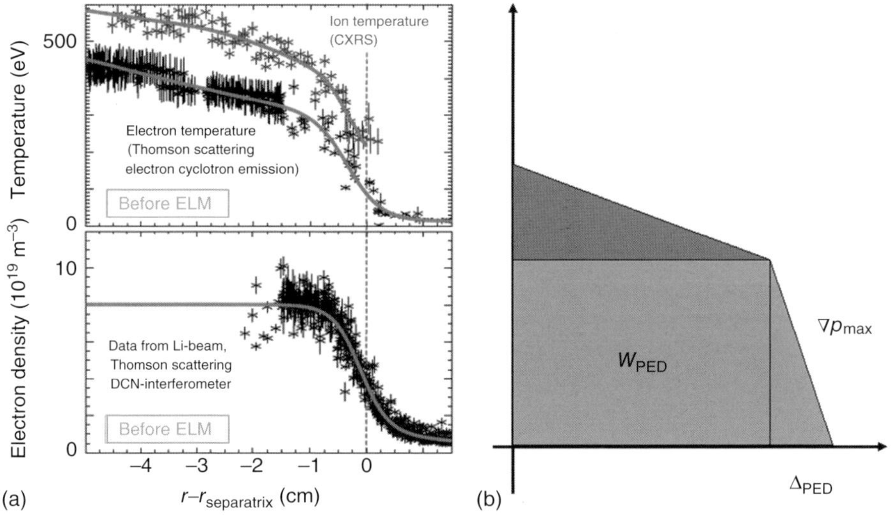

Figure 6.1 Edge kinetic profiles measured during an H-mode discharge on the ASDEX Upgrade tokamak (a) and sketch of the pressure pedestal (b). *Source (a):* Pütterich *et al.* 2009 [18], reproduced with permission of AIP. (Please find a color version of this figure on the color plates.)

been developed to explain different aspects of ELMs The important topic of active ELM control will be treated at the end of the chapter.

6.1
ELM Phenomenology

Figure 6.2 shows typical traces of the temporal evolution of the edge electron temperature during the ELM cycle (Figure 6.2b). As can be seen in Figure 6.2a, the global plasma parametres are not strongly affected by these edge cycles and in fact, ELMs lead to quasi-stationarity of H-mode discharges as they flush out impurities, which otherwards might pollute the plasma core and lead to unacceptable radiation losses there.

While the onset conditions in terms of ∇p_{max} and Δ_{ped} may be accessible by linear MHD stability analysis, the energy loss δW_{ELM} induced by the ELM crash depends crucially on the non-linear evolution and must therefore be assessed using non-linear MHD simulations. Similarly, the prediction of the ELM repetition frequency, ν_{ELM} will ultimately require an understanding of the pedestal dynamics that is highly non-linear. This has been addressed using simple limit cycle models as will be discussed in Section 6.3.1 but will certainly need refined modelling once the transport properties of the pedestal region can be understood in a predictive manner.

The energy loss per ELM is very important as for large devices of the scale of ITER, it can be so large that the impact of energy on the surface of the plasma

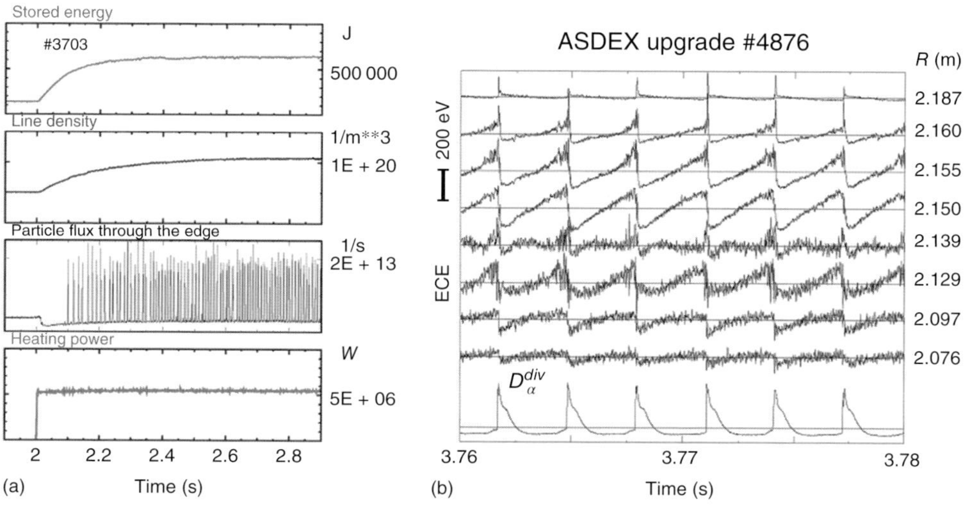

Figure 6.2 Temporal evolution of plasma parametres in an H-mode discharge in ASDEX Upgrade. While the integrated density and stored energy are quasi-stationary on the timescale of the global energy confinement time (a), the edge temperature is modulated by the occurrence of ELMs (b, the outer plasma edge is at $R = 2.17$ m). ELMs can also be seen in (a) as sharp spikes of D_α light, indicating the particle outflux during each ELM event. (Please find a color version of this figure on the color plates.)

facing material leads to melting or local destruction. Hence, active means to influence the size of the ELM loss are an important area of research.

While the principal nature of an ELM is defined by a rapid edge localized MHD instability that expels heat and particles from the edge, there are ELMs of different nature. Owing to the incomplete physics understanding of ELMs, it has so far not been possible to give a unified description of ELMs based on their physical nature. However, a phenomenological classification exists that is based on common features of different ELM types observed on several tokamaks worldwide. A way to distinguish different ELM types is by looking at the dependence of their repetition frequency ν_{ELM} with respect to the power crossing the separatrix, $P_{sep} = P_{tot} - P_{rad} - dW/dt$, where P_{rad} is the power radiated from the plasma inside the separatrix and W is the kinetic energy of the plasma. This criterion is mainly used to distinguish between type I and type III ELMs:

- Type I ELMs are characterized by $d\nu_{ELM}/dP_{sep} > 0$ and can occur over a large range of the tokamak operational space. Owing to their usually large δW_{ELM}, they are sometimes also referred to as *giant ELMs.*
- Type III ELMs are characterized by $d\nu_{ELM}/dP_{sep} < 0$ and usually occur close to the L–H threshold power[1]. In a situation of increasing P_{sep}, their δW_{ELM} will increase as ν_{ELM} decreases.

1) It is found experimentally that the H-mode operational regime is only accessible if the total heating power exceeds a threshold value.

Type II ELMs do not show a clear dependence of ν_{ELM} with P_{sep}. They usually occur in highly shaped plasmas, especially in configurations close to double-null, that is with two active magnetic X-points and have been discussed connected with access of the edge to second stability for ballooning modes [19]. Their δW_{ELM} is usually smaller than that of type I ELMs. In comparison to type III ELMs, they usually do not show any coherent magnetic precursor activity, whereas for type III ELMs, such activity is regularly observed with mode numbers in the range $n = 5-10$. We note that these three ELM types have been identified on several tokamak devices according to the characteristics pointed out earlier. There is a number of other ELM types that have so far not uniquely been identified as a type that would exist in several machines of different size and plasma parametres; hence, these are not treated here.

6.2
Linear Stability of the Pedestal

With ELMs being a violent MHD instability, their onset should be to some extent described by linear ideal MHD stability analysis. Hence, right after the discovery of ELMs, efforts were undertaken to see which criterion could be violated at the ELM onset.

A first analysis showed that the pressure gradient at the type I ELM onset was roughly consistent with the infinite-n ideal ballooning limit, whereas type III ELMs can also be found at pressure gradients substantially below that limit. Supporting evidence came from the observation that experimental values of T_e and n_e at the pedestal top before atype I ELM crash lie on an isobar over a large range of values, indicating that the pedestal top pressure p_{ped} might indeed be limited by an ideal MHD instability. Conversely, type III ELMs rather occur up to a certain absolute value of T_{ped} with a variation in p_{ped}, which indicates that they may rather be resistive instabilities. This diagram is shown in Figure 6.3.

However, there are some experimental observations concerning type I ELMs that cannot be explained by the infinite-n ideal ballooning stability. One is the fact that ELMs are associated with an abrupt crash of the pedestal pressure profile, whereas high-n modes would rather be expected to be a soft 'transport' limit. In addition, it was found that the pressure gradient can often be at the ideal ballooning limit for many ideal MHD timescales before an ELM crash occurs, making it unlikely that ideal ballooning alone can explain the crash itself. Hence, another ingredient in linear ELM stability analysis was added by considering the large edge bootstrap current driven by the steep pedestal pressure gradient[2]. As outlined in Chapter 4, this will lead to enhanced susceptibility against external kink modes. In fact, as was mentioned there, in a cylinder, finite edge current density will always lead to instability against high-n external kinks, the so-called peeling modes. In order to understand the relevance of the peeling mode for the ELM crash, one therefore has to include the effects of pressure and toroidicity. Following a

2) The bootstrap current j_{bs} is a thermoelectric toroidal electric current driven by finite pressure gradient in a toroidal magnetic confinement device, $j_{bs} \propto \sqrt{r/R_0}\nabla p/B_{\text{pol}}$.

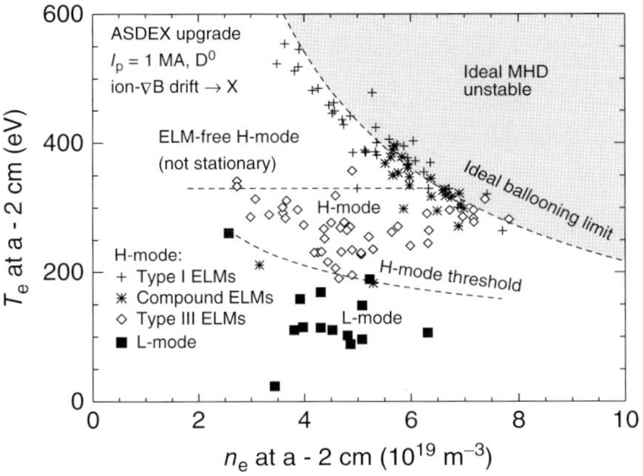

Figure 6.3 Occurrence of different ELM types in terms of pedestal top density and temperature. While type I ELMs follow an isobar, type III ELMs are rather occurring below a certain pedestal top temperature. The power threshold for the L–H transition is also shown. *Source*: Adapted from W. Suttrop *et al.* 1997 [20] with permission from IOP.

similar procedure as outlined for the localized pressure driven modes in Chapter 5, but this time keeping the parallel current, one arrives at an expression for δW for peeling modes under the assumption that the plasma is only affected in a small region Δ at the edge and the vacuum term δW_V does not play a role due to the strong localization of the mode [21]:

$$\delta W = \frac{\Delta^2}{2}\int_{\Delta}^{\infty} dx \left(Px^2 \left(\frac{d\xi}{dx}\right)^2 + Q\xi^2 + S\frac{d}{dx}(x\xi^2) \right) \tag{6.1}$$

where x is a radial coordinate starting at the small layer Δ affected by the peeling mode. The coefficients are again averages over the flux surface under consideration

$$P = 2\pi(q')^2 \left(\oint \frac{JB^2}{R^2B_p^2} d\chi \right)^{-1} \tag{6.2}$$

$$Q = \frac{p'}{2\pi} \left(\oint \frac{\partial J}{\partial \Psi} d\chi - p' \oint JB_p^2 d\chi + \mu_0 I_{\text{pol}} \oint JR^2B_p^2 d\chi \right.$$
$$\left. \times \left(\mu_0 I_{\text{pol}} p' \oint \frac{J}{R^2B_p^2} d\chi - 2\pi q' \right) \left(\oint \frac{JB^2}{R^2B_p^2} d\chi \right)^{-1} \right) \tag{6.3}$$

$$S = P + q' \oint \frac{j_{\parallel}BJ}{R^2B_p^2} d\chi \left(\oint \frac{JB^2}{R^2B_p^2} d\chi \right)^{-1} \tag{6.4}$$

In this form, one can recognize the first two terms as equivalent to the terms contained in the cylindrical expression for δW (Eq. (3.35)), but now the coefficients contain appropriate toroidal averaging. The term containing S is the equivalent of the fluid term that has been partially integrated and added to the boundary term in Eq. (3.35).

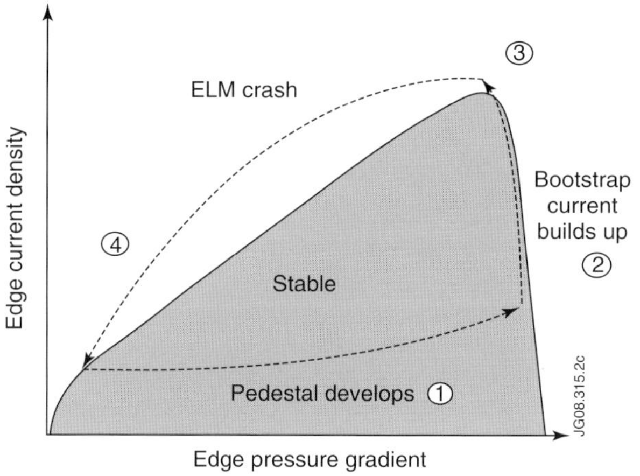

Figure 6.4 Schematic of the peeling ballooning diagram for tokamak edge stability. *Source:* Connor *et al.* 1998 [21], reproduced with permission of AIP.

Minimization of this energy functional using the Euler–Lagrange equation leads to a condition for peeling stability in toroidal geometry:

$$\alpha\left(\frac{r}{R}\left(1-\frac{1}{q^2}\right)+s\Delta'-f_t\frac{Rs}{2r}\right) > Rqs\left(\frac{j_{\|,\text{driven}}}{B}\right) \tag{6.5}$$

where the normalized pressure α has been defined previously in the context of ballooning modes, Eq. (5.32), the normalized shear is $s=(r/q)(dq/dr)$ and f_t is the fraction of trapped particles. Δ is the radial derivative of the Shafranov shift given by Eq. (2.51). The term $j_{\|,\text{driven}}$ represents the parallel current except the Pfirsch-Schlüter[3)] and the bootstrap current which, owing to their dependence on the pressure, have been separated and appear explicitly on the left-hand side of Eq. (6.5).

In Eq. (6.5), the first term on the left-hand side represents the Mercier contribution, the second one the stabilizing Pfirsch–Schlüter term and the third one the destabilizing bootstrap contribution. For finite pressure, there is obviously a window of stability against peeling modes even at finite edge current density that increases with α. We expect, however, that at some point, ballooning modes will limit the achievable α. Carrying out a finite n ballooning mode analysis for the edge region using the same assumptions as for the peeling analysis above [21], one finds that stability for finite n is somewhat reduced with respect to the infinite-n result but roughly follows the same trends[4)]. Combining this analysis with the peeling mode stability analysis leads to the stability diagram shown schematically in Figure 6.4.

3) The Pfirsch–Schlüter current results from the effect that in a torus, the divergence of the diamagnetic current is not automatically zero on a flux surface and additional currents flow along the field lines to guarantee this condition.
4) As mentioned in Chapter 5, a different expansion of the energy functional has also to be applied, namely in $n^{1/3}$ rather than $n^{1/2}$.

It can be seen that the stable window is limited by both edge current and pressure. Analysis of the limiting instabilities shows that the ballooning limit occurs at higher n numbers of the order of 15–20, whereas the peeling limit is usually connected with lower $n \leq 5$. It is hence conjectured that the limitation of the pressure is due to the ballooning limit while the ELM crash itself occurs when ballooning and peeling modes couple in the upper right corner of the stability diagram. In addition, it offers the possibility of discharge trajectories at constant edge pressure gradient limited by the ballooning mode, as indicated by the dashed line, provided the build-up of the bootstrap current is slower than that of the pressure gradient, as one might expect for hot edge plasmas where resistive diffusion should be slow.

This model is very successful in describing a number of experimental observations relating to type I ELMs. In particular, the evolution of the edge parametres often lies within the stability boundaries given by the peeling-ballooning theory and ELMs occur when the experimentally determined edge parametres are reasonably close to the boundary, albeit counter examples can also be found[5]. However, there are also experimental observations that cannot be explained by the peeling-ballooning model, such as the fact that in medium-sized devices at low edge temperature, the bootstrap current seems to be fully developed for a relatively long time interval before the ELM crash, so that the phase of saturated gradients without ELM still awaits a rigorous explanation. In summary, the peeling-ballooning model as described earlier captures many of the essentials of the ELM onset but cannot be considered complete. This is not surprising as potentially important physics elements such as sheared edge rotation, finite resistivity, two-fluid effects or an influence of fast particles are not rigorously included. Moreover, in the edge pedestal, the typical banana orbit widths are comparable to the gradient lengths, challenging one of the basic assumptions for use of a fluid theory (see Chapter 1). Finally, we note that attempts to explain other ELM types than type I using the standard peeling- ballooning model have not yet been fully conclusive.

While the analysis so far has focused on the explanation of the ELM onset, another important question is if linear stability can explain the pedestal width Δ_{ped}. Parametric studies of the peeling- ballooning stability in which the pedestal parameters are systematically varied to see if a limit to Δ_{ped} can be found reveal that in principle, there is no such limit within the model. When extending the steep gradient zone, it is found that the maximum allowable pressure gradient is somewhat reduced, but a scaling of the pedestal top pressure $p_{\text{ped}} \sim \Delta_{\text{ped}}^{3/4}$ is found, indicating that the pedestal can in principle be arbitrarily wide if only limited by peeling-ballooning modes. Hence, the model needs to be extended by another constraint. Such a constraint may be found in analysing the residual turbulent transport in the pedestal and finding a relation between the pressure gradient length and the pedestal width. Motivated by the experimental finding that $\Delta_{\text{ped}} \sim \sqrt{\beta_p}$, it has been conjectured that the pedestal top pressure could scale as $p_{\text{ped}} \sim \Delta^2$ due to the onset of kinetic ballooning modes. Owing to the different

5) We note here that the determination of the edge gradients is a particularly challenging diagnostics task.

scalings of p_{ped} with Δ_{ped}, there is usually an intersection between the two constraints and in the so-called EPED model [22], this intersection is identified as the pedestal width. While this model has had some success in explaining experimentally observed pedestal widths, the transport constraint needs to be justified more rigorously before the model can be considered truly predictive. In particular, there are indications that the density and the temperature pedestal vary differently with plasma parametres, which is beyond the scope of one-fluid MHD.

6.3 Non-linear Evolution

The non-linear evolution of the ELM cycle includes three elements

- The timescale of the full ELM cycle consisting of pedestal breakdown during the ELM crash and the following recovery phase determines the ELM repetition frequency ν_{ELM}. As the ELM crash itself is very short compared to the recovery phase, it is the latter that mainly determines ν_{ELM}, together with the magnitude of the energy loss during the ELM. Simple, limit cycle models of this cycle capture some of the observed properties but are not detailed enough to provide predictive capability. For a full description of this phase, understanding of the pedestal transport is required, which clearly is beyond the scope of this book. However, we will show in the framework of a simple limit cycle model how non-linear cycles can occur in a system with disparate timescales.
- The amount of energy lost during an ELM crash, δW_{ELM}, could be so large that it might endanger the first wall components in future devices, and so predictive capability is crucial. This will be discussed further in the following.
- The duration of the ELM crash, τ_{crash}, varies between 100 s of micro-seconds and several milli-seconds in present day devices. As the energy impact on the first wall also depends on τ_{crash} (see Section 10.2), predictive capability for this number is as important as for δW_{ELM}.

6.3.1 Non-linear Cycles

As mentioned earlier, a detailed modelling of the whole ELM cycle involves different physics aspects which are presently not known comprehensively. However, we will discuss here briefly how non-linear cycles can arise from a system in which the driving force is continuously supplied and an instability that acts to remove the drive is triggered above a certain threshold. Such a system can have different stationary solutions: one possibility is a state in which all time derivatives vanish and an equilibrium value is reached where the system just sits at marginal stability. Another, more interesting case is the possibility of limit cycles, that is periodic growth of an instability which removes the driving force such that it needs some time to be restored. Obviously, such a limit cycle describes the global characteristics of the ELM cycle as outlined earlier. A famous example for a limit cycle occurring in nature is the 'predator–prey' system known from biology. However, the system of two non-linearly coupled equations describing this system (the so-called Lotka–Volterra equations) has to be modified in order to describe our situation.

In the situation under consideration, limit cycles can be obtained from a set of two non-linearly coupled differential equations representing the temporal evolution of the instability drive and the MHD mode. If we assume, for simplicity, that the drive for the ELM instability is the edge pressure gradient, we can write the temporal evolution of the MHD mode amplitude $\xi(t)$ as

$$\frac{d^2\xi}{dt^2} = (p' - 1)\xi - \delta\frac{d\xi}{dt} \tag{6.6}$$

where ξ is the amplitude normalized to a typical length scale and the pressure gradient p' is normalized to the critical pressure gradient for onset of the instability such that it occurs at $p' > 1$, mimicking for example a linear stability criterion such as the peeling-ballooning analysis in the case of ELMs. The time is normalized to the linear growth rate γ of the instability under consideration and δ describes the damping of the amplitude, for example by resistivity or viscosity depending on the assumption if the MHD instability is resistive or ideal. If we assume that this dissipative damping can be described by a diffusive process with diffusivity χ, the coefficient δ is the normalized inverse diffusion time, that is $\delta \propto \chi/(\gamma L^2)$ where L is the length over which diffusion takes place. In the absence of damping, we just recover the harmonic oscillator for the stable case $p' < 1$ (stable oscillation around the equilibrium value) and exponential growth for $p' > 1$. Damping will act to dissipate the energy put into the MHD mode and hence lead to an exponential decay of mode amplitude once the drive is removed. The timescale introduced by the damping sets the timescale of enhanced losses due to finite MHD amplitude, that is determines the magnitude of the ELM crash in the regime where the solution of the system of equations yields a crash-like structure.

The second equation is concerned with the temporal evolution of p' as a balance of normalized heating h and transport losses characterized, for example, by an energy confinement time τ_E

$$\frac{dp'}{dt} = \eta(h - p' - \chi_{\text{MHD}}\xi^2 p') \tag{6.7}$$

where the last term on the RHS provides a coupling between the two equations. It mimics the additional losses due to the MHD mode by a normalized transport coefficient χ_{MHD} and η is the ratio of the transport timescale in the absence of the MHD mode and the timescale of the growth of the MHD mode, $\eta = \gamma\tau_E$. In the absence of MHD losses (i.e. $\xi = 0$), this equation is not coupled to Eq. (6.6) and it just describes the temporal evolution towards the stationary operation point $p' = h$ according to $p'(t) = \eta h(1 - \exp(-\eta t))$.

As shown in Figure 6.5, this system of coupled equations exhibits a rich non-linear behaviour in $\eta - \delta$ space. In the regime where the MHD timescale is much shorter than the confinement timescale, that is $1 \times 10^{-4} < \eta < 1 \times 10^{-2}$, indeed a regime of non-linear oscillations is found which reproduces the essential characteristics of the ELM cycle. An example is shown in Figure 6.5b, which shows that

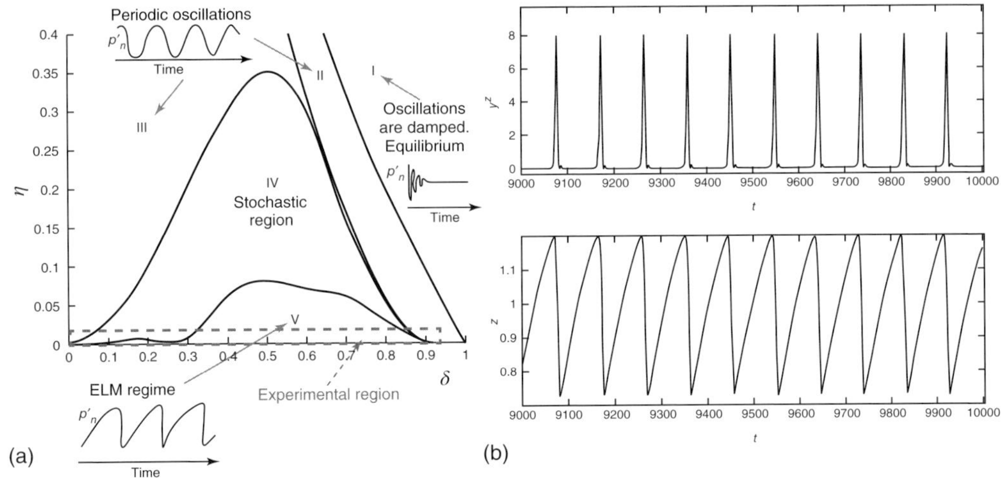

Figure 6.5 Occurrence of different dynamics in parametre space (a) for a simple non-linear model of coupled differential equations describing the interaction between MHD modes and a driving force that is restored on a much slower timescale. In the relevant regime of timescales ($\eta \ll 1$) non-linear cycles are observed that resemble the dynamics of ELMs (time traces of ξ^2 and p', b). *Source*: Adapted from D. Constantinescu *et al.* 2011 [23] with permission from AIP. (Please find a color version of this figure on the color plates.)

the time traces of the pressure gradient exhibit the typical sawtooth-like behaviour and the crashes are correlated to short bursts of MHD activity.

While being quite simple, this model can be thought of as a model system for several non-linear cycles treated in this book, such as the ELMs discussed in this section, or FIR NTMs treated in Section 12.4, where the driving force is the pressure gradient at the resonant surface of interest. Similarly, the sawtooth oscillations treated in Section 11.1 are a limit cycle driven by the current gradient at the $q = 1$ surface and the fishbones treated in Section 11.2 are driven by the fast particle pressure.

6.3.2 Magnitude of the ELM Crash

An early attempt to explain the variation of $\delta W_{\rm ELM}$ with plasma parameters and machine size assumed that $\delta W_{\rm ELM}$ is determined by the parallel heat conduction in the SOL, which limits the energy loss for given ELM duration. This reproduces some global trends but cannot account for the variety of ELM sizes that occur experimentally for similar SOL parameters. Hence, $\delta W_{\rm ELM}$ must also be linked to the detailed MHD characteristics of ELMs to refine this approach.

Within linear stability analysis, an attempt has been made to link the radial extent affected by ELMs to the width of the linearly unstable eigenfunction from the peeling-ballooning model as shown in Figure 5.6. While this has been successful in explaining some of the experimentally observed changes in $\delta W_{\rm ELM}$,

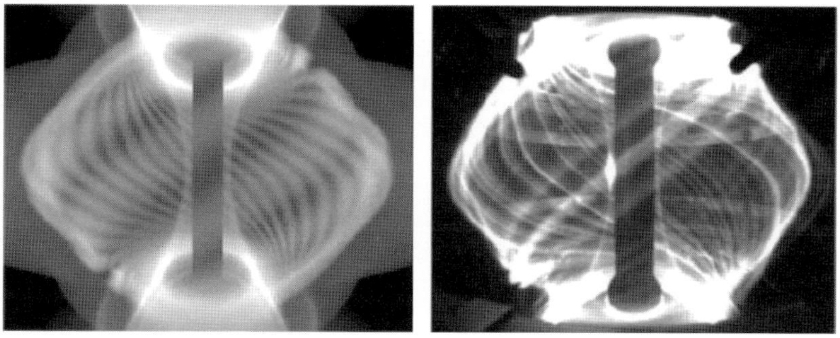

(a) (b)

Figure 6.6 Filamentary structure observed during the ELM crash on MAST (b) and in a non-linear MHD model (a). *Source*: Adapted from A. Kirk *et al.* 2005 [24] with permission of IOP.

for example the decrease when moving towards type II ELMs, many features remain unexplained and this approach is far from being predictive. In particular, the presence of co-existence of small and large ELMs, as is observed experimentally for example during the transition from type I to type II ELMs, is hard to explain as the edge conditions are virtually identical and yet δW_{ELM} can differ a lot. The fact that for very similar linear stability, the evolution of the ELM crash can lead to very different δW_{ELM} has therefore motivated experimental and theoretical studies of the non-linear nature of the ELM crash.

Experimentally, it is observed that the rapid ELM crash consists of the ejection of a series of filaments[6] that carry away part of the pedestal energy. An example is shown in Figure 6.6, together with the output of a non-linear MHD simulation of an ELM crash. While the result of the simulation in Figure 6.6 visually suggests a good agreement between model and experiment, these simulations presently suffer from the fact that they need a very high spatial and temporal resolution and hence can only be run on large computers. Owing to these limitations, it is presently not possible to access the same parametre regime as in the experiment, for example in the magnetic Reynolds number (see Chapter 8). The non-linear models reproduce the linear peeling- ballooning boundary and show a large crash developing due to the non-linear coupling of the linearly uncoupled modes, involving also low-n modes through which the higher n-modes couple, thus explaining qualitatively the large impact on the pedestal. In addition, the simulations reproduce the trend of filament formation due to the non-linear coupling of several linearly unstable modes[7]. Future development will show if they can also quantitatively reproduce the experimental results and can be used in a predictive manner.

We also note here that, experimentally, the energy loss due to filaments does not correspond to the full δW_{ELM}. A possible explanation for this observation is

6) The term filament here corresponds to a poloidally and toroidally localized structure elongated along the field lines.
7) This trend is also observed in analytical non-linear ballooning theory.

that due to the experimentally observed breakdown of the sheared rotation in the edge pedestal during the ELM, also the underlying anomalous transport will be greatly enhanced and can lead to significant energy loss.

6.3.3 Timescale of the ELM Crash

Concerning the ELM crash time, the rapid onset of the crash on the order of 10 s of micro-seconds is clearly linked to an ideal timescale. It is however not clear that this is adequately described by the linear growth rate because of the non-linear nature of the crash. For the crash duration, τ_{crash}, a large variation of experimental observations exists, making clear that it is necessary to understand the details of the non-linear phase in order to be able to predict it adequately. In particular, the observation of significantly enhanced duration of the ELM crash when the first wall material is changed from C to W in the ASDEX Upgrade and JET tokamaks points to a role of the scrape-off layer and divertor conditions in the dynamics of the ELM crash, indicating that also this region will have to be described adequately in future non-linear modelling to obtain a comprehensive predicative capability.

Finally, we note that it is not clear that the non-linear evolution of the linearly unstable mode(s) will always lead to an ELM crash. Non-linear theory also provides the possibility of saturated states in which modes of finite constant amplitude exist. The occurrence of the so-called quiescent H-mode (QH-mode), which is a state without ELMs, but with pedestal pressure profiles similar to those at which type I ELMs occur, has been linked to such a saturated state. Here, a stationary, 3D perturbation of the plasma edge, the so-called edge harmonic oscillation (EHO) is found that provides enough particle transport to prevent the impurity accumulation usually observed in ELM-free H-modes. An example of a QH-mode phase is shown together with a magnetic spectrogram of the EHO in Figure 6.7.

As indicated by its name and evident from Figure 6.7, the typical frequency spectrum of the EHO consists of multiple harmonics that are phase locked, leading to a poloidal and toroidal localization of the mode amplitude resembling that of a localized current filament inside the separatrix. This suggests that the EHO could be a saturated non-linear state of the modes usually leading to the ELM crash. Such a saturation at finite amplitude could for example be provided by the strongly sheared rotation experimentally observed in experiments during QH-mode. We will come back to QH-mode in Section 6.4.1 where we address the mitigation or control of ELMs.

6.4 ELM Control

As outlined earlier, large ELM crashes can potentially harm the first wall components of future large devices. According to Section 10.2.1, Eq. (10.2), the energy impact on the first wall is proportional to $\delta W_{\text{ELM}}/(A\sqrt{\tau_{\text{crash}}})$ and for a crash time

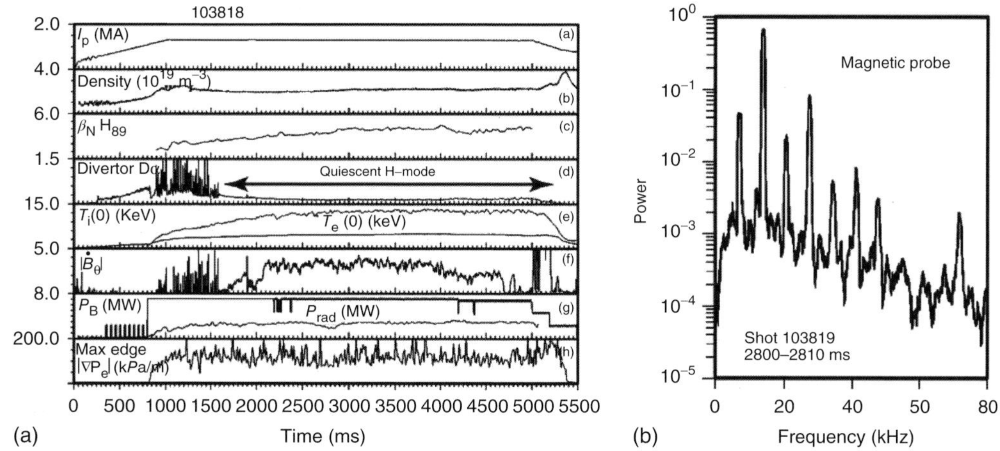

Figure 6.7 Temporal evolution of a QH-mode discharge (a) and magnetic spectrum of the edge harmonic oscillation (EHO) observed during a QH-phase (b) in DIII-D. *Source:* K.H. Burrell *et al.* 2001 [25], reproduced with permission of AIP.

of 1 ms and a wetted area of 1 m^2, representative for present-day experiments, the critical number given there is exceeded if δW_{ELM} exceeds an energy of 1 MJ, which is of the order of the total stored energy in medium-sized tokamaks and hence clearly not an issue. For JET, this can already be problematic for the largest ELMs with $\delta W_{\mathrm{ELM}}/W_{\mathrm{plasma}}$ of the order of several 10% at the highest achievable energy. For ITER, however, it is clear that while the wetted area only increases with machine size R, the ejected energy is roughly proportional to R^3 and hence unmitigated ELMs will exceed the threshold for surface melting. It is therefore very important to either mitigate or suppress ELMs. In the following, we first discuss operation scenarios where ELMs with smaller $\delta W_{\mathrm{ELM}}/W_{\mathrm{tot}}$ than typical for type I ELMs are observed and then active control schemes where additional control measures are applied to mitigate or completely suppress ELMs.

6.4.1
Small ELM Regimes

Usual ELM-free H-mode is not a stationary scenario since due to the absence of the ELMs there is a loss of (impurity and main ion) density control that leads to impurity accumulation and radiative collapse. However, there exists a number of operational modes in which ELMs can be substantially smaller than regular type I ELMs, but the edge particle transport is high enough to avoid the problems of ELM-free H-mode. The above-mentioned QH-mode is an example for such a scenario with no ELMs at all and the transport given by the EHO. In terms of a physics description, the reasons for the occurrence of these smaller ELMs are usually not understood to a degree that would allow predictions of how these scenarios will extrapolate to future devices of bigger size and different plasma parametres. We will hence only briefly give a phenomenological description of these regimes[8]:

- *Type II ELMs:* The properties of type II ELMs have been described earlier. The conditions for their occurrence are mainly strong shaping (close to a double null configuration), and a combination of high β_p, high q_{95} and high collisionality. It has been speculated that the effect of the strong shaping is to squeeze radially the most unstable Eigenfunctions, leading to a smaller radial extent of the ELM affected zone, but the occurrence of 'mixed' phases in which type I and type II ELMs can occur together seems to contradict such a 'continuous' transition from type I to type II ELMs. It has also been noted that the conditions mentioned earlier tend to decrease, for fixed β_p, the bootstrap current, and this might move the operational point closer to the ballooning stability boundary where higher mode numbers and smaller radial extent are expected. Some analysis also suggests that type II ELMs occur in the transition region between first and second ballooning stability which occurs with strong shaping (see Figure 5.5). However, the theoretical description is by no means predictive such that it remains unclear if the type II regime can be achieved in future low collisionality devices.
- *Type III ELMs:* As described earlier, type III ELMs have been observed to occur below a certain pedestal temperature, usually closer to the L–H threshold than type I ELMs. In present-day experiments, this restricts their occurrence to regimes of low to moderate heating power, and the confinement, as for example characterized by the H-factor is somewhat reduced with respect to type I ELMy discharges ($H \approx 0.8$). However, in discharges applying impurity seeding to reduce the power flux into the scrape-off layer and divertor, type III ELMs are observed to occur also at high total heating power but still consistent with the lowered P_{sep} by radiation losses that also leads to lower pedestal temperature and brings the pedestal closer to the H–L backtransition. Under these conditions, values around $H = 1$ have also been observed. Since at present, no first principles physics model exists to predict the maximum edge temperature for type III ELMs, the applicability of the regime to future devices is unclear.
- *EDA Mode:* This regime derives its name from the enhanced D_α radiation observed during its occurrence, indicating that ELMs are replaced by a quasi-coherent mode (QCM) that provides enough particle transport for stationary conditions. It is presently only observed at high collisionality, and it has in fact been conjectured that the absence of ELMs is due to the decrease of bootstrap current with increasing collsionality, but a generally accepted explanation for the QCM does not yet exist, limiting the predictive capability for extrapolation of the operation space for this mode as well.
- *I mode:* This essentially ELM-free mode of operation makes use of the observation that close to the L–H transition, there can be a regime exhibiting a temperature pedestal but not a density pedestal. This mode of operation hence provides

8) Here, we only list scenarios that have been demonstrated to exist in several devices of different size.

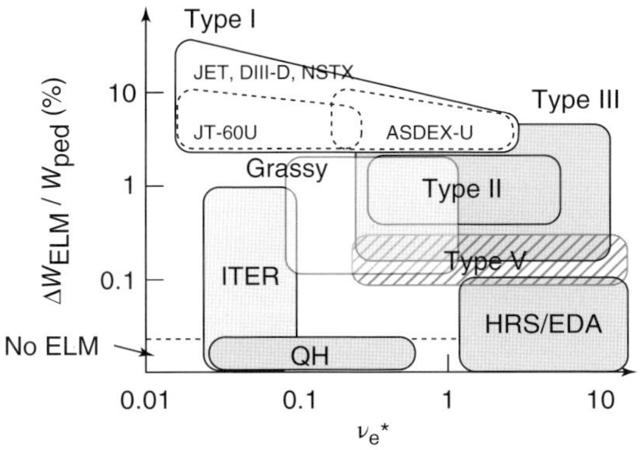

Figure 6.8 Overview of the occurrence of different ELM regimes, ordered by relative energy loss $\delta W_{\mathrm{ELM}}/W_{\mathrm{plasma}}$, as function of collisionality ν^*. *Source:* P.T. Lang *et al.* 2013 [26], reproduced with permission of the IAEA. (Please find a color version of this figure on the color plates.)

H-mode like energy confinement ($0.6 \leq H \leq 1$) with a reduced pedestal pressure gradient. The reduction of the pressure gradient can explain the absence of ELMs in terms of the linear peeling-ballooning stability, and the absence of impurity accumulation may be related to the missing strong neoclassical inward pinch that is usually observed due to the steep density gradient in the H-mode pedestal. Again, theoretical understanding is not mature enough to predict the operational space of this mode for future devices.

- *QH mode:* The QH mode was described earlier as a possible saturated state of the MHD mode usually responsible for type I ELMs. Energy confinement is of the order of that of type I ELMy H-modes ($H \approx 1$). Its occurrence is presently limited to regimes of low collisionality, which in present-day devices means low normalized density n/n_{GW}[9)], but as future large devices such as ITER and DEMO will operate simultaneously at high n/n_{GW} and low ν^*, it is presently not clear if QH-mode is a viable scenario for these devices.

The occurrence of the different regimes is shown in Figure 6.8. It can be seen that in the parameter space achieved so far, only QH mode has some overlap with the ITER requirements. However, as pointed out earlier, present-day devices cannot simultaneously achieve ITER collisionality and normalized density n/n_{GW}, and hence we still need better understanding of the underlying physics in order to safely extrapolate the operational space for these regimes.

6.4.2
Active ELM Control

We now discuss active ELM control, that is means to directly influence the ELM properties externally. A first class of ELM control schemes makes use of the fact

9) The maximum attainable density in a tokamak is approximately given by the so-called Greenwald density, n_{GW}, see Section 10.1.1.

that it has been found experimentally that the fraction of energy transported by ELMs is roughly constant, that is

$$\frac{P_{\mathrm{ELM}}}{P_{\mathrm{tot}}} = \frac{W_{\mathrm{ELM}}}{W_{\mathrm{plasma}}} \nu_{\mathrm{ELM}} \tau_E \approx const. \tag{6.8}$$

Consequently, if ELMs are frequently triggered by external perturbations such that ν_{ELM} increases with respect to the unperturbed value, δW_{ELM} deceases[10]. There are presently two methods that have shown to reliably trigger ELMs:

- Type I ELMs can be triggered by small ($\delta z/a$ of the order of few percentage) and fast (order of millisecond in present-day experiments) vertical 'kicks' exerted by the vertical position control system (PF coils)[11]. If ν_{ELM} is not too far from the kick frequency, a frequency locking appears and an increase of ν_{ELM} over the 'natural' value (i.e. without vertical kicks) of up to a factor of 3 has been demonstrated. An initial attempt to explain this behaviour was to assume that the vertical motion of the plasma in the nonuniform poloidal field induces an additional toroidal edge current density, which would change the peeling mode stability and hence the ELM behaviour. However, a more detailed analysis showed that also the change in the plasma shape occurring during the vertical motion can have an effect on the stability and hence the interpretation is more difficult. At present, it is not clear what the required $\delta z/\delta t$ will be to reliably trigger ELMs at a certain rate. This number is important as it sets the requirements for the PF coil system. In superconducting tokamaks, such a modulation will introduce additional AC losses in the coils and this may be one of the limitations to the applicability of vertical kicks as ELM pacing scheme in ITER or a reactor. Another potential problem lies in the fact that the changing plasma position may also introduce an unwanted modulation of the heat flow to the first wall[12] and challenge the vertical position control system, reducing the margin against vertical displacement events (see Section 4.4).
- A different method to trigger ELMs is the injection of frozen pellets of hydrogenic fuel into the edge plasma. Experimentally, it is found that in type I ELMy H-modes, this triggers additional type I ELMs such that δW_{ELM} on average decreases. We note here that during ablation, the pellet initially creates a local perturbation in the form of a dense 'plasmoid' on the flux surface, as the local cooling is compensated by parallel heat flux occurring roughly at electron thermal speed while the increased density will only equilibrate on the timescale given by the ion thermal velocity. Usually, the ELM is already triggered in this phase and hence, pellet triggering of ELMs is most likely not adequately described by the usual peeling-ballooning analysis that assumes that the pressure is equilibrated on the flux surfaces. In fact, non-linear simulations

10) We note that it is presently not clear if the energy impact decreases linearly with δW_{ELM} as this requires constant wetted area
11) We note that this method also triggers type III ELMs, but this is of less relevance in the context studied here.
12) On the other hand, such a modulation may be useful to reduce the average load on the diverter plates.

of pellet triggering clearly show the development of localized structures at the pellet ablation location at the ELM onset. In experiments in ASDEX Upgrade und JET employing a carbon wall, pellets were shown to reliably trigger ELMs virtually throughout the whole ELM cycle. However, owing to the above-mentioned change in the non-linear ELM phase when operating with a W wall, there can be conditions in which the recovery phase after the longer ELM crash is different and there is a time interval after the ELM in which no ELM can be triggered by a pellet. Under circumstances where this behaviour prevails, it might set an upper limit to the applicable pellet pacing frequency. Hence, we need to understand the non-linear phase of the ELM cycle better in order to be able to extrapolate the applicability of this scheme to future devices. A drawback of the method is that there can be a moderate loss in confinement quality during pellet pacing, depending on the pellet size. Hence, it is also needed to understand the physics behind pellet pacing in order to predict the minimum pellet size that reliably triggers ELMs. Minimizing the pellet size will also minimize the amount of additional fuel introduced that does not penetrate into the core and hence will be a burden for the tokamak fuel cycle system, especially concerning the throughput of the pumping system. Figure 6.9 shows an example for pellet triggered ELMs. It can be seen that while increasing the pacing frequency above the 'natural' ELM frequency (about 30 Hz in this example), more and more of the ELMs are pellet triggered leading to the occurrence of full pacing at around 60 Hz pellet frequency.

We note here that it may also be possible to influence the ELM repetition frequency by local heating or current drive, using for example ECRH/ECCD,

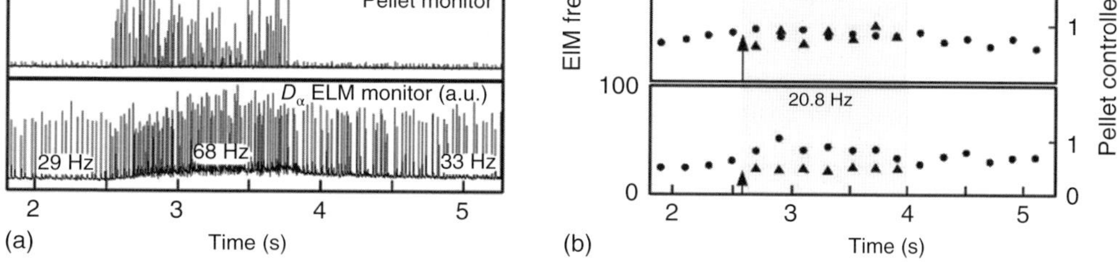

Figure 6.9 Pellet pacing of ELMs: the ELM repetition frequency locks to the pellet frequency, leading to smaller ELMs. The fraction of pellet triggered ELMs, indicated in (b), approaches 1 for sufficiently high $\nu_{pell}/\nu_{ELM,natural}$. *Source:* P.T. Lang *et al.* [27], reproduced with permission from the IAEA. (Please find a color version of this figure on the color plates.)

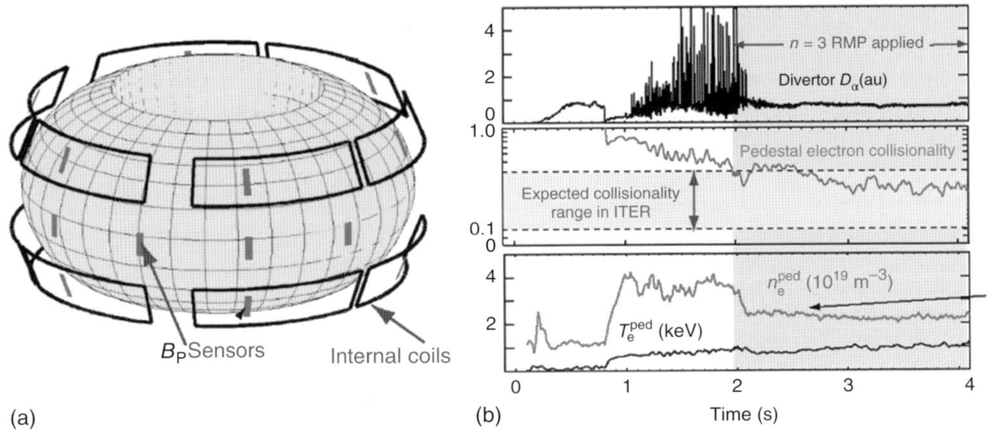

Figure 6.10 Suppression of type I ELMs by application of RMPs in DIII-D. (a) A schematic drawing of the coil set used to produce the RMP; (b) Time traces giving evidence of the complete ELM suppression. *Source*: Adapted from T. Evans 2008 [28], reproduced with permission from the IAEA. (Please find a color version of this figure on the color plates.)

but experiments are inconclusive yet as to the possibility of increasing the ELM frequency further than just according to the increased total heating power that will increase the type I ELM frequency according to $\nu_{\mathrm{ELM}} \propto P_{\mathrm{tot}}$.

Finally, we discuss a different way of active control of ELMs, namely the application of non-axisymmetric fields using suited control coils that generate a spectrum of (m, n) perturbations[13] which has been shown to influence the plasma edge. While an influence on the L–H threshold and type III ELM frequency were already discovered in the early 1990s on JFT 2-M and COMPASS-D, it was found later on DIII-D that application of RMPs to type I ELMy discharges can completely suppress type I ELMs. In low collisionality conditions (roughly for pedestal top collisionality $\nu^*_{\mathrm{ped}} \leq 0.4$ in DIII-D), it is observed that the edge gradients are reduced when RMPs are applied, leading to ELM suppression, consistent with peeling-ballooning analysis of the edge at this reduced pressure. Owing to the reduction in pedestal top pressure, this method also leads to a moderate reduction of confinement quality with respect to the 'natural' type I ELM phase.

While the ELM suppression itself may be described by the peeling-ballooning stability analysis, the explanation of the observed reduction of the pedestal pressure is more difficult. In the low collisionality regime, suppression is only observed in narrow windows of q_{95}, indicating a resonance between plasma and RMP helicity. Early attempts to explain this by ergodization[14] of the edge failed to explain why the temperature seems to be less affected than the density profile and also faced the problem to calculate the shielding of the applied RMPs by the plasma

13) Such fields are often called resonant magnetic perturbations (RMPs).
14) The effect of ergodization of magnetic fields by resonant perturbations is described in more detail in Section 9.1.

(see also Section 9.3.3). More advanced theories take this into account but still require ad-hoc assumptions about the nature of the additional edge transport introduced by the RMPs.

Experiments on RMP suppression of ELMs on a number of tokamaks have shown that there exists a regime at high edge collisionality with type I ELM suppression as well. In this regime, type I ELMs are replaced by high-frequency, small amplitude transport events resembling a small ELM regime. On ASDEX Upgrade, the transition into this regime occurs above a certain pedestal top density that scales roughly linearly with current, indicating a scaling rather with Greenwald fraction n/n_{GW} than with collisionality. The regime can be introduced by different RMP coil configurations and is also not particularly sensitive to the value of q_{95}, so that it does not seem to be a resonant phenomenon. Comparing with type III ELMs on ASDEX Upgrade, this regime does not suffer from confinement degradation with respect to the type I ELM regime at equal parameters and it seems that the RMPs open up a window in which the existence regime of small ELMs is extended to higher pedestal temperatures than for the usual type III ELMs. Clearly, a better understanding is required to predict the applicability of this regime in future devices.

In summary, RMP suppression of ELMs has been shown at ITER edge collisionality and at ITER values of n/n_{GW} but obviously in two different regimes that are not continuously connected in parameter space. Hence, the situation is similar to the small ELM regimes such that predictive capability is not yet established to assess the applicability of these methods to future large devices.

7
Combined Pressure and Current Driven Modes: The Ideal β-Limit

In Chapter 5, we analysed pressure driven MHD modes in a local approach for high (m, n) modes. Conversely, in Chapter 4, the stability of low (m, n) modes against current driven instabilities was analysed in the limit $\beta \rightarrow 0$. In this chapter, we analyse the remaining case, that is the effect of finite pressure on global modes with low (m, n). More specifically, the external kink mode stability is substantially modified by finite pressure and toroidal mode coupling. These modes then lead to a global limitation of the achievable β in tokamak discharges, the so-called β-limit. This will be treated in the last section of the chapter.

7.1
Tokamak Operational Scenarios

As we will see below, the onset of β-limiting ideal MHD modes depends sensitively on the shape of the radial profiles of current and pressure. While these can vary substantially depending on how the tokamak discharge is evolving, tokamak operational scenarios can in general be classified by the characteristics of these profiles. Hence, we briefly introduce here typical tokamak operational scenarios before analysing in detail the stability.

One class of tokamak operational scenarios is based on current profiles that essentially follow the electrical conductivity profile, as would be the case for a purely ohmically driven current. As the electrical conductivity is linked to the temperature profile via $\sigma \propto T_e^{3/2}$, the resulting current profile will be peaked in the centre, resulting in a monotonically increasing q-profile[1]. Owing to the constraint of internal kink mode stability (see Section 4.3), the central q value in this case is usually close to 1. Typical operational scenarios based on these current profiles are the so-called L-(low confinement) mode or the H-(high confinement) mode. The latter is characterized by a steep edge pressure gradient ('pedestal') that has been described in more detail in Chapter 6. Tokamak operational scenarios with a large fraction of ohmic current and a monotonically increasing q-profile are usually developed for pulsed operation and are called *conventional* scenarios.

1) If $dj/dr < 0$, $I_p(r)$ increases less than quadratically which means that $q \propto r^2/I_p(r)$ will increase with radius.

Magnetohydrodynamic Stability of Tokamaks, First Edition. Hartmut Zohm.

On the other hand, in the so-called advanced tokamak scenarios, emphasis is not only put on optimizing the current and pressure profiles for maximum β and τ_E and hence maximum fusion power but also on obtaining conditions under which the toroidal current is largely driven noninductively. As the efficiency of typical auxiliary current drive systems is generally not high enough to drive the full plasma current, advanced tokamak scenarios aim at a high fraction of bootstrap current.

The bootstrap current fraction f_{bs} can be expressed as

$$f_{bs} \propto \frac{I_{bs}}{I_p} \propto \sqrt{\frac{a}{R_0}} \frac{a^2 \nabla p}{B_{\text{pol}} I_p} \propto \sqrt{\frac{a}{R_0}} \frac{p}{B_{\text{pol}}^2} \propto \sqrt{\frac{a}{R_0}} \beta_{\text{pol}} \tag{7.1}$$

where we have assumed $\nabla p \approx p/a$. This equation points to the need of high β_{pol} in advanced tokamak scenarios, that is reduced I_p which tends to decrease τ_E and hence introduces a need for optimization of the operational point. Moreover, the proportionality constant will depend on the profile shapes, and as high-pressure gradient at low poloidal field will maximize the local bootstrap current, advanced tokamak scenarios tend to have flat or even reversed, elevated q-profiles in the radial region where ∇p is already significant. These q-profiles result from current profiles which are peaked off-axis, that is are broader than those of conventional scenarios. Under these conditions, pressure profiles can also exhibit a transport barrier inside the plasma, a so-called internal transport barrier (ITB), which reinforces the magnitude of the bootstrap current at this radius. In this case, the profile of j_{bs} can largely be aligned with the total current profile, allowing a self-consistent solution with large f_{bs}.

The typical pressure and safety factor profiles in these scenarios are shown schematically in Figure 7.1. In praxi, the two classes of scenarios described earlier

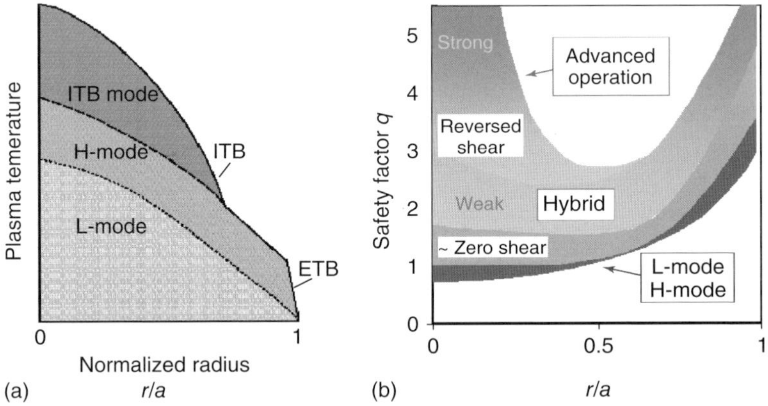

Figure 7.1 Typical pressure (a) and safety factor (b) profiles of different tokamak operational scenarios. Adapted from ITER Physics Basis Update 2007 [29], reproduced with permission of the IAEA. (Please find a color version of this figure on the color plates.)

are connected by a continuum of profile shapes[2] and optimization of both profiles for maximum stability and confinement under the constraint of maximizing pulse length is an area of ongoing research in tokamak physics. In the following, we will discuss in more detail the stability aspects of this problem.

7.2
External Kink Modes in a Tokamak with Finite β

In the analysis presented in Chapter 4 we found that external kinks with $m/n < q_a$ are always stable in a cylinder at negligible β. However, as pointed out in Chapter 5, the assumption of negligible β is violated in the vicinity of resonant surfaces. In Chapter 5, we therefore argued that pressure driven modes should be quite localized around the resonant surface and applied a local expansion to determine stability for high n pressure driven modes, namely interchange and ballooning modes. However, for low magnetic shear, this may be different as the region in which the stabilizing contribution of field line bending is small, may be quite extended. According to the expansion (Eq. (4.1)) and (Eq. (4.2)), we have to inspect the condition

$$\beta' r \ll \left(\frac{m}{nq} - 1\right)^2 (m^2 - 1) \tag{7.2}$$

over the plasma radius to determine the validity of the assumption.

Figure 7.2 shows a comparison for the (2, 1) mode for two different model q-profiles with same total current (resulting in $q_a = 4$ for both cases) according to

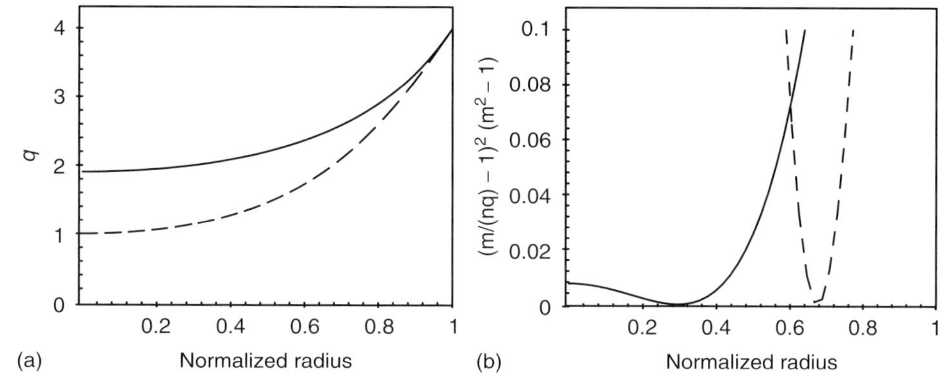

Figure 7.2 (a) Evaluation of criterion (Eq. (7.2)) for two different q-profiles shown. (b) While for the profile with $q(0) = 1$ (dashed lines), the criterion is violated only in a narrow region around $q = 2$ for typical values of $\beta' r$ of few percentage, this region is much larger for the elevated q-profile (solid lines).

2) In particular, scenarios with flat elevated q-profiles that have improved performance but do not necessarily provide steady-state conditions are sometimes called *hybrid* scenarios.

Eq. (2.12), a typical 'conventional' one using $\mu = 3$ and one with a flatter and elevated q-profile in the core ($\mu = 1.1$), more typical of 'advanced' tokamak operation. As can be seen, the conventional q-profile has only a small region in which violation of condition (Eq. (7.2)) occurs at typical $\beta' r$-values in the percentage range, validating the assumption of localized Eigenfunctions: for interchange and ballooning analysis. However, the 'advanced' case shows a large region in which the destabilizing pressure must be considered, opening up the possibility of low n, global MHD modes with $r_s < a$ being destabilized by the pressure gradient.

While quite simplified, this is the basic physics picture of the so-called infernal mode, that is an external kink with resonant surface inside the plasma in a region where the magnetic shear is low and the pressure gradient is large around the resonant surface. As outlined earlier, these conditions coincide with those required for advanced tokamak operation. In practical applications, advanced tokamak scenarios are hence often characterized by a relatively low ideal β-limit due to ideal kink modes. Furthermore, typical ideal MHD modes limiting advanced tokamak scenarios show a coupling of several external kink components with identical n but on different rational surfaces, indicating that cylindrical stability analysis, which essentially deals with modes at single resonant surfaces, is insufficient to describe the stability. Figure 7.3 shows an example of an experimentally determined Eigenfunction of a β-limiting MHD mode in an advanced tokamak scenario (Figure 7.3b) and a comparison to an Eigenfunction resulting from a linear stability analysis employing realistic experimental profiles and geometry (Figure 7.3a) using a numerical approach.

It is clear from the figure that, when decomposed into poloidal harmonics, the Eigenfunction consists of several components. This can be understood as follows: as discussed in Chapter 2, in a torus, the magnetic field components B_{pol} and B_t vary on a flux surface due to shaping and toroidicity, and the field line pitch varies along the poloidal circumference according to Eq. (2.86). Thus, the geometrical poloidal angle θ is no longer an ignorable coordinate and m is not a 'good' quantum number any more while n still is because the toroidal direction remains an ignorable coordinate in a tokamak.

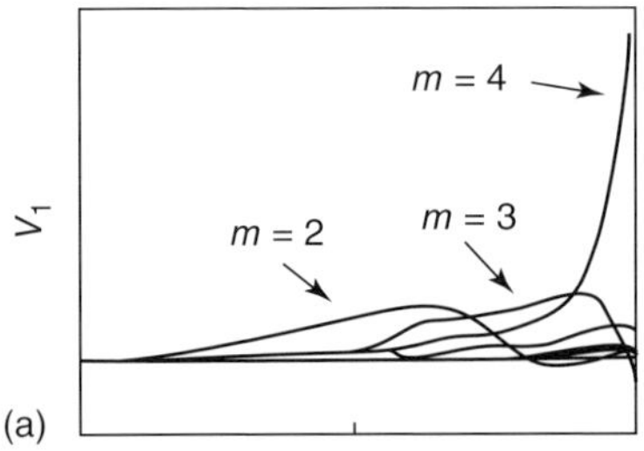

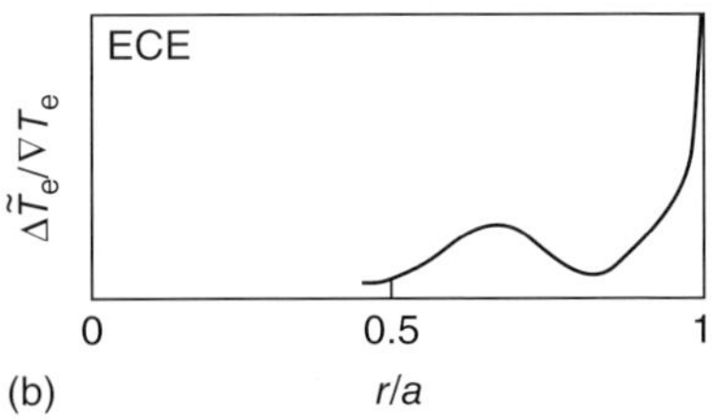

Figure 7.3 Experimentally determined Eigenfunction of an MHD mode limiting β in an advanced tokamak discharge (b) and Eigentfunction obtained from linear numerical stability analysis of the discharge (a). *Source:* S. Günter *et al.* 2000 [30], reproduced with permission of the IAEA.

Using the straight field line angle θ^* derived in Section 2.2.1.4, the phase variation of a mode with constant phase along field lines can be expressed as $\xi \propto \exp\,(im\theta^* - n\phi)$. Assuming for simplicity just the Merezhkin correction given in Eq. (2.76), $\theta^*(\theta) = \theta - \lambda \sin\theta$, Fourier decomposition of a component with poloidal mode number m in the straight field line angle yields for the Fourier component with mode number m' in poloidal angle

$$\xi_{m'} = \frac{\xi_m}{2\pi} \int_{\theta=0}^{2\pi} d\theta e^{i((m-m')\theta - m\lambda \sin\theta)} = \xi_m J_{m-m'}(m\lambda) \tag{7.3}$$

where $J_{m-m'}(m\lambda)$ is the Bessel function of the first kind. Hence, for $\lambda \neq 0$, there is toroidal mode coupling to neighbouring surfaces of same n but different m'. For $m\lambda \ll 1$, the Bessel function can be expanded and the amplitude of the $m - m' = 1$ component is $m\lambda/2$, that is of order (r/R_0). Higher order shape moments will give additional sidebands, for example $m \pm 2$ due to ellipticity and $m \pm 3$ due to triangularity. It is due to these sidebands that components with different m but same n can couple in a torus, whereas they are uncoupled in a cylinder[3)], leading to the above-mentioned low β-limit in advanced scenarios with peaked pressure and broad current profiles. The situation is relaxed when the pressure peaking is decreased, that is for an ITB that has its footpoint at large minor radius. However, it may still be desired to run advanced tokamak discharges at β-values that are prone to the external kink instability. In Section 7.3, we hence analyse ways to overcome this limitation.

7.3 The Effect of a Conducting Wall on External Kink Modes

In the previous analysis, we found that external kink modes can be a serious limitation to β in scenarios with broad current profiles. So far, we have neglected the stabilizing effect of a conducting wall at $r = r_w$ in the analysis. We will include it in this section, considering first the case of an ideally conducting wall, which can be seen as the limit where the eddy currents induced in the wall decay on a timescale longer than the typical growth time. Then, we treat a wall with finite conductivity, also known as the *resistive wall* case.

7.3.1 Ideally Conducting Wall

The stabilizing effect of an ideally conducting wall enters through Λ, given by Eq. (3.38), into δW. For small argument $nr_w/R_0 \ll 1$, the Bessel functions can be

3) We stress here, however, that there is only one Eigenfunction for a particular n, which has a poloidal structure that can be thought of as decomposed in different ms in the geometrical poloidal angle coordinate.

expanded and we obtain

$$\Lambda \approx \frac{1 + \left(\frac{a}{r_w}\right)^{2m}}{1 - \left(\frac{a}{r_w}\right)^{2m}} \tag{7.4}$$

as stabilizing contribution from the vacuum term. While the exact derivation of this term from the energy principle requires some algebra, physical insight can be gained by recalling that an ideally conducting wall has the property of shielding magnetic fields generated inside it to the outside. This can be expressed by the boundary conditions

$$\mathbf{B}_{\text{out}} = 0 \tag{7.5}$$

$$\mathbf{n} \cdot \mathbf{B}|_{\text{wall}} = 0, \tag{7.6}$$

where $\mathbf{n}$ is the wall surface normal vector. Equation (7.6) forces B_r to zero on the wall and implies a jump of the tangential component of $\mathbf{B}$ across the wall which, in the limit of a 'thin' wall (meaning that the current density in the wall is assumed to be constant over the wall thickness d), is determined by

$$\mathbf{n} \times \mathbf{B}|_{\text{out}} - \mathbf{n} \times \mathbf{B}|_{\text{in}} = \mu_0 \, \mathbf{J}_w \tag{7.7}$$

where $J_w = j_w d$ is the surface current density flowing in the wall. Relation 7.7 holds for any surface current and will be used frequently in the remainder of the book. For the case of a conducting wall that shields the field generated inside it, we have to use Eq. (7.5) in addition.

We will now assume a screw pinch with circular cross section in which the distortion of the plasma surface due to a kink mode is described by a surface current with constant amplitude $\bar{J}_s$ located at $r = r_s$

$$J_s(r, \theta) = \bar{J}_s \delta(r - r_s) e^{im\theta} \tag{7.8}$$

Note that as the unit of $\delta(r)$ is 1/m, the amplitude $\bar{J}$ has the unit A, that is it is a current. Here, we have assumed that the current is flowing in the z-direction and the sinusoidal modulation due to the MHD mode is dominant in the poloidal direction[4]. Then, the magnetic field is given by the vacuum solution of the magnetostatic equations $\nabla \cdot \mathbf{B} = 0$ and $\nabla \times \mathbf{B} = 0$. As already discussed in Chapter 2, if an ignorable coordinate exists (such as the z coordinate here), it is convenient to derive the magnetic field from a scalar flux function. For our problem, we define this flux function by

$$\mathbf{B} = \nabla \Psi^* \times \mathbf{e}_z \text{ with } \Psi^*(r, \theta) = \Psi^*(r) e^{im}\theta \tag{7.9}$$

so that the magnetic field perturbation can be calculated by

$$B_r = \frac{1}{r}\frac{\partial \Psi^*}{\partial \theta} = \frac{\text{im}}{r}\Psi^* \tag{7.10}$$

$$B_\theta = -\frac{d\Psi^*}{dr} \tag{7.11}$$

4) Neglecting the toroidal variation here is equivalent to expanding the Bessel functions above, that is $nr/R_0 \ll 1$.

This definition is closely related to the 'helical flux function' that will be introduced in Chapter 8. The z-component of Ampère's law then reads $\Delta\Psi^* = -\mu_0 j_z$ which, for $j_z = 0$, leads to the general vacuum solution

$$\Psi^* = (c_1 r^m + c_2 \frac{1}{r^m})e^{im\theta} \tag{7.12}$$

Applying the condition (7.7) at r_s for the surface current defined by Eq. (7.8) and chosing the solution such that it is finite both at $r = 0$ and $r \to \infty$, we obtain

$$\Psi^* = \overline{\Psi}^* \left(\frac{r}{r_s}\right)^m e^{im\theta} \text{ for } r < r_s \tag{7.13}$$

$$\Psi^* = \overline{\Psi}^* \left(\frac{r_s}{r}\right)^m e^{im\theta} \text{ for } r > r_s \tag{7.14}$$

with the constant

$$\overline{\Psi}^* = \mu_0 \bar{J}_s r_s/(2m) \tag{7.15}$$

If we assume that the surface current is located close to the plasma boundary at $r_s \approx a$, the radial magnetic field decaying like $B_r \propto r^{-(m+1)}$ will induce a current of magnitude $J_w \approx J_s(a/r_w)^{m+1}$ there. This in turn will lead to a magnetic field at the plasma surface $B_r(a) \propto -J_w(a/r_w)^{m-1}$, so that we finally obtain a reduction by the current induced in the wall according to

$$B_r^{w/wall}(a) \approx B_r^{w/o\ wall}(a)\left(1 - \left(\frac{a}{r_w}\right)^{2m}\right) \tag{7.16}$$

motivating the form of Eq. (7.4).

Depending on how close the wall is, it can have a large effect on the β-limit. Owing to the m-dependence, the stabilizing effect will be most pronounced for low mode numbers and stronger for broad current profiles than for peaked ones. This is borne out by numerical calculations of the onset of external kinks for 'conventional' and 'advanced' tokamak scenarios as shown in Figure 7.4.

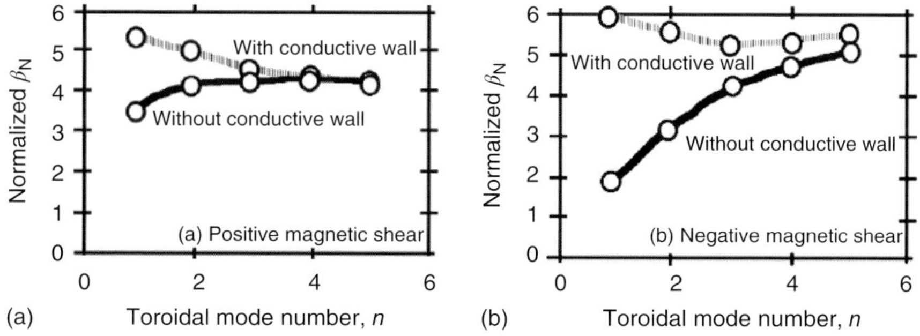

Figure 7.4 Numerical analysis of the effect of an ideally conducting wall close to the plasma for peaked (a, representing a conventional scenario) and broad (b, representing an advanced scenario) current profiles as function of mode number. Adapted from: J. Manickam *et al.* 1994 [31], reprinted with permission of AIP.

For the peaked current profile case, shown in Figure 7.4a, one can see that the 'no-wall limit' (i.e. the onset β[5] of the external kink with $r_{\text{wall}} \to \infty$) is already appreciably high[6]. In this case, while a wall close to the plasma (defining the 'ideal wall limit') has a noticeable effect, it is not too large. Conversely, the broad current profile case shown in Figure 7.4b shows a reduced no-wall limit and a large gain with the ideal wall. We remind the reader that the gain for the broad current profile will depend on the peakedness of the pressure profile, which has been assumed to be quite peaked in the case shown here. For both cases, the effect is pronounced for low mode numbers but vanishes for higher ones due to the stronger decay of the field in the vacuum region between plasma surface and wall, as shown earlier. As a consequence, conducting wall elements play an important role in the design of advanced tokamaks which target operation at β-values above the ideal no-wall limit.

7.3.2
Resistive Wall

The above-mentioned analysis assumed an ideally conducting wall, meaning that the eddy current pattern induced in the wall by the MHD perturbation would not decay resistively. This is the case for example in RFPs with thick shells and short discharge times, or for quickly rotating modes in tokamaks, but for instabilities that are stationary in the rest frame of the wall, the eddy currents will eventually decay due to finite resistivity, weakening the stabilizing effect and leading to a penetration of the magnetic field through the wall.

Mathematically, including the finite resistivity leads to a change in the boundary condition at the wall. In an ideally conducting wall, currents flow without electrical resistance and hence $E_w = 0$. For finite resistivity, there will be an electric field according to Faraday's law. In the approximation of surface currents and vacuum fields used in Section 7.3.1, it is possible to construct an analytical solution of the time-dependent problem that gives insight into the effect of the resistive wall as follows: as before, we represent the MHD mode by a surface current located at $r = r_s$, Eq. (7.8), but allow for a temporal variation[7] of both surface current and wall current according to

$$J_{s,w}(r,\theta,t) = j_{s,w} d\delta(r - r_{s,w}) e^{i(m\theta - \omega t)} \tag{7.17}$$

The analysis is carried out by constructing the solution from the general vacuum solution in three separate areas as indicated in Figure 7.5a. Inside the resonant surface, denoted as region I in Figure 7.5, Ψ^* has to vanish for $r = 0$ (i.e. will be of the form (7.13)). Outside the wall radius, denoted as region III in Figure 7.5, Ψ^* has to vanish for $r \to \infty$ (i.e. will be of the form (7.14)). Between the two surfaces,

5) Here, β is expressed in terms of the normalized beta β_N that will be introduced in Section 7.5.
6) For comparison, the ITER Q=10 operational point is at $\beta_N = 1.8$.
7) The parameter ω is in general complex and can hence refer to mode growth (imaginary part of ω) and rotation in the laboratory frame (real part of ω).

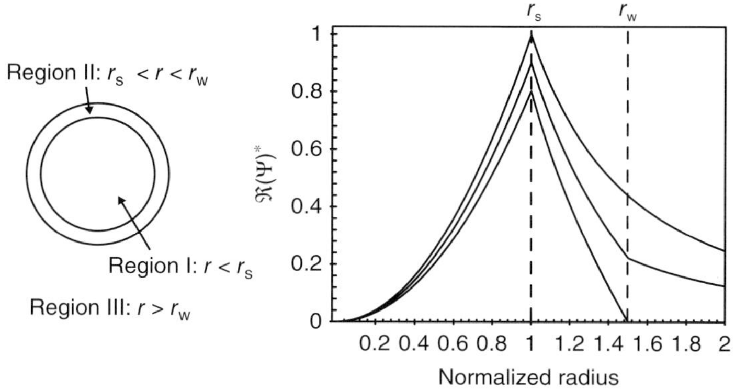

Figure 7.5 Geometry used to analyse the effect of the resistive wall (a) and solutions $\Re(\Psi)^*$ for $\omega\tau_w =$ 0.1 (upper curve, corresponding to the absence of the wall), $\omega\tau = 100$ (lower curve, corresponding to an ideally conducting wall) and a case in between (curve in the middle, $\omega\tau_w = 1$). (b) In this plot, $r_s = 1$ and $r_w = 1.5$.

denoted as region II in Figure 7.5, Ψ^* consists of both terms from Eq. 7.12. Furthermore, we allow for a temporal variation of Ψ^* by a Fourier ansatz similar to that for the wall current (Eq. (7.17)). The geometry is shown in Figure 7.5 together with three typical solutions.

In total, the problem is hence described by the four coefficients in front of the known radial and angular dependence of Ψ^* in the different regions. The different parts of the solution are connected at the boundaries by the conditions of continuity of Ψ^* (equivalent to B_r being continuous at both interfaces) and equaling the jump of the first derivative of Ψ^* to $\mu_0 J$ (equivalent to Eq. (7.7)). While for the resonant surface, this condition is the same as in the analysis of the ideally conducting wall, the condition at the wall radius is now modified to allow for finite resistivity. The r-component of Faraday's law yields

$$-i\omega B_r = \omega \frac{m}{r_w}\Psi^* = -\frac{1}{r}\frac{\partial E_{w,z}}{\partial \theta} = -\frac{\mathrm{im}}{r_w \sigma} j_w \tag{7.18}$$

where σ is the electrical conductivity of the wall. Relating this to the jump in B_θ across the wall by Eq. (7.7), we arrive at

$$\frac{m}{r_w}\Psi^* = -\frac{i}{2\omega\tau_w}\left(\left.\frac{\partial \Psi^*}{\partial r}\right|_{\text{out}} - \left.\frac{\partial \Psi^*}{\partial r}\right|_{\text{in}} \right) \tag{7.19}$$

where we have introduced the resistive timescale of the wall, $\tau_w = \mu_0 \sigma d r_w/(2m)$. Mathematically, the four conditions across the two interfaces lead to four linear equations for the four coefficients which can be solved in a straightforward way. Thus, the full analytical solution is obtained as

$$\Psi_I^* = \overline{\Psi}^* \left(1 - \frac{\omega\tau_w}{i + \omega\tau_w}\left(\frac{r_s}{r_w}\right)^{2m}\right)\left(\frac{r}{r_s}\right)^m e^{i(m\theta-\omega t)} \tag{7.20}$$

$$\Psi_{II}^* = \overline{\Psi}^* \left(\left(\frac{r_s}{r}\right)^m - \frac{\omega\tau_w}{i + \omega\tau_w}\left(\frac{r_s}{r_w}\right)^{2m}\left(\frac{r}{r_s}\right)^m\right) e^{i(\theta-\omega t)} \tag{7.21}$$

$$\Psi_{III}^* = \overline{\Psi}^* \frac{i}{i + \omega\tau_w}\left(\frac{r_s}{r}\right)^m e^{i(m\theta-\omega t)} \tag{7.22}$$

where $\overline{\Psi}^*$ is given by Eq. (7.15). Figure (7.5) shows three typical solutions, one with fast temporal variation with respect to the wall timescale, $\omega\tau_w \gg 1$, one with slow variation ($\omega\tau_w \ll 1$) and one with $\omega\tau_w = 1$. It can be seen that the two limiting cases correspond to an ideal wall ($\Psi^* = 0$ for $r > r_w$) and the no-wall case (Ψ^* goes smoothly through r_w). The reduction of Ψ^* for given $\overline{J}_s$ due to the presence of the wall leads to the stabilizing effect according to Eq. (7.4).

This solution allows the following conclusions:

- For a purely growing mode, $\omega = i\gamma$, where γ is the growth rate, the wall acts as a perfect conductor if $\gamma\tau_w \gg 1$, as in this limit, the ideal wall solution from above is recovered: Ψ^* goes to zero outside the wall and the radial component is attenuated by roughly a factor of $(r_s/r_w)^{2m}$. The stability properties will be the same as in the presence of a perfectly conducting wall.
- Conversely, a purely growing mode can penetrate the wall if $\gamma\tau_w \ll 1$, leading to an instability on this timescale. For typical fusion experiments, this timescale will be on the order of 10 s of ms, which is much slower than the Alfvén timescale on which an ideal kink mode would grow in the absence of the wall, meaning that in the sense of ideal MHD, the mode goes through a series of MHD equilibria (formally $\gamma\tau_A \to 0$). This mode is therefore called the *resistive wall mode* (RWM) and will be discussed in more detail in Section 7.4.
- An interesting case occurs if the mode is rotating with angular frequency ω_{mode} while it grows. Then, $\omega = i\gamma + \omega_{\text{mode}}$, and if $\omega_{\text{mode}}\tau_w \gg 1$, the wall acts as perfect conductor independently of the magnitude of γ. It follows that an ideal kink can be stabilized by rotation of the mode structure with respect to the resistive wall. We will come back to this in Section 7.4.
- Finally, the radial field is in general not in phase with the current induced in the wall and the phase difference, which can be calculated from the ratio of real and imaginary part, varies with $\omega\tau_w$. This will lead to a force on a rotating mode that acts to slow it down, potentially towards the region in which the wall loses its stabilizing character. This process will be treated in detail when the locking of tearing modes to the vessel wall is discussed in Section 9.3.2.

7.4
The Resistive Wall Mode (RWM)

In Section 7.3, it was found that for MHD modes that grow slowly compared to the timescale of the resistive wall, the stabilizing effect of the wall vanishes and hence a new branch of ideal[8] MHD instability, the RWM can develop on the timescale τ_w. However, the analysis given in the previous section has included the effect of the eddy currents in the wall but not the drive for the instability itself, which was introduced by just assuming a positive growth rate γ. In order to arrive at a dispersion relation for the RWM, one can match the vacuum field to the plasma solution, evaluating $\partial_r(rB_r)$ by relating it to δW_F across the plasma surface. That way, the stability of the plasma enters into the problem in the usual way by the potential energy δW. It can be shown [1], that neglecting the toroidal mode variation and assuming $\gamma\tau_A \to 0$ as pointed out earlier, this yields

$$\gamma\tau_{\text{wall}} = -\frac{2m}{1-\left(\frac{a}{r_w}\right)^{2m}}\frac{\delta W_{\text{no wall}}}{\delta W_{\text{id wall}}} \tag{7.23}$$

where $\delta W_{\text{no wall}}$ is evaluated with $r_w \to \infty$ while $\delta W_{\text{id wall}}$ is evaluated assuming an ideally conducting wall at $r = r_w$ as treated in Section 7.3. Owing to the assumption $\gamma\tau_A \to 0$, the resistive timescale of the wall is the only timescale appearing in the problem, while keeping τ_A finite, we expect a transition of the growth rate from resistive to Alfvénic at the ideal wall limit similar to the analysis for the VDE in Section 4.4.

The RWM plays an important role in the RFP which is always unstable to purely current gradient driven external kinks without wall due to $q < 1$. Experimentally, the stabilizing effect of the wall in RFPs has been clearly demonstrated by showing that replacing a shell with $\tau_{\text{wall}} \gg \tau_{\text{discharge}}$ by one with $\tau_{\text{wall}} \leq \tau_{\text{discharge}}$, a stable RFP discharge converts into one with an RWM growing with $\gamma \approx \tau_{\text{wall}}^{-1}$. An example for the occurrence of a current gradient driven RWM in a tokamak is the growth of the external (2,1) kink mode when approaching the $q_a = 2$ limit shown in the left part of Figure 4.1. Clearly, the timescale of 10 s of milliseconds relates to the resistive timescale of the wall and not the ideal timescale $\gamma \approx \tau_{\text{A}}^{-1}$.

However, the most important role that the RWM plays in tokamaks is for advanced scenarios where it limits both the maximum bootstrap current density (which require broad current profiles) and the maximum achievable β as shown in Figure 7.4. To demonstrate the role of the RWM in limiting β, we note that $\delta W_{\text{no wall}}$ becomes negative when the no-wall β-limit is exceeded while $\delta W_{\text{id wall}}$ will become negative for β above $\beta_{\text{id wall}}$. Assuming that the dependencies can be approximated linearly, the dispersion relation becomes

$$\gamma\tau_{\text{wall}} \propto -\frac{\beta_{\text{no wall}} - \beta}{\beta_{\text{id wall}} - \beta} \tag{7.24}$$

This function is plotted in Figure 7.6. Above $\beta_{\text{no wall}}$, the RWM becomes unstable, with growth rate of the order of τ_{wall}^{-1}. Approaching the ideal wall limit, the growth rate diverges. As mentioned earlier, this is due to the fact that we neglected the Alfvén timescale in the derivation of (7.23), while including it will set an upper limit to the growth rate of that order. We note that the physics of slowing down the timescale from the ideal timescale to the resistive timescale of the wall is similar

8) Ideal refers to the fact that the plasma is still treated using ideal MHD while the wall is resistive.

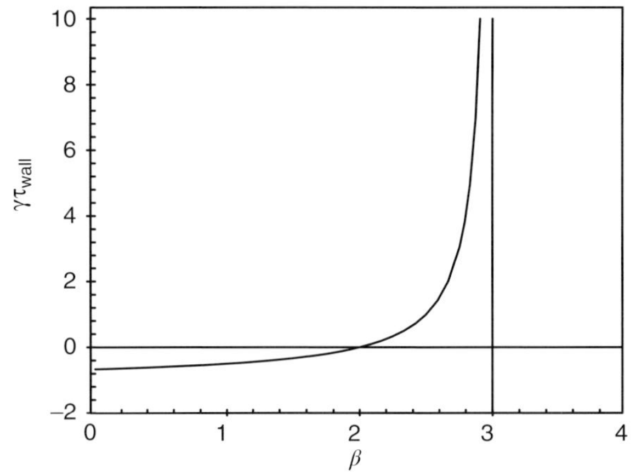

Figure 7.6 Idealized dispersion relation for the resistive wall mode, indicating the window of RWM instability between $\beta_{\text{no wall}}$ and $\beta_{\text{id wall}}$ according to Eq. (7.23). The singularity comes from neglecting the ideal MHD timescale in the calculation ($\gamma\tau_A \to 0$, see also Figure 4.7).

to that treated for the $n = 0$ vertical instability in Section 4.4, but with the stable region for the VDE, which is the equivalent of the 'no-wall' limit for this case, occurring at control parameter $n < 0$.

For practical application, it is important to note that, also similar to the control strategy for the $n = 0$ vertical instability, slowing down the growth of the ideal external kink to the wall time scale opens up the possibility to apply active feedback control using coils that produce helical fields counteracting those of the mode, effectively replacing the response of the ideally conducting wall[9]. For a current gradient driven RWM, an example of such an active control was already shown for the case of the (2,1) external kink occurring when $q_a \to 2$ in Figure 4.1. For an RWM occurring at the β-limit, the response time of the feedback system will determine to what fraction of the ideal wall limit the discharge can be stabilized, but it is clear that this scheme will be especially beneficial in cases where the difference between no-wall and ideal limit is large, such as in Figure 7.4b. However, in experimental applications, the picture turns out to be more complex than described by the ideal MHD analysis shown earlier. This is due to the effects of rotation and fast particles, both of which are not included in the description earlier. Although a rigorous treatment of these effects is beyond the theoretical framework used in this book, we will briefly describe them below.

The RWM should not occur in tokamak plasmas that rotate fast enough so that the ideal kink is stabilized. Experimentally, it has been observed that the onset of RWMs is preceded by a reduction in plasma rotation. Figure 7.7 shows an example where a tokamak discharge in the DIII-D tokamak exceeds the no-wall β-limit

9) Other feedback strategies are also possible, for example aiming at zeroing the radial field at the plasma surface rather than the wall.

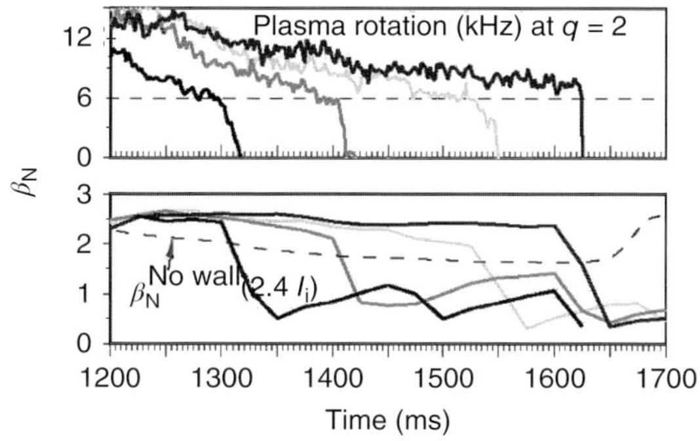

Figure 7.7 β-limit experiments in the DIII-D tokamak: the discharges significantly exceed the no-wall limit as long as the $q = 2$ surface rotates at a rate much faster than the inverse wall time, but as soon as the plasma rotation slows down and locks to the wall, an RWM grows to large amplitude and terminates the high performance phase. *Source:* E. J. Strait *et al.* 2003 [32], reproduced with permission of the IAEA.

for an appreciable duration, corresponding to many Alfvén times, as long as the rotation of the $q = 2$ surface is much higher than the inverse wall time but becomes unstable to an RWM as soon as the rotation falls below a value associated with a bifurcation in the torque balance. This in turn leads to disruptive termination of the discharge. Note that the final slowing down occurs abruptly, implying that the growth of the RWM contributes to slowing down the plasma rotation. We note that while the picture described here only leads to the final state of a locked RWM when $\beta > \beta_{\text{no wall}}$, there is also an amplification of external error fields by a stable RWM at $\beta < \beta_{\text{no wall}}$ which leads to the β-dependence of error field penetration mentioned in Section 9.3.3.

A quantitative assessment of the minimum torque input for avoidance of the RWM is not straightforward as in ideal MHD, a rotating plasma does not exert a force on a mode[10] and the calculation of the effect of rotation on RWM stability hence needs to include non-ideal effects such as neoclassical toroidal viscosity [33], sound wave excitation or ion Landau damping to provide finite viscosity due to the presence of the mode. Calculations based on the latter indicated that there can be a stability window in terms of the wall location at which the wall is close enough to suppress effectively the ideal kink mode due to rotation and on the other hand far enough so that the RWM is suppressed by the plasma rotating with respect to the mode structure [34]. A further ingredient that has to be considered is the presence of an intrinsic helical field due to the unavoidable finite tolerance in shape and positioning of the tokamak coil system which, as mentioned earlier, play a role in the locking process as well and will be further discussed in Section 9.3.2.

The question remains how the RWM is stabilized by the plasma flow with respect to the mode structure. It has been conjectured that this is mainly due to

10) This is different for resistive modes as treated in Section 9.3.2.

another important ingredient of RWM stability that occurs in present tokamak experiments, namely the kinetic resonance with thermal or fast particles. As will also be discussed in Chapter 11 in the context of the stability of the internal kink mode, these can have both stabilizing and destabilizing effects as on the one hand, work must be done to displace fast particle orbits, on the other hand, releasing the free energy stored in the gradient of the fast particles at the resonant surface by the MHD mode will lower the total δW. Hence, kinetic effects must be included in a quantitative way to correctly describe RWM stability. As these effects are resonant and hence crucially depend on the rotation of the bulk plasma, they can significantly modify the rotation dependence of RWM stability. In the frequency region of interest for RWM stability, the precession drift or the bounce motion of trapped particles on banana orbits are possible candidates for kinetic stabilization of the RWM, making it necessary to treat in detail the kinetic corrections to RWM stability analysis. Figure 7.8 shows an example for a detailed analysis of these effects for discharge parameters from the NSTX spherical tokamak. It can be seen that both effects are significant in the region of interest (typical values in tokamaks are of the order of $\omega_{\text{mode}}/(v_A/R)$ several percentages), thus making it necessary to include these when analysing in detail the critical rotation velocity for rotational stabilization of RWMs.

Hence, experiments that aim at demonstrating active RWM stabilization using non-axisymmetric control coils above the nominal no-wall β-limit must address stabilization of RWMs at low plasma rotation to unambiguously prove that they do not just rely on rotational stabilization of the external kink, effectively

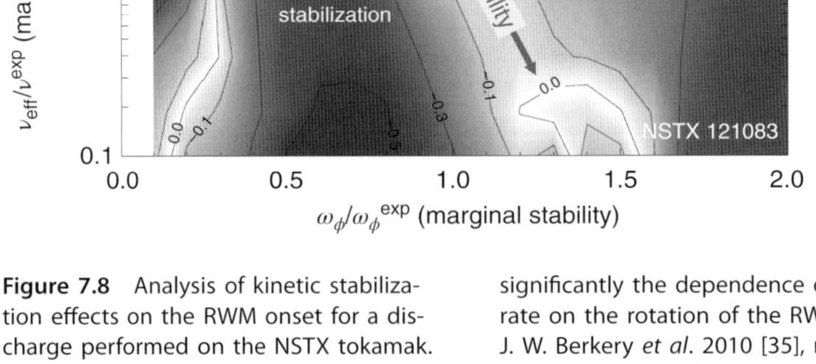

Figure 7.8 Analysis of kinetic stabilization effects on the RWM onset for a discharge performed on the NSTX tokamak. Two regions of increased stability due to kinetic effects are obtained, modifying significantly the dependence of the growth rate on the rotation of the RWM. *Source:* J. W. Berkery *et al.* 2010 [35], reproduced with permission of AIP. (Please find a color version of this figure on the color plates.)

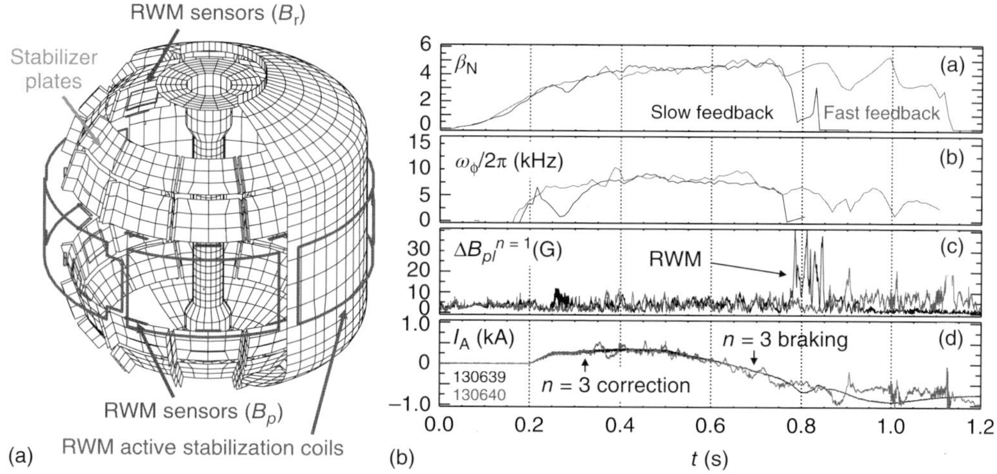

Figure 7.9 (A) Active stabilization of an RWM at low rotation using a set of control coils located on the outside of the vacuum vessel of the NSTX tokamak shown. (B) With additional direct feedback on the $n = 1$ mode (fast feedback), the amplitude of the RWM is controlled and β_N (trace a) and rotation frequency (trace b) are maintained at a high level while with error field correction only (slow feedback), the mode grows (the mode amplitude is shown in trace (c) and causes a disruption. Trace (d) shows the current in one of the control coils. *Source:* S. A. Sabbagh *et al*. 2010 [36], reproduced with permission of the IAEA. (Please find a color version of this figure on the color plates.)

removing the RWM branch by this effect. In fact, plasmas run at β_N-values significantly above the no-wall limit and at much lower plasma rotation than shown in Figure 7.7 have been run through the region stabilized by ion precession drift resonance shown in Figure 7.8. When the plasma rotation profile drops to low values, active feedback control is generally required to stabilize the RWM. Such an experimental demonstration is shown in Figure 7.9 where active control using the coil system shown in the Figure 7.9A is applied to control the RWM. The traces labelled 'slow feedback' show a case where the coils are only used for error field correction and a disruption occurs at reduced plasma rotation. The traces labelled 'fast feedback' show a case where the applied control field directly counteracts the $n = 1$ RWM growth. Both cases use non-resonant $n = 3$ braking. Clearly, the action of the active control prolongs the stable phase and a β-collapse is observed with only error field correction. It is expected that this situation will be relevant in future devices where the plasma rotation induced by NBI, which is quite strong in present-day experiments, will be negligible.

Thus, while a treatment of kinetic effects is beyond the framework for MHD stability developed in this book, we note that only when these are included it can be expected that predictive capability for the requirements of RWM stabilization in future large devices can be obtained.

7.5 The Troyon Limit

Concluding this chapter, we ask the question if, considering the combination of all ideal MHD limits that we have discussed so far, it is possible to give a simple estimate of the ideal β-limit for given machine parameters, that is cross section, aspect ratio, plasma current and toroidal field.

This question has been studied numerically, based on linear ideal MHD stability considering the main limiting factors discussed so far, namely Mercier stability, ballooning stability and stability to external kinks [37]. In this optimization, current and pressure profiles are varied to obtain the limiting β for series of equilibria and it is found that Mercier modes mainly limit the central current and pressure gradients[11], ballooning modes limit the plasma pressure in the outer half of the plasma radius[12], but the most restrictive condition is usually the onset of the $n = 1$ external kink mode. The variation of $\beta_{t,\max}$ with I_p, B_t, R_0 and the aspect ratio $A = R_0/a$ can be approximated by a surprisingly simple formula

$$\beta_{t,\max} \propto \frac{AI_p}{R_0 B_t} \tag{7.25}$$

or, when expressed in dimensional quantities for circular cross section

$$\beta_{t,\max}[\%] = 2.8\frac{I_p[\mathrm{m}]}{a[\mathrm{m}]B_t[\mathrm{T}]} \tag{7.26}$$

This so-called Troyon-limit gives a very robust description of the global ideal β-limit in tokamaks, which has led to the definition of the 'normalized β'

$$\beta_N = \frac{\beta_{t,\max}[\%]}{I_p[\mathrm{m}]/(a[\mathrm{m}]B_t[\mathrm{T}])} \tag{7.27}$$

Figure 7.10 shows the envelope of a number of data points from a variety of tokamaks. It can be seen that the upper limit to β_t is well described by the scaling 7.25. The proportionality constant $\beta_{N,\max}$ is higher than for circular cross section, which will be addressed below. Note in particular that in Figure 7.10b, the data points for spherical tokamaks which have a high value of $I_p/(aB_t)$ (which is basically the ratio B_{pol}/B_t), extend the conventional tokamak data range by a factor of more than 2[13].

The wide range of validity of the Troyon limit indicates that there might be a fundamental principle behind it which gives rise to the scaling. While such a scaling is not easily derived for the $n = 1$ external kink mode when finite pressure and toroidicity are involved (Section 7.2), an argument based on the first stability ballooning boundary can reproduce the trend [40]. To see that, we write β as

$$\beta = \frac{4\mu_0 \int_0^a p(r) r\, dr}{a^2 B^2} = \frac{\int_0^a \alpha r^2/q^2\, dr}{Ra^2} = -0.3\frac{1}{a^2 R}\int_0^a \frac{d}{dr}\left(\frac{1}{q^2}\right) r^3\, dr \tag{7.28}$$

11) Most of the cases studied in the original paper were limited to $q_0 \geq 1$ by that condition so that the internal kink did not play a significant role.
12) The pressure profile is not assumed to exhibit a pedestal structure, so ELMs are not an issue.
13) More recent experiments in NSTX have extended further these parameters, also in β_N [36].

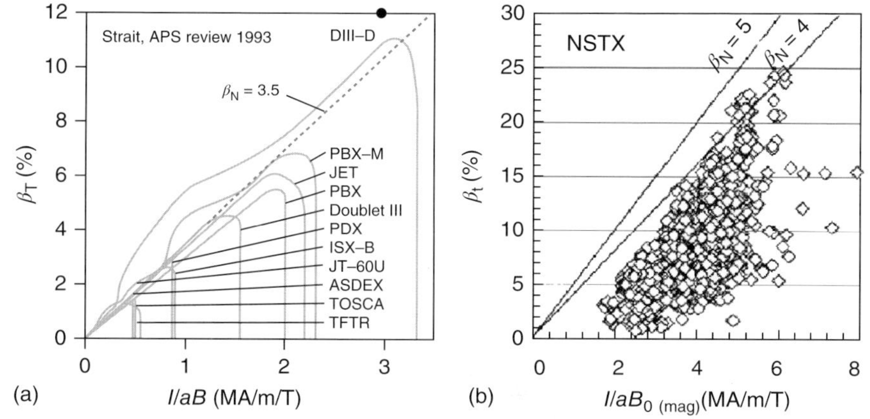

Figure 7.10 (a) Envelope of data points from various tokamaks indicating the validity of the Troyon limit. (b) Including a small aspect ratio tokamak, here NSTX, increases the region in which the scaling is validated significantly. *Source (a):* E. J. Strait 1994 [38] and (right) S. A. Sabbagh *et al.* 2002 [39], both reproduced with permission of the AIP. (Please find a color version of this figure on the color plates.)

where we have approximated the first stability ballooning boundary as $\alpha = 0.6s$. It is clear that this expression will be optimized for given q_a by maximizing the shear in the outermost region of the plasma, leading to a large pressure gradient there. Hence, a suited model current profile has a constant current density inside a certain radius r_j, leading to zero shear in this region and is zero outside, leading to a linear increase of q. The value of the current density is limited by the internal kink or the Mercier limit to $q_0 = 1$. Using the formula for q_0 (Eq. (4.13)), we obtain for the radius at which the current density drops to zero

$$r_j = \sqrt{\frac{\mu_0 I_p R}{2\pi B_t}} \quad \rightarrow \quad \frac{r_j}{a} = \frac{1}{\sqrt{q(a)}} \tag{7.29}$$

The corresponding profiles are shown in Figure 7.11.

Evaluating β leads to $\beta = 1.2\frac{a}{Rq_a^2}(\sqrt{q_a} - 1)$ which, for $q_a \geq 2$, can be approximated by

$$\beta[\%] = 5.6\frac{I_p[\mathrm{MA}]}{a[\mathrm{m}]B[\mathrm{T}]} \tag{7.30}$$

While this limit is numerically more optimistic than that found by Troyon (consistent with the fact that in the Troyon calculation, it is the $n = 1$ external kink that sets the ultimate limit), it reproduces the typical scaling.

As evident from Figure 7.10, there is a variation in $\beta_{N,\max}$. This is mainly due to a dependence on the shape of the poloidal cross section and on the variation in current profile. As for the shape dependence, we observe that according to Eq. (7.26), $\beta_{t,\max}$ will increase with I_p but, for given shape, this will lead to lower q_a and ultimately be limited by the low-q limit. However, if the cross section is

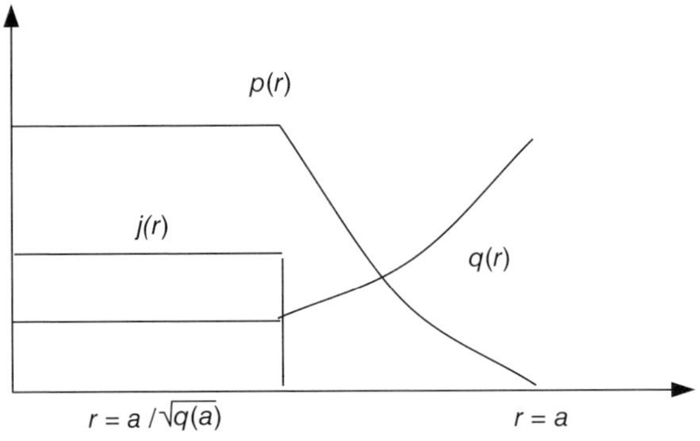

Figure 7.11 Profiles used in the optimization for maximum β using the first stability regime ballooning stability boundary.

shaped appropriately, I_p can be increased at constant q_a with respect to the circular case, leading to higher $\beta_{t,\max}$ [14]. For example, if the cross section is elongated, the poloidal circumference increases approximately by $\sqrt{(1+\kappa^2)/2}$, and hence q_a will be

$$q_a = q_{\mathrm{cyl}} \frac{1+\kappa^2}{2} \tag{7.31}$$

leading to an increase of $\beta_{t,\max}$ of the same factor. While this argument is in principle in agreement with the experimentally observed trends, it over-estimates the increase with κ, and it has also been found that higher moments of shape will enter in a more subtle way, changing the achievable $\beta_{t,\max}$ in a way not directly related to the poloidal circumference but by increasing the magnetic well as discussed in the context of Mercier stability in Section 5.2.1. (see also Figure 5.2).

As for the effect of the current profile, we already found that the external kink is more unstable with a broader current profile and may hence expect an increase of $\beta_{N,\max}$ with increased peaking of the current profile. Such an increase is clearly seen in experiments and has led to a modification of the original Troyon scaling by including the internal inductance ℓ_i which, according to Eq. (2.55) is a measure of the peakedness of the current profile.

Figure 7.12 shows how the quality of the fit can be improved by including ℓ_i, leading to the scaling

$$\beta[\%] = 4\ell_i \frac{I_p[\mathrm{MA}]}{a[\mathrm{m}]B[\mathrm{T}]} \tag{7.32}$$

Again, it is not possible to directly derive this formula from $n = 1$ external kink stability, but similar to the derivation shown earlier, an analytical analysis based on infinite-n ideal ballooning mode stability qualitatively reproduces this dependence although the numerical factor is not reproduced quantitatively [42].

14) We note that this argument implies that the Troyon limit is actually a limit at given q_a, not given I_p.

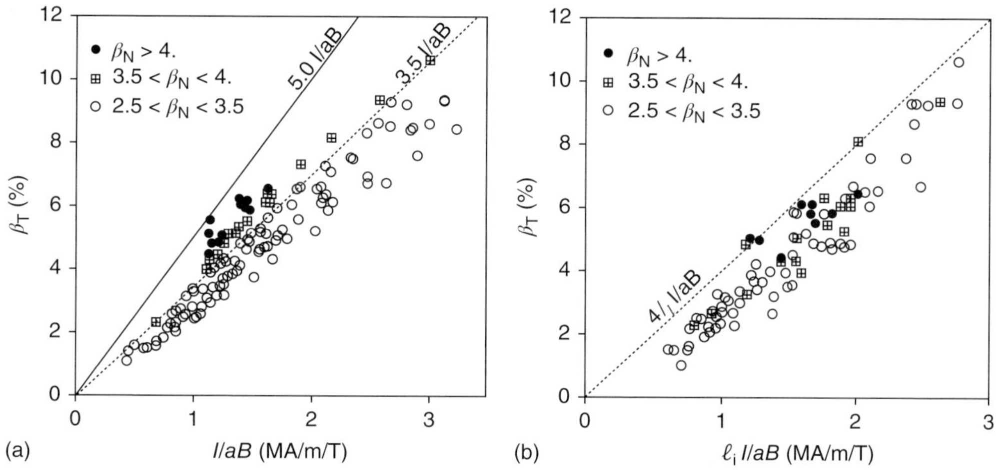

Figure 7.12 Plot of β-limit experiments in DIII-D against the Troyon scaling without (a) and with an ℓ_i-dependence b). The scatter is reduced if the ℓ_i-effect is included, confirming the role of the current profile in the β-limit. *Source:* T. S. Taylor [41], reproduced with permission of the author.

Finally, we note that while this scaling describes the essential properties of the β-limit, a practical limitation to β can arise due to the occurrence of neoclassical tearing modes, a resistive MHD instability not included in the analysis earlier, which is based on ideal MHD. Resistive MHD instabilities will be studied in the following chapters of this book.

8
Resistive MHD Stability

So far, we have analysed ideal MHD stability, that is the motion $\mathbf{v}_1$ was assumed to be flux conserving and hence not changing the topology of the nested flux surfaces due to the assumption of infinite electrical conductivity[1]. It was already mentioned in Chapter 1 that finite electrical conductivity changes this picture and allows to generate or annihilate magnetic flux. This breaks the frozen-in flux constraints and allows changes of the flux surface topology, hence giving rise to a new class of MHD instabilities not accessible in ideal MHD. This can be seen from the equation for the perturbed magnetic field

$$\frac{\partial \mathbf{B}}{\partial t} = \nabla \times (\mathbf{v} \times \mathbf{B} - \frac{1}{\sigma}\mathbf{j}) = \nabla \times (\mathbf{v} \times \mathbf{B} - \frac{1}{\mu_0 \sigma}\nabla \times \mathbf{B}) \tag{8.1}$$

where we have assumed for the sake of simplicity that the electrical conductivity is constant in space. In contrast to Eq. (3.6), this equation not only allows the generation of a perturbed magnetic field by streaming across field lines but also by a current parallel to the magnetic field which, for finite conductivity, generates an electric field $\mathbf{E}_1$.

The timescale of resistive MHD can be estimated by evaluating the second term on the right-hand side of Eq. (8.1) to yield

$$\nabla \times (\nabla \times \mathbf{B}) = -\Delta \mathbf{B} = -\mu_0 \sigma \frac{\partial \mathbf{B}}{\partial t} \tag{8.2}$$

Equation (8.2) has the mathematical structure of a diffusion equation with diffusion coefficient $D_{\mathrm{mag}} = 1/(\mu_0 \sigma)$. In classical electrodynamics, this diffusion process is known as the *skin effect*, that is the penetration of current into a conductor when a voltage is applied rapidly at its surface. This allows to derive the characteristic timescale over which the magnetic field diffuses into a plasma of typical cross section L^2 and electrical conductivity σ:

$$\tau_R = \mu_0 \sigma L^2 \tag{8.3}$$

which was already introduced in Eq. (1.28). Hence, magnetic flux in a plasma changes on this timescale; for $\tau \ll \tau_R$, flux is conserved and ideal MHD is valid. The diffusion of magnetic field in a plasma on the resistive timescale can also be

1) Note that this is different from superconductivity as there, magnetic flux inside an object is zero rather than conserved at the initial value.

Magnetohydrodynamic Stability of Tokamaks, First Edition. Hartmut Zohm.

seen as a 'slipping' of field lines through the plasma as opposed to the frozen-in condition. Obviously, resistive effects become important if the ratio

$$\frac{|\mathbf{v} \times \mathbf{B}|}{1/(\mu_0\sigma)|\nabla \times \mathbf{B}|} \approx \mu_0 \sigma L v = Re_M \tag{8.4}$$

is small. Owing to the similarity to the Reynolds number in ordinary hydrodynamics, this number is called the *magnetic Reynolds number*.

An upper bound for Re_M can be obtained if v is assumed to be the timescale of ideal MHD, that is the Alfvèn velocity v_A. Following this definition, ideal MHD is valid if the Lundquist number

$$S = \frac{\tau_R}{\tau_A} \approx \mu_0 \sigma L v_A \tag{8.5}$$

is much bigger than 1. For a hot fusion plasma, this is usually the case as S may range anywhere between 10^6 and 10^9 for typical values of T, n, B and L. However, we remind that finite conductivity allows perturbations that would not be possible in the framework of ideal MHD so that there is no competition between ideal and resistive MHD instabilities for these modes. Furthermore, as we will see in detail in the following, the actual change in topology introduced by a resistive MHD instability originates from a narrow layer around the resonant surface on which it grows while the flux surfaces further away from this layer are deformed according to ideal MHD motion. Then, the resistive timescale will drop by the square of the ratio of layer width and typical length scale. Assuming, for example, a typical layer width of 1 cm and a macroscopic length scale of 1 m, τ_R drops from 10 s for current redistribution to 1 ms for growth of resistive instabilities. Hence, resistive MHD instabilities will also play a role in limiting the tokamak operational space.

We note here that often, resistive MHD instabilities are found to grow on a faster timescale than predicted by the simple one-fluid estimate given earlier. This is true not only in fusion plasmas but also in astrophysical phenomena. In these cases, Ohm's law has to be reconciled to consider effects that can effectively decrease the conductivity and hence increase the growth rate. While this is an active field of research in high-temperature plasma physics in general, it is beyond the theoretical framework developed in this book and will hence not be treated in detail in the following.

8.1 Stability of Current Sheets

In this section, we treat the stability of a current sheet against reconnection. Reconnection of field lines is the process by which the topology of a flux surface structure in a plasma can change. It occurs in situation in which magnetic field lines of opposing direction occur close to each other, indicating the presence of a current sheet. While this current sheet persists infinitely long in ideal MHD, it decays on the resistive timescale in resistive MHD. By forming topologically new objects, the so-called magnetic islands, the free energy of the system can be

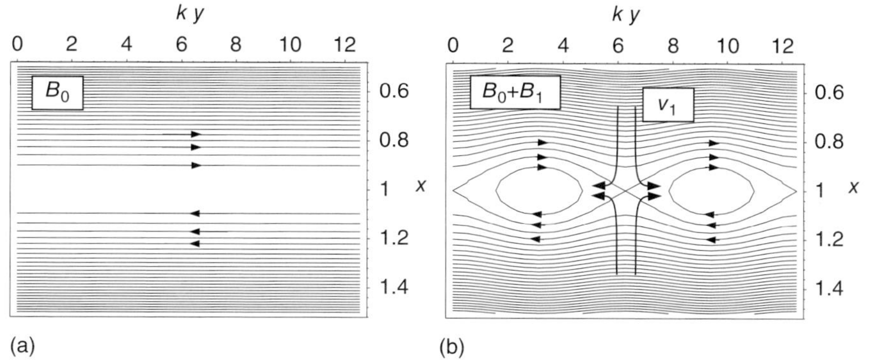

Figure 8.1 (a and b) Reconnection of field lines in an area of opposing field. During the process, magnetic islands are formed. Also indicated is the flow pattern of plasma into the island during formation.

reduced. This instability, which tears and reconnects field lines, is called a *tearing mode*.

Figure 8.1 shows the basic geometry. In Figure 8.1a, the initial situation is shown, assuming a current sheet at $x = 1$. During the formation of the magnetic islands, the sheet breaks up into areas of currents with periodically changing sign, forming the so-called X-points through which the plasma streams into the newly formed island flux surfaces. We will examine the magnetic island in more detail in Section 8.3. In the following, we study the stability of the current sheet using a simple argument following [43].

Let us assume that the system is characterized by a length L in the x-direction perpendicular to the initial magnetic field B_0 which we assume to vary linearly in this direction and change sign at the surface $x = x_s$:

$$B_0 = B_{0y}\frac{x - x_s}{L} \tag{8.6}$$

In y-direction, the characteristic length is the wavelength λ of the perturbation, that is the wave vector is $k = 2\pi/\lambda$. These conditions are sketched in Figure 8.1a.

Now we assume that an external force F_{ext} acts to compress the plasma homogeneously in x-direction towards the $x = x_s$ surface. Far away from x_s, the motion will be governed by ideal MHD and the field lines move with the plasma. This can be understood as follows: assume that the plasma would move across the field lines without changing B_0 (i.e. violate ideal MHD). Then, according to Faraday's law, there is an induced electric field and using Ohm's law with finite conductivity, we obtain a current

$$j_{1z} = \sigma v_{1x} B_0 \tag{8.7}$$

This current will lead to a restoring $\mathbf{j} \times \mathbf{B}$ force

$$F_x = \sigma v_{1x} B_0^2 \tag{8.8}$$

Then, applying the limit of ideal MHD ($\sigma \to \infty$), this force becomes arbitrarily large and hence a decoupling of plasma and field is not possible[2]. However, close to the $\mathbf{B} = 0$-line ($x = x_s$), this argument becomes invalid as the restoring force approaches 0 as $B_0 \to 0$. Hence, there will be a point at $x = x_s \pm \delta/2$ at which the restoring force equals the initial external force and inside this point, the plasma decouples from the field. We can determine the thickness δ of the resistive layer by

$$F_{\text{ext}} = F_x = \sigma v_{1x} B_0^2(x_s \pm \delta) = \sigma v_{1x} B_{0y}^2 \left(\frac{\delta}{L}\right)^2 \tag{8.9}$$

Outside of the layer, the plasma can be treated using ideal MHD, while inside, finite conductivity is essential. Examining the energetics of this process, the power dissipated in the layer, $P = vF$, must equal the power flowing into the system due to the external force

$$P = F_x v_{1x} = \sigma v_{1x}^2 B_{0y}^2 \left(\frac{\delta}{L}\right)^2 \tag{8.10}$$

This power must be equal to the rate of kinetic energy change during the acceleration of the plasma in the ideal region by the external force. If we assume an incompressible flow of plasma into the island, that is

$$\nabla \cdot \mathbf{v} = 0 \;\; \to \;\; \frac{1}{\delta} v_{1x} + k v_{1y} = 0 \tag{8.11}$$

we see that because of $\delta k \ll 1$ (layer width much smaller than the wavelength of the perturbation) the velocity is determined by the flow into the island, $v_{1y} \gg v_{1x}$. Effectively, the flow through the X-point resembles that through a nozzle as the width in x-direction is much smaller than in y-direction.

If we now assume the typical timescale of this process to be $1/\gamma$, the power needed to accelerate the plasma becomes

$$\gamma \rho v_{1y}^2 = \gamma \rho \frac{v_{1x}^2}{(\delta k)^2} \tag{8.12}$$

and equating Eqs (8.10) and (8.12) allows to obtain an equation for the layer width

$$\delta = \left(\frac{\gamma \rho L^2}{\sigma k^2 B_{0y}^2}\right)^{1/4} \tag{8.13}$$

where γ is still unknown. In order to obtain an equation for this quantity, we examine the process of island formation in more detail. In particular, the formation of the island structure implies the growth of a field B_1 in x-direction according to

$$\gamma B_{1x} = k E_{1z} = \frac{k}{\sigma} j_{1z} \tag{8.14}$$

where we have neglected the term $\mathbf{v} \times \mathbf{B}$ as inside the layer, the resistive effect should dominate. As already discussed in Section 7.3, this current can be expressed by the jump of the tangential component of B_1 across the layer, that is

$$\mu_o j_{1z} = \frac{B_{1y}^+ - B_{1y}^-}{\delta} \tag{8.15}$$

2) This argument is essentially similar to the derivation of the frozen-in flux in Chapter 1.

where the indices + and − denote the boundary of the layer at $x_s \pm \delta/2$.

$$\gamma = \frac{1}{\mu_o \sigma \delta} \frac{k(B_{1y}^+ - B_{1y}^-)}{B_{1x}} = \frac{1}{\mu_o \sigma \delta} \Delta' \tag{8.16}$$

where in the last step, we have introduced the stability parametre Δ' which determines the stability of the system since its sign will determine the sign of γ. Inserting Eq. (8.16) into Eq. (8.13), we obtain an equation for δ

$$\delta = \left(\frac{\Delta' \rho L^2}{\mu_o \sigma^2 k^2 B_{0y}^2} \right)^{1/5} \tag{8.17}$$

which, using the characteristic timescales $\tau_R = \mu o \sigma L^2$ and $\tau_A = L/v_A = L/(B_{0y}/\sqrt{\mu o \rho})$ can be expressed as

$$\delta = L \left(\frac{\Delta' L \tau_A^2}{(kL)^2 \tau_R^2} \right)^{1/5} \tag{8.18}$$

which finally gives an equation for the growth rate γ:

$$\gamma = \tau_R^{-3/5} \tau_A^{-2/5} (\Delta' L)^{4/5} (kL)^{2/5} \tag{8.19}$$

Thus, the linear growth rate of the tearing mode is given by a hybrid timescale involving both τ_R and τ_A. For $\Delta' > 0$, an arbitrarily small external force will lead to the growth of a tearing mode, that is the situation is unstable, while for $\Delta' < 0$, the system opposes tearing. Still, in this case, reconnection can happen, driven by the external force, but in this case, the flow pattern will be approximately stationary ($\partial \mathbf{B}/\partial t = 0$) and the timescale is different. In Astrophysics, this process is known as *Sweet–Parker reconnection* and we will come back to it when we treat the sawtooth reconnection in Chapter 11.

As pointed out earlier, this simple picture often fails to describe the experimentally observed timescales, which can be much faster, indicating that more physics ingredients have to be incorporated into Ohm's law. For example, two-fluid effects or turbulence may lead to a broadening of the resistive layer and hence to a larger mass flow into the island. Turbulence can also give rise to 'anomalous' resistivity, decreasing the effective electrical conductivity in the layer. However, a detailed treatment of these effects is beyond the scope of this book.

8.2 Reconnection in the Presence of a Guide Field

In this section, we describe the analysis of the tearing mode in a tokamak plasma. While in the previous example, we assumed that the equilibrium field was entirely due to the current sheet, the situation is different in a tokamak where the toroidal field component is created externally and does not take part directly in the reconnection process. In fact, even the poloidal field does not change sign over the radius of a usual tokamak discharge. However, a situation comparable to that in Figure 8.1 is due to the shear in the magnetic field. If we follow a field line on a resonant surface with $q = q_s$, the field lines on neighbouring surfaces will have different helicity and hence lag behind or advance the field line on the resonant surface in poloidal angle. For example, after two turns around the torus on the $q = 2$ surface, the field line closes in itself, whereas, for positive shear, the neighbouring field line at $r < r_s$ (i.e. $q = q_s - \epsilon$) has advanced already to a poloidal angle greater than 2π. Similarly, the field line on the $q > 2$ surface is lagging behind. Thus, there is a field component relative to the $q = q_s$ surface that changes sign across $q = q_s$. This field, denoted as helical field B^*, can now be prone to a tearing instability in the same way it was discussed in Section 8.1.

The field B^* can be described by a flux function Ψ^* known as *helical flux*, where the flux integral is carried out through a surface extending in radial direction along the equilibrium field line. This is shown schematically in Figure 8.2.

The helical co-ordinate system is obtained from the cylinder co-ordinate system which describes the screw pinch by

$$\hat{\mathbf{e}}_r = \hat{\mathbf{e}}_r \tag{8.20}$$

$$\hat{\mathbf{e}}_\mu = \frac{1}{\sqrt{1 + (\frac{r}{R_0}\frac{1}{q_s})^2}} \left(\hat{\mathbf{e}}_\theta - \frac{r}{R_0}\frac{1}{q_s}\hat{\mathbf{e}}_z \right) \tag{8.21}$$

$$\hat{\mathbf{e}}_\eta = \frac{1}{\sqrt{1 + (\frac{r}{R_0}\frac{1}{q_s})^2}} \left(\hat{\mathbf{e}}_z + \frac{r}{R_0}\frac{1}{q_s}\hat{\mathbf{e}}_\theta \right) \tag{8.22}$$

where the (constant) helicity of the co-ordinate system is that of the field lines on the resonant surface. Hence, $\hat{\mathbf{e}}_\eta$ is parallel to the equilibrium field B_0 on the resonant surface, whereas $\hat{\mathbf{e}}_\mu$ is perpendicular to both $\hat{\mathbf{e}}_\eta$ and $\hat{\mathbf{e}}_r$ (both coordinate

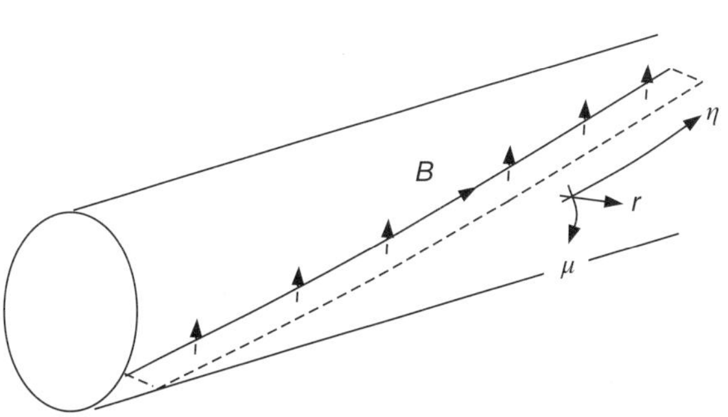

Figure 8.2 Definition of a helical co-ordinate system relative to the field lines on a resonant surface. The helical flux Ψ^* is obtained by the integral of $\mathbf{B}$ over the surface indicated with dashed lines.

systems share the radial coordinate). By definition, $B^*_{0\mu}$ vanishes at the resonant surface

$$B^*_{0\mu} = \mathbf{B}^*_{\mathbf{0}} \cdot \hat{\mathbf{e}}_\mu = \frac{1}{\sqrt{1+(\frac{r}{R_0}\frac{1}{q_s})^2}} \left(B_{0\theta}(r) - B_{0\theta}(r_s)\, \frac{r}{r_s} \right)$$
$$\approx B_{0\theta}(r) - B_{0\theta}(r_s)\, \frac{r}{r_s} = B_{0\theta}(r) \left(1 - \frac{q(r)}{q_s} \right) \tag{8.23}$$

From the definition of the equilibrium helical flux Ψ^*_0, we obtain a relation between Ψ^*_0 and $B^*_{0\mu}$

$$\mathbf{B}^*_0 = \nabla\Psi^*_0 \times \hat{\mathbf{e}}_\eta \approx \nabla\Psi^*_0 \times \hat{\mathbf{e}}_z \;\rightarrow\; \frac{d\Psi^*_0}{dr} = -B^*_{0\mu} \tag{8.24}$$

Note that here, Ψ^* is a flux per unit length in helical direction and we have used the fact that $B_\phi \gg B_\theta$, that is $\hat{\mathbf{e}}_\eta$ and $\hat{\mathbf{e}}_z$ are approximately parallel.

Using Eq. (8.24) together with Eq. (8.23), we obtain

$$\left.\frac{d^2\Psi^*_0}{dr^2}\right|_{r_s} = -\left.\frac{dB^*_{0\mu}}{dr}\right|_{r_s} = B_{0\theta}\left.\frac{q'}{q}\right|_{r_s} \tag{8.25}$$

From Eq. (8.25), it follows that Ψ^*_0 can be approximated by a parabola close to the resonant surface where the sign of the curvature is determined by q'. Specifically, for $q' > 0$, $B^*_{0\mu}$ is positive for $r < r_s$ and negative for $r > r_s$. Figure 8.3 shows, for a typical monotonically increasing q-profile, the quantities $B_{0\theta}$ and $B^*_{0\mu}$.

We will now derive an equation for the perturbed helical flux Ψ^*_1. Assuming a periodic screw pinch, we decompose Ψ^*_1 in Fourier components in the two independent variables θ and z/R_0

$$\Psi^*_1 = \Psi^*_1(r)e^{i(m\theta - n\frac{z}{R_0})} \tag{8.26}$$

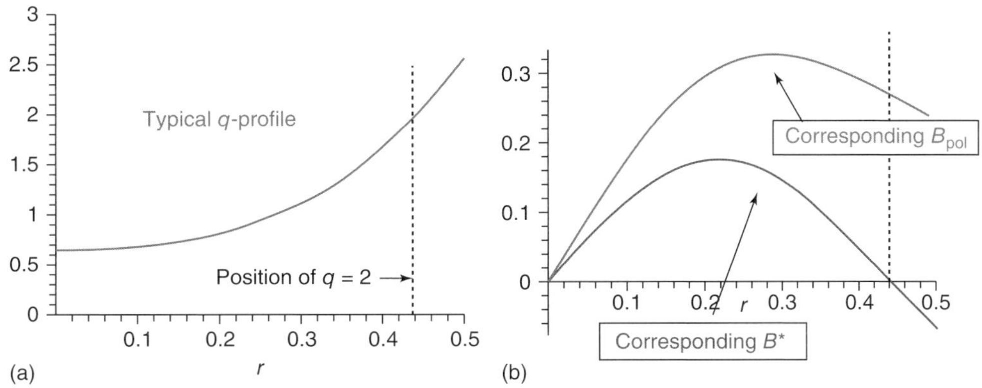

Figure 8.3 (a and b) Visualization of the equilibrium poloidal field B_{pol} and the corresponding $q = 2$ helical field $B^*_{0\mu}$ for a typical q-profile shown. The helical field changes sign at the $q = 2$ surface. (Please find a color version of this figure on the color plates.)

According to Section 8.1, outside the resistive sheet, the helical deformation of the flux surfaces can be described by ideal MHD. As the ideal MHD timescale is much shorter than the resistive timescale on which the instability develops, we can assume that the plasma goes through a sequence of ideal MHD equilibria, that is according to $\mathbf{j} \times \mathbf{B} = \nabla p$. For the ideal MHD energy functional, this means the marginal case $\delta W = 0$. As we are dealing with current driven modes, we use the low β form of δW according to Eq. (4.6) neglecting the external part. In analogy to the procedure used when deriving the Suydam criterion (Eq. (5.6)), we obtain an Euler–Lagrange equation that reads

$$r(m^2-1)\left(\frac{n}{m}-\frac{1}{q}\right)^2 \xi - \frac{d}{dr}\left(r^3\left(\frac{n}{m}-\frac{1}{q}\right)^2\frac{d\xi}{dr}\right) = 0 \tag{8.27}$$

for the radial component of the displacement vector, $\xi = \xi_r$. We relate ξ and B_1^* using the linearized MHD kinematic equation (Eq. (3.6)) and relate B_1^* to Ψ_1^* using Eq. (8.26) to obtain

$$B_{1r} = \frac{B_{0\theta}}{r}\frac{\partial \xi_r}{\partial \theta} + B_{0z}\frac{\partial \xi_r}{\partial z} = \frac{i}{r}(m-nq)B_{0\theta}\xi_r \;\rightarrow\; \Psi_1^* = (1-\frac{nq}{m})B_{0\theta}\xi_r \tag{8.28}$$

This relation can be inserted into the Euler–Lagrange equation (Eq. (8.27)) to express it in the perturbed helical flux. After some algebra, where also we make use of Ampère's law, we obtain

$$\Delta\Psi_1^* - \frac{\mu_0(dj_{0z}/dr)}{B_{0\theta}(r)(1-q(r)n/m)}\Psi_1^* = 0 \tag{8.29}$$

This equation is known as *tearing mode equation* [3]. Again, we have made use of the fact that $\hat{\mathbf{e}}_\eta \approx \hat{\mathbf{e}}_z$.

The tearing mode equation describes the ideal deformation of the plasma according to the helicity given by (m, n) (in Eq. (8.26)) in a linearized approximation. As we have not assumed anything about the region inside the resistive layer, we cannot obtain information on the growth rate of a possible tearing instability. However, we can infer the stability itself from the tearing mode equation in the linearized sense, that is for $\delta \rightarrow 0$ by integrating it, noting that it becomes singular at the resonant surface. Practically, this means that we can integrate it, for example starting from $r = 0$ with the appropriate boundary condition $dj/dr = 0$[4], up to the resonant surface. The same procedure can be done from the other side, for example assuming a perfectly conducting wall at $r = r_{\text{wall}}$. As we are dealing with a linear equation where the amplitude is arbitrary, the two solutions can always be scaled by a constant factor to continuously connect across r_s. However, this will in general lead to a jump of $d\Psi_1^*/dr = B_{1\mu}$ at $r = r_s$, indicating the existence of a surface current there. Figure 8.4 shows an example of a numerical solution of the tearing mode equation for a current profile $j \propto (1-(r/a)^2)^3$.

3) We note that due to the force-free condition, this equation is the linearized form of $(\mathbf{B}\cdot\nabla)j_z = 0$, that is the perturbed current is constant on the perturbed flux surfaces. We will come back to this when we derive the nonlinear evolution equation.
4) This means that close to $r = 0$, the solution is essentially the vacuum solution (Eq. (7.13)).

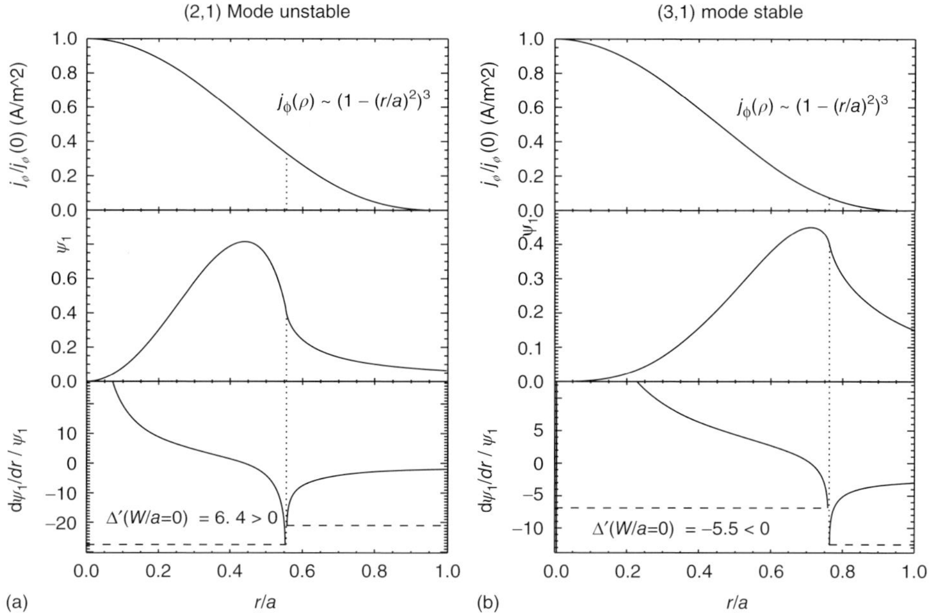

Figure 8.4 The perturbed helical flux function Ψ_1^*, calculated using the tearing mode equation based on the current profile $j \propto (1-(r/a)^2)^3$. The discontinuity in $d\Psi_1^*/dr$ at r_s determines stability against tearing. In this particular case, (a) the (2,1) mode is unstable, whereas (b) the (3,1) mode is stable, as can be seen from the jump in the first derivative which determines the sign of Δ'.

The direction of the surface current generated at r_s by the deformation of the flux surfaces will determine the stability of the system: if it acts to reinforce the ideal perturbation, the situation is unstable while an opposing current means that the plasma resists tearing. The sign of the surface current is determined by the sign of the jump of $B_{1\mu}^*$ across r_s, that is when normalized to $\Psi_1^*(r_s)$, by

$$\lim_{\epsilon\to 0} \frac{\Psi_1^{*\prime}(r_s+\epsilon) - \Psi_1^{*\prime}(r_s-\epsilon)}{\Psi_1^*(r_s)} = \Delta' \tag{8.30}$$

which is equivalent to the definition of Δ' in Section 8.1 (Eq. (8.16)). For positive Δ', the current will reinforce the initial perturbation and hence the stability criterion for the tearing mode in this case is the same as above, namely

$$\Delta' < 0 \leftrightarrow \text{system is stable} \tag{8.31}$$

The role of Δ' can also be seen from the energetics of the tearing mode equation. The energy density of the magnetic field is given by

$$W_{\text{mag}} = \frac{1}{2\mu_0}\left(\left(\frac{d\Psi_1^*}{dr}\right)^2 + \frac{m^2}{r^2}\Psi_1^{*2}\right) \tag{8.32}$$

On the other hand, writing out the tearing mode equation (Eq. (8.29)) and multiplying it by Ψ_1^* gives

$$\frac{d^2\Psi_1^*}{dr^2}\Psi_1^* + \frac{1}{r}\frac{d\Psi_1^*}{dr}\Psi_1^* - \frac{m^2}{r^2}\Psi_1^{*2} - \frac{\mu_0 dj_{0z}/dr}{B_{0\theta}(1-n/mq(r))}\Psi_1^{*2} = 0 \tag{8.33}$$

which can be rearranged to become

$$-2\mu_0 W_{\text{mag}} + \frac{1}{r}\frac{\partial}{\partial r}\left(r\Psi_1^*\frac{d\Psi_1^*}{dr}\right) = \frac{\mu_0 dj_{0z}/dr}{B_{0\theta}(1-n/mq(r))}\Psi_1^{*2} \tag{8.34}$$

which can be interpreted as a time-integrated rate equation for the radial energy flux: the first term is the field energy, the second one the radial divergence of the quantity $\Psi_1^*\Psi_1^{*\prime}$ and the term on the right-hand side can be considered as a local source term, coming from the current gradient which provides the free energy for the classical tearing mode. The quantity $\Psi_1^*\Psi_1^{*\prime}$ can hence be identified as a (time-integrated) flux of electromagnetic energy. Using the definition of the Poynting flux, we see that

$$\mathbf{S} = \frac{1}{\mu_0}\mathbf{E}\times\mathbf{B} \rightarrow S_r = \frac{1}{\mu_0}E_z B_\theta = \frac{1}{\mu_0}\frac{\partial\Psi_1^*}{\partial t}\frac{\partial\Psi_1^*}{\partial r}\cos^2 m\theta \tag{8.35}$$

so that for an exponentially growing perturbation with growth rate γ, the quantity $\gamma\Psi_1^*\Psi_1^{*\prime}/(2\mu_0)$ is the radial component of the Poynting flux, averaged over the poloidal angle (the average over $\cos^2 m\theta$ just contributes the factor 1/2). Using the definition of Δ' (Eq. (8.30)), the jump of the Poynting flux gives the relation

$$\frac{1}{\gamma}\Delta S_r = \lim_{\epsilon\to 0}\frac{1}{2\mu_0}\Psi_1^*(\Psi_1^{*\prime}(r_s+\epsilon) - \Psi_1^{*\prime}(r_s-\epsilon)) = \frac{1}{2\mu_0}\Psi_1^{*2}\Delta' \tag{8.36}$$

and we see that Δ' can be interpreted as the sink or source of a Poynting flux across the ideal MHD region into or out of the rational surface, depending on the sign of Δ'.

While usually, the tearing mode equation has to be integrated numerically, for a vacuum field, produced by a surface current at $r = r_s$, the perturbed flux is given by Eqs (7.13) and (7.14). From this, Δ' can easily be derived to be $\Delta' = -2m/r_s$, that is as expected, a vacuum field is stable. In fact, this value is sometimes used as an approximation for Δ' in cases where classical tearing modes are stable. We also see that for higher m, the stabilizing energy is larger so that the lowest resonant (m, n) instability is usually favoured on a given resonant surface, the physical reason being that the curvature of field lines is minimized.

Further insight can be gained by analysing the linear tearing instability of the 'top hat' current profile already introduced in Section 7.5 when treating the Troyon limit and shown in Figure 7.10 together with the resulting q-profile. As this profile assumes $j = j_0$ to be constant inside radius r_j and to be 0 outside, we have $dj_{0z}/dr = 0$ everywhere except at $r = r_j$, where it is represented by a δ-function

$$\frac{dj_{0z}}{dr} = -\frac{2B_{0\theta}(r)}{\mu_0 r}\delta(r - r_j) \tag{8.37}$$

and we have used the fact that for the current profile under consideration, $B_\theta = \mu_0 j_{0z} r/2$ for $r < r_j$.

The tearing mode equation then reads

$$\frac{1}{r}\frac{d}{dr}\left(r\frac{d\Psi_1^*}{dr}\right) - \frac{m^2}{r^2}\Psi_1^* + \frac{2}{r}\frac{\delta(r-r_j)}{1-nq/m}\Psi_1^* = 0 \tag{8.38}$$

Integrating this over a small region across r_j, only the singular terms survive and we obtain

$$\lim_{\epsilon\to 0}\left(r\frac{d\Psi_1^*}{dr}\right)_{r_j-\epsilon}^{r_j+\epsilon} + \frac{2}{1-nq(r_j)/m}\Psi_1^*(r_j) = 0 \tag{8.39}$$

As $\Delta\Psi_1^* = 0$ in the rest of the region, the solutions are the vacuum solutions given by Eqs (7.13) and (7.14). The solution has to be finite at $r = 0$, so that for $r < r_j$, it will be of the form $\propto r^m$. On the other hand, if the solution also has to be finite for $r \to \infty$, it will have to be of the form $\propto r^{-m}$ there. However, different from the jump condition for a surface current (Eq. (7.7)), condition (8.39) can in general not be fulfilled simultaneously with the condition of a continuous Ψ_1^* across r_j by matching these two solutions. Hence, we also have to include the resonant surface at $r = r_s$ in the analysis. Assuming $r_j < r_s$, we have three separate regions in which Ψ_1^* takes the following form:

$$\Psi_1^* = c_1 r^m \text{ for } r < r_j \tag{8.40}$$

$$\Psi_1^* = c_2 r^m + c_3 r^{-m} \text{ for } r_j < r < r_s \tag{8.41}$$

$$\Psi_1^* = c_4 r^{-m} \text{ for } r > r_s \tag{8.42}$$

From the continuity of Ψ_1^* across r_j and r_s, we obtain the conditions

$$c_1 r_j^m = c_2 r_j^m + c_3 r_j^{-m} \tag{8.43}$$

$$c_4 r_s^{-m} = c_2 r_s^m + c_3 r_s^{-m} \tag{8.44}$$

In addition, condition (Eq. (8.39)) gives another relation between the coefficients at r_j:

$$m(c_2 r_j^m - c_3 r_j^{-m} - c_1 r_j^m) + \frac{2}{1-nq(r_j)}(c_2 r_j^m + c_3 r_j^{-m}) = 0 \tag{8.45}$$

hence, we have a system of three linear equations for the four coefficients as, different from the shielding problem treated in Section 7.3, one free coefficient remains in this problem because the amplitude of Ψ_1^* as determined from the tearing mode equation is arbitrary. If, for example, we chose c_1 such that Ψ_1^* is dimensionless and normalized to 1 at the maximum, we get

$$c_1 = \frac{1}{r_j^m} \tag{8.46}$$

$$c_2 = \frac{1}{r_j^m}\frac{m - nq(r_j) - 1}{m - nq(r_j)} \tag{8.47}$$

$$c_3 = \frac{r_j^m}{m - nq(r_j)} \tag{8.48}$$

$$c_4 = \frac{(m - nq(r_j) - 1)(r_s/r_j)^{2m} + 1}{m - nq(r_j)} r_j^m \tag{8.49}$$

In any case, since Δ' is a ratio involving the flux function in a linear way, it can be calculated to yield

$$\Delta' = -\frac{2m}{r_s}\frac{m - nq(r_j) - 1}{m - nq(r_j) - 1 + \left(\frac{r_j}{r_s}\right)^{2m}}$$
$$= -\frac{2m}{r_s}\frac{m(1-\left(\frac{r_j}{r_s}\right)^2) - 1}{m(1-\left(\frac{r_j}{r_s}\right)^2) - 1 + \left(\frac{r_j}{r_s}\right)^{2m}} \quad (8.50)$$

where in the last step, we have made use of the fact that for the top hat current profile, we have $q(r) = q(a)(r/a)^2$ for $r \geq r_j$.

While not suited for a quantitative statement about the tearing stability of realistic tokamak profiles, this result is quite instructive to discuss the basics of tearing stability of a tokamak plasma. First, we note that if the resonant surface is far away from the equilibrium current gradient, that is $r_j/r_q \to \infty$, the result coincides with the (stable) vacuum case $\Delta' = -2m/r_s$. Then, if $m(1-(r_j/r_s)^2) - 1 > 0$, the stability index Δ' will always be negative, that is the system is stable. Conversely, if

$$m\left(1-\left(\frac{r_j}{r_s}\right)^2\right) - 1 < 0 \to \frac{r_j}{r_s} > \sqrt{1-\frac{1}{m}} \quad (8.51)$$

the system will always be unstable[5]. For $m = 1$, there is no stable window ($\Delta' \to \infty$ for all values of r_j/r_s). For $m > 1$, there are windows of instability in the range $r_j/r_s = 0$ (resonant surface at infinity) to $r_j/r_s = 1$ (resonant surface coincides with the surface at which the equilibrium current gradient occurs). Physically, it means that moving the resonant surface closer to the current gradient will increase the (initially negative) value of Δ' until the system becomes unstable when the condition (Eq. (8.51)) is fulfilled. In Figure 8.5a, two solutions are plotted, indicating the typical shape for a stable and an unstable situation. In Figure 8.5b, the stability index Δ', normalized to the vacuum value, is plotted as function of r_j/r_s for different m. We see that for higher m, the unstable region becomes smaller, consistent with our previous conclusion that low-mode numbers tend to be the most unstable ones.

8.3 Magnetic Islands in Tokamaks

Equation (8.31) is the linear stability criterion, valid for infinitesimal layer width ($\epsilon \to 0$ in Eq. (8.30)). In the nonlinear case, we have to treat the formation of a magnetic island as shown in Figure (8.1) explicitly and match the ideal solution at the island separatrix, which will be done in Section 8.4. Here, we examine the

5) To prove this statement, one can show that under this condition, the denominator in Eq. (8.50) is always positive as for $m > 1$, it falls monotonically from $m - 1$ at $r_j/r_s = 0$ to 0 at $r_j/r_s = 1$.

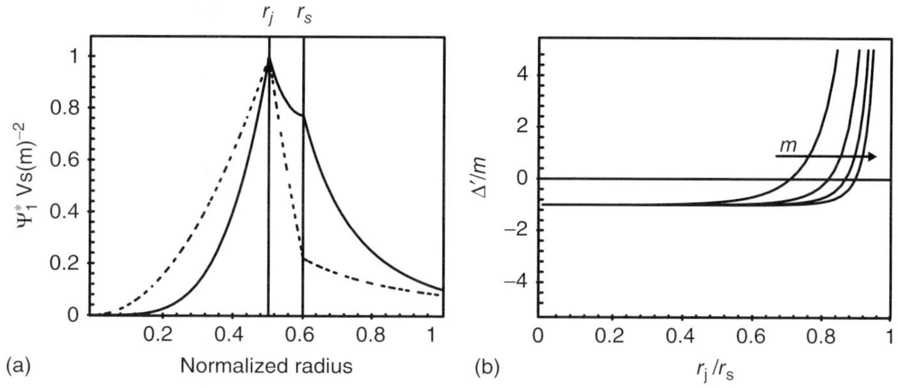

Figure 8.5 (a) Perturbed flux function Ψ_1^* for the top hat current profile, for a case with $\Delta' < 0$ ($r_j/r_s = 5/6, m = 4$, solid line) and one with $\Delta' > 0$ ($r_j/r_s = 5/6, m = 2$, dashed line). (b) Δ'/m as function of r_j/r_s for $m = 2, 3, 4$ and 5.

structure of magnetic islands due to the superposition of helical equilibrium flux Ψ_0^* and perturbed flux Ψ_1^*.

To start with, we analyse how the topology of the flux surfaces changes in the presence of a sinusoidally modulated sheet current generated by the tearing mode instability. To do so, we superpose the helical equilibrium flux Ψ_0^* with the perturbed component Ψ_1^*. These two quantities have to be calculated for a given equilibrium current profile according to the definition of the equilibrium helical flux (Eq. (8.24)) and the tearing mode equation for the perturbed flux (Eq. (8.29)). It is clear that in general, this can only be done numerically. An analytically tractable case is however obtained making use of the above-mentioned fact that close to the resonant surface, Ψ_0^* can be approximated by a parabola. If furthermore, we assume that Ψ_1^* has constant amplitude $\overline{\Psi}_1^*$ over the island[6], we can add the fluxes and obtain

$$\Psi^* = \Psi_0^*(r_s) + \frac{1}{2}\Psi_0^{*\prime\prime}(r - r_s)^2 + \overline{\Psi}_1^* \cos(m\zeta) \tag{8.52}$$

where the coordinate $\zeta = m\theta - nz/R_0$ is the 'helical angle' which is constant along the equilibrium field lines. This function can be inverted to obtain the flux surfaces as the contours $\Psi^* = $ const:

$$r - r_s = \sqrt{\frac{2}{\Psi_0^{*\prime\prime}}(\Psi^* - \Psi_0^*(r_s) - \overline{\Psi}_1^* \cos(m\zeta))} \tag{8.53}$$

These contours are shown in Figure 8.6 in the $r - \zeta$-plain for $m = 2$

The formation of magnetic islands is evident from Figure 8.6. At the helical angle where the perturbed flux has its minimum, the island structure attains its maximum radial width, the island width W. This point is usually called the *O-point* of

6) This is often called the *constant-psi* approximation. It is reasonable for not too big islands and $m > 1$ while for the $m = 1$ case, it is generally not a good approximation, see Chapter 11.

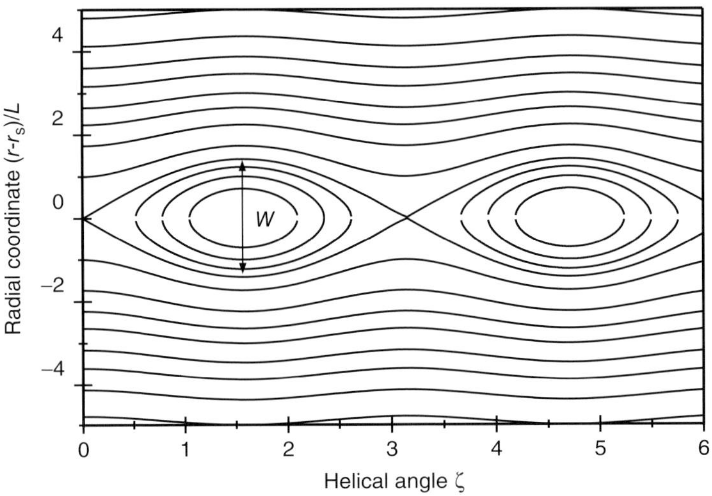

Figure 8.6 Adding a radially constant helical flux Ψ_1^* that varies sinusoidally in helical angle to a parabolic equilibrium flux Ψ_0^*, the resulting flux surfaces are islands of width W (parametres for $m = 2$).

the island. Conversely, at $\zeta = 0$, the perturbed flux has a maximum and the island separatrix has its X-point. These conditions apply for a parabola with positive curvature; if we change the sign of the curvature, they will revert. According to Eq. (8.24), positive curvature means $q' > 0$, that is positive magnetic shear.

For this situation, a simple formula for the island width can be obtained by analysing the equation for the island separatrix. Without loss of generality, we can set $\Psi_0^*(r_s) = 0$[7] and as at the X-point, we have $\zeta = 0$, and the separatrix passes through $r = 0$, the flux on the separatrix is $\Psi_{\text{sep}}^* = \overline{\Psi}_1^*$. At the O-point, we have $\zeta = \pi$ and the radial excursion of the separatrix is

$$\Psi_{\text{sep}}^* = \frac{1}{2}\Psi_0^{*\prime\prime}\left(\frac{W}{2}\right)^2 - \overline{\Psi}_1^* \tag{8.54}$$

where we have used the fact that in this symmetric geometry, the island width W is twice the radial excursion at the O-point. Inserting the relation for the X-point, we obtain a formula for the island width:

$$W = 4\sqrt{\frac{\overline{\Psi}_1^*}{\Psi_0^{*\prime\prime}}} = 4\sqrt{\frac{B_{1r} r_s q}{m q' B_{0\theta}}} \tag{8.55}$$

where in the last step, we have expressed the perturbed flux by the radial field and used Eq. (8.25) for the equilibrium flux.

According to Eq. (8.55), the island width is proportional to the square root of the perturbed current and decreases with square root of magnetic shear. While these dependencies also hold for more realistic geometries, magnetic islands occurring

7) Addition of a constant flux does not change the B-field derived from it.

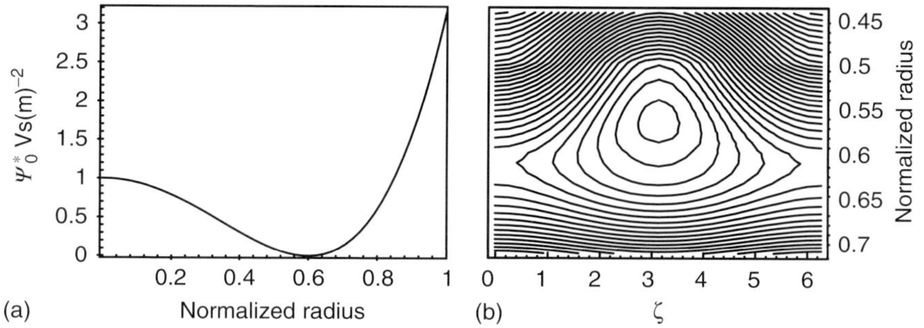

Figure 8.7 Helical equilibrium flux for the model current profile $j = j_0(1-(r/a)^2)$ (a) and (b) resulting asymmetric island structure when adding the perturbed flux $\overline{\Psi}_1^*(r)\cos\zeta$.

in tokamak discharges are usually not symmetric with respect to the resonant surface. This is due to both a deviation of the equilibrium flux from the parabolic form and the variation of the perturbed flux with radius. These effects can be visualized in a simple extension of the assumptions made earlier. For the equilibrium flux, the curve will increase less than parabolic when moving away from the resonant surface towards the magnetic axis as the first derivative of Ψ_0^* has to go to zero there due to $B^*(r=0)=0$. For example, assuming that the current profile is of the form $j = j_0(1-(r/a)^2)$, that is using Eq. (2.12) with $\mu = 1$, we obtain

$$\Psi_0^*(r) = \frac{\mu_0 I_p}{8\pi}\left(\left(\frac{r}{a}\right)^2 - \left(\frac{r_s}{a}\right)^2\right)^2 \tag{8.56}$$

Using this together with a constant $\overline{\Psi}_1^*$, we obtain the flux surface contours shown in Figure 8.7.

It can be seen that the island is asymmetric, extending further towards the magnetic axis than towards the plasma edge. This asymmetry is also observed experimentally (see Chapter 9, Figure 9.1) and the island shape hence contains information about Δ'.

8.4
The Rutherford Equation

In the final section of this chapter, we derive an equation for the nonlinear evolution of a magnetic island. While the linearized evolution treated in Section 8.1 assumed that the perturbed current flows in a sheet of width δ and is treated as surface current $J = j\delta$, for $W > \delta$, we now have to consider the equilibration of current on the perturbed flux surfaces which leads to a modification of the spatial distribution of current density which in turn is responsible for the formation of the island. This problem can be treated self-consistently analytically in the limit of constant amplitude $\overline{\Psi}_1^*$ and parabolic Ψ_0^* as introduced earlier.

We start by noting that for typical fusion plasma parametres, δ as evaluated according to Eq. (8.13) will be very small[8)], of the order of an ion Larmor radius, and hence the regime $W > \delta$ is the usual case. Here, inertial effects do not play a role any more and the evolution is governed by the resistive diffusion. Consequently, Ohm's law can be written as

$$\langle E_{1z}\rangle_\Omega = \frac{d\overline{\Psi}_1^*}{dt}\langle\cos\zeta\rangle_\Omega = \frac{1}{\sigma}(j_1(\Omega) - j_{1,ni}(\Omega)) = \frac{1}{\sigma}\langle j_{1z} - j_{1z,ni}\rangle_\Omega \tag{8.57}$$

where $\langle ...\rangle_\Omega$ denotes an average on an island flux surface parameterized by the flux surface label Ω and $j_{1z,ni}$ is the non-inductive part of the perturbed current, that is any current that will not give rise to a loop voltage, such as for example bootstrap current or auxiliary current driven by an external system. As indicated earlier, the averaging of the current over the flux surfaces will give the correct current $j(\Omega)$ as flux function.

For the island geometry under consideration, a parameterization of the flux surfaces can be found by inserting Eq. (8.55) into Eq. (8.52) and introducing the normalized flux label

$$\Omega = \frac{\Psi^*}{\overline{\Psi}_1^*} = 8\frac{x^2}{W^2} + \cos\zeta \tag{8.58}$$

where $x = r - r_s$ and only one island is treated, but the results are valid for arbitrary $m > 1$. With this definition, the O-point corresponds to $\Omega = -1$ and the island separatrix is described by $\Omega = +1$. The flux surface average of a function $f(\Omega, \zeta)$ is calculated by evaluating the volume integral

$$\langle f\rangle_\Omega = \frac{\int d^3x' f(\mathbf{x}')\delta(\Omega - \Omega')}{\int d^3x'\,\delta(\Omega - \Omega')} = \frac{\int \frac{d\zeta}{J} f}{\int \frac{d\zeta}{J}} \tag{8.59}$$

where in the second step, we have assumed that the z-direction is an ignorable coordinate such that we are only integrating in the r, ζ plane and J is the Jacobian for the transformation from r, ζ to Ω, ζ. Evaluating the Jacobian using Eq. (8.58), the average over an island surface can then be calculated by

$$\langle f\rangle_\Omega = \frac{\oint \frac{d\zeta}{2\pi}\frac{f(\Omega,\zeta)}{\sqrt{\Omega - \cos\zeta}}}{\oint \frac{d\zeta}{2\pi}\frac{1}{\sqrt{\Omega - \cos\zeta}}} \tag{8.60}$$

where the integration inside the island ranges over the definition range of ζ, that is from $\arccos\Omega$ to $2\pi - \arccos\Omega$, while outside, it is over the full range from 0 to 2π.

In analogy to the treatment in the case of the current sheet, Δ' may be defined as the jump of the derivative of Ψ_1^* outside the tearing layer, which is related to the $\cos\zeta$ component of the current flowing inside the layer. While in the linear case, the layer is arbitrarily small and hence the matching is done at $r_s \pm \epsilon$, for a

8) Inserting $n_i = 1 \times 10^{20}\mathrm{m}^{-3}$, $m = m_H$, $L = 1\mathrm{m}$, $k = 2\pi/L$, $\Delta' = 1\mathrm{m}^{-1}$, $\sigma = \sigma(T_e = 1\mathrm{keV})$ and $B_{0y} = 0.1$ T into (8.17) gives $\delta \approx 1$ mm.

finite island we have to consider that tearing mode theory is a theory of asymptotic matching and hence the matching to the outer solution should be done at a radius far enough from the island that the detailed nonlinear physics determining the perturbed flux there is not important[9]. For the case treated here, there is rapid convergence of the radial integral when extending the integration region across the island separatrix so that integration can be expanded to $\pm\infty$, avoiding the discussion where exactly the matching is done[10]. Then, Δ' can be calculated according to

$$\Delta' = \frac{2}{\overline{\Psi}_1^*} \int_{r=-\infty}^{\infty} dr \oint \frac{d\zeta}{2\pi} j_{1,z} \cos\zeta \tag{8.61}$$

where the factor 2 accounts for the fact that for a pure cos component, the cosine Fourier integral will just yield 1/2.

Using the parameterization (Eq. (8.58)) of the flux surfaces, this integral can be rewritten as an integral over the perturbed flux surfaces

$$\Delta' = \frac{W}{\sqrt{2}\overline{\Psi}_1^*} \int_{\Omega=-1}^{\infty} d\Omega \oint \frac{d\zeta}{2\pi} \frac{j_1(\Omega)\cos\zeta}{\sqrt{\Omega - \cos\zeta}} \tag{8.62}$$

where we have made use of $dx = W/(4\sqrt{2}\sqrt{\Omega - \cos\zeta})\, d\Omega$ and integration is only over half a radial width which due to the symmetry just results in a factor of 2.

Inserting $j_1(\Omega)$ from Eq. (8.57) and using the relation between $\overline{\Psi}_1^*$ and W (Eq. (8.55)) leads to

$$\begin{aligned}\Delta' = \mu_0\sigma \frac{dW}{dt} \sqrt{2} \int_{\Omega=-1}^{1} d\Omega \oint \frac{d\zeta}{2\pi} \frac{\langle\cos\zeta\rangle_\Omega \cos\zeta}{\sqrt{\Omega - \cos\zeta}} \\ + \frac{16q}{\sqrt{2}Wq'B_{0\theta}}\Big|_{r_s} \int_{\Omega=-1}^{1} d\Omega \oint \frac{d\zeta}{2\pi} \frac{\langle\mu_0 j_{1z,ni}\rangle_\Omega \cos\zeta}{\sqrt{\Omega - \cos\zeta}}\end{aligned} \tag{8.63}$$

The integral on the first part of the RHS can be carried out analytically in ζ, leading to an expression involving Elliptic functions in Ω. The Ω integration has to be done numerically, and it yields

$$\int_{\Omega=-1}^{\infty} d\Omega \oint \frac{d\zeta}{2\pi} \frac{\langle\cos\zeta\rangle_\Omega \cos\zeta}{\sqrt{\Omega - \cos\zeta}} = 0.582 \tag{8.64}$$

We remind of the above discussion about the limit of integration and state here that integrating just up to the island separatrix will lead to a numerical value of

9) This procedure is consistent with the assumption of a radially constant $\overline{\Psi}_1^*$ which also means that the detailed nonlinear physics within the island and, even more important, around the island separatrix, is neglected.

10) Clearly, also the approximation made when using a slab geometry will break down at finite r.

0.5, which validates a posteriori the extension of the integration up to ∞ in the sense discussed earlier[11].

Finally, neglecting the noninductive part of Eq. (8.63), the nonlinear island evolution is given by

$$\frac{\tau_R}{r_s}\frac{dW}{dt} = r_s\Delta' \tag{8.65}$$

where we have defined the resistive timescale including the numerical factor

$$\tau_R = 0.82\ \mu_0\sigma r_s^2 \tag{8.66}$$

This is the evolution equation for the island width if only the inductive currents are considered. It is known as the Rutherford equation [44]. Similar to the linear case, the stability criterion is $\Delta' > 0$ for instability, but now Δ' is given by the nonlinear definition (Eq. (8.61)). If there are other noninductive currents, the second term on the right-hand side of Eq. (8.63) has to be evaluated accordingly and it is usually called the *modified* Rutherford equation. We will come back to this when we treat finite β-effects and also when external current drive by ECCD as a method to control tearing modes is discussed.

11) A pragmatic way to summarize this discussion is that the precise numerical value of the integral is only of mathematical interest.

9
Current Driven ('classical') Tearing Modes in Tokamaks

In Chapter 8, we have derived the Rutherford equation that governs the nonlinear resistive growth of magnetic islands in tokamaks. This equation was derived for cylindrical geometry and describes the evolution of a single helicity mode. In this chapter, we first examine the effects of tearing modes on the kinetic profiles of tokamak discharges. It will be shown that large magnetic islands change the equilibrium profiles, modifying the stability properties of the mode in turn. The consequences of this nonlinear evolution will be described in the following sections, which deal with the experimentally important cases of the disruptive instability and the sawtooth instability.

9.1
Effect of Tearing Modes on Kinetic Profiles

Owing to the change in topology introduced by tearing modes, they can have an important effect on the kinetic profiles of tokamak discharges. Mainly, magnetic islands connect different radial regions of the plasma along the island field lines. As the transport coefficients in a hot magnetized plasma are in general much larger along the field lines than across them, this provides an effective shortcut for radial heat and particle fluxes. We can estimate the island size above which this effect becomes important using the example of the heat flux $\mathbf{q}$ in a plasma with a magnetic island of width W[1]. If we assume that there is no local heat source, the heat flux obeys the condition $\nabla \cdot \mathbf{q} = 0$. Furthermore, assuming a conductive heat transport $\mathbf{q} = -\kappa \nabla T$, where κ is the heat conductivity, the transport equation can be written as

$$-\kappa_{\parallel} \nabla_{\parallel}^2 T - \kappa_{\perp} \nabla_{\perp}^2 T = 0 \tag{9.1}$$

which for $\kappa_{\parallel} \gg \kappa_{\perp}$ indicates that the parallel gradients are indeed small for the usual case of nested flux surfaces.

However, a radial temperature gradient across a magnetic island will also give rise to a heat flux around the island as it is equivalent to a parallel gradient of order

$$\nabla_{\parallel} \approx \frac{B_r}{B_z} \nabla_{\perp} = \left(\frac{W}{4}\right)^2 \frac{n s_s}{R_0 r_s} \nabla_{\perp} \tag{9.2}$$

1) A similar argument can be made for the particle flux.

Magnetohydrodynamic Stability of Tokamaks, First Edition. Hartmut Zohm.

where we have made use of Eq. (8.55) for the island width. Equation (9.1) can then be written as

$$(\kappa_{\|}\left(\frac{W}{4}\right)^4\left(\frac{ns_s}{R_0 r_s}\right)^2 + \kappa_{\perp})\nabla_{\perp}^2 T = 0 \tag{9.3}$$

One can see that if the first term in the bracket is small compared to the second one, the gradient across the island will be of order of the radial gradients in the island vicinity, while for the first term dominating, the gradient will be much smaller and the temperature gradient across the island is essentially flat. This can be rearranged as a condition for the island width W_0 at which the temperature gradient across the island vanishes. Considering the exact island geometry, we obtain a slightly different numerical factor compared to the simple argument presented earlier and the criterion for flattening becomes [45]

$$W > W_0 = 5.1\left(\frac{\kappa_{\perp}}{\kappa_{\|}}\right)^{1/4}\left(\frac{R_0 r_s}{ns_s}\right)^{1/2} \tag{9.4}$$

For typical ratios of $\kappa_{\perp}/\kappa_{\|}$ in hot fusion plasmas, W_0 is smaller than a saturated island so that the profile flattening due to a magnetic island can be observed experimentally. Figure 9.1 shows an example of an experimentally measured temperature distribution across a magnetic island.

For large single islands, the flattening of the kinetic profiles leads to a reduction of the stored energy. Assuming that the transport coefficients of the plasma outside the island are unchanged by the presence of the mode, this loss can be modelled by representing the island as a 'belt' in the poloidal cross section with

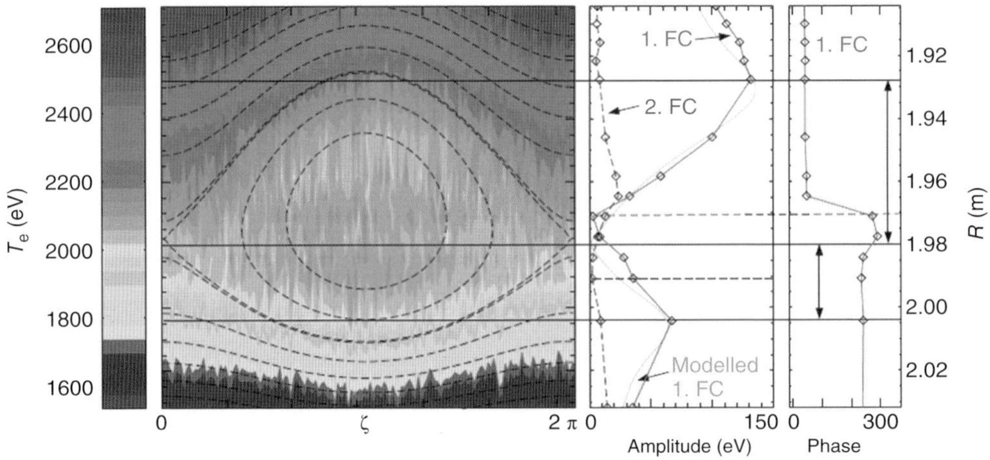

Figure 9.1 Measured temperature distribution across a large island. The contour plot shows the temperature with the modelled asymmetric island overlaid. The panels on the right show how the island structure can be determined from the profiles of the perturbed temperature in the first and second Fourier component (FC) with the location of r_s given by the characteristic phase jump. *Source:* Meskat 2001 [46]. Reproduced with permission of IOP. (Please find a color version of this figure on the color plates.)

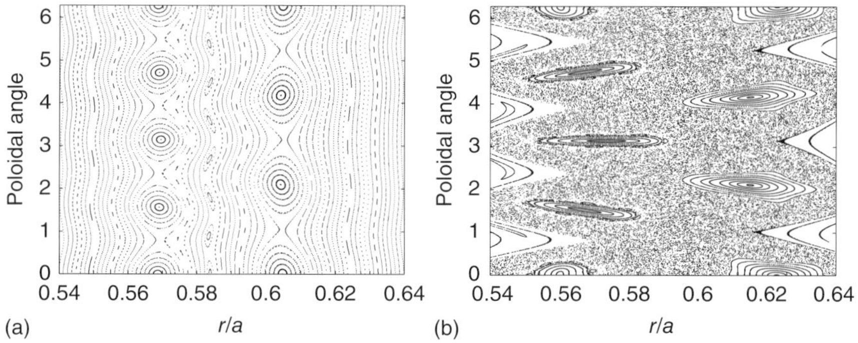

Figure 9.2 Poincaré plot of field lines in the presence of (4,3) and (3,2) island chains. (a) At small perturbation amplitude, the flux surfaces are regular and due to non-linear coupling, a (7,5) island can be seen in between the two chains. (b) At large perturbation amplitude, the field becomes stochastic, with some KAM islands remaining. *Source:* Yu *et al.* 2006 [48], reproduced with permission of AIP.

flat temperature distribution. In the case of a flat density profile, constant heat source and heat conductivity, the reduction of stored kinetic energy δW_{kin} due to the presence of a saturated island of width W_{sat} can be calculated to be approximately [47]

$$\delta W_{\mathrm{kin}} \approx -4 W_{\mathrm{kin},0} \frac{W_{\mathrm{sat}}}{a} \left(\frac{r_s}{a} \right)^3 \tag{9.5}$$

It can be seen that the reduction is more pronounced for islands that are located further outside, which is generally observed in experiments where the (2,1) mode has a larger impact than more central modes as for example the (3,2) mode.

If there is more than one mode present, one might expect a flattening at several radial locations and a reduction of stored energy according to several 'belts'. However, when the islands are of a size that they would overlap in radial direction, a new phenomenon sets in, the so-called stochastization of field lines[2)]. In this case, field lines do in general no longer form closed flux surfaces but rather fill the region in between the two resonant surfaces in a stochastic manner[3)]. A Poincaré plot of field lines resulting from two island chains, a (4,3) and a (3,2) perturbation, is shown in Figure 9.2. In Figure 9.2a, the amplitude is small and nested flux surfaces can be seen, whereas in Figure 9.2b, the amplitude is so large that the local field becomes stochastic.

As can be seen in Figure 9.2, in the case of stochastization, the field lines 'diffuse' in the radial direction, and appropriate averaging leads to an effective radial component δB_r. As in a collisionless plasma, the condition $\lambda_{\mathrm{mfp}} \gg L$ holds (see also Chapter 1), the electrons can essentially move freely along the field lines around the torus and the effective perpendicular heat diffusivity χ_{eff} can be

2) The overlap criterion is also known as the *Chirikov criterion*.
3) It can be shown that some islands remain according to the KAM (Kolmogorov, Arnold, Moser) theorem, but these do not provide an effective radial heat insulation.

estimated to be [49] (for a discussion of the different transport regimes, see [48])

$$\chi_{\mathrm{eff}} \approx \pi R_0 \left(\frac{\delta B_r}{B_0} \right)^2 v_{\mathrm{th,e}} \tag{9.6}$$

Owing to the large values of $v_{\mathrm{th,e}}$, this can be substantial already for small values of δB_r. For stochastization due to large islands, χ_{eff} will become much larger than the usual value, leading to complete loss of energy in the affected region. An experimental example is shown in Figure 9.3 where two island chains, the (2,1) and the (3,1) island, rotate in phase and induce a sudden loss of thermal insulation as can be seen from the sudden flattening of the temperature profile. This process is believed to be responsible for the sudden energy loss observed at the beginning of the disruptive instability (Chapter 10).

Finally, we note that stochastization will only occur if the interacting magnetic islands are phase locked with respect to each other, that is differential rotation will effectively suppress stochastic transport. This is important in the discussion of disruptions in Chapter 10 where cases with locked modes are most prone to disruptions and for the FIR NTMs discussed in Section 12.4.

9.2
Nonlinear Saturation

In Chapter 8, we derived the Rutherford equation and saw that it will lead to a linear growth of the island width due to a current gradient driven tearing mode. For sufficient island width, we may expect an effect of the tearing mode on the

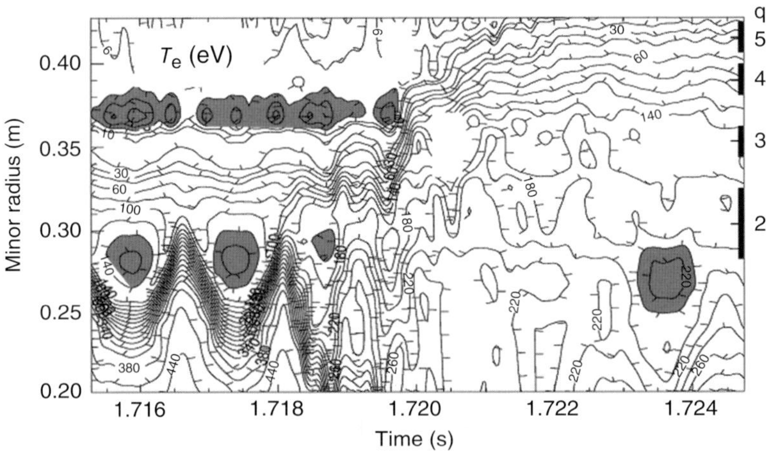

Figure 9.3 Experimentally observed rapid energy loss due to the interaction of two phase locked island chains on ASDEX Upgrade, leading to stochastization. Here, rapid edge cooling when approaching the density limit (see Section 10.1.1) has led to the occurrence of (2,1) and (3,1) current gradient driven tearing modes. Isothermals as function of time and radius are shown. The modes rotate in the laboratory frame. *Source:* Suttrop *et al.* 1997 [50], reproduced with permission of the IAEA. (Please find a color version of this figure on the color plates.)

equilibrium parameters due to the flattening of kinetic profiles as discussed in the previous section. This will lead to a deviation from the linear growth, as the flattening of the temperature profile will reduce the driving gradient and hence can finally lead to a saturation. This effect should thus result in a reduction of $\Delta'(W)$ with increasing W. In a simplified picture, $\Psi_1^*(r)$, the solution of the tearing mode equation for $W \to 0$ can be thought to be matched to a constant $\overline{\Psi}_1^*$ at the left and right boundary of an island of finite width, that is at $r_s \pm W/2$, and $\Delta'(W)$ can then approximately be calculated as

$$\Delta(W) = \frac{\Psi_1^{*\prime}(r_s + W/2) - \Psi_1^{*\prime}(r_s - W/2)}{\Psi_1^*(r_s)} \tag{9.7}$$

although, as pointed out in Chapter 8, this matching should in principle be done further away from the island boundary. Figure (9.4) illustrates this procedure for the example given in Figure 8.4.

It can be seen that $\Delta'(W)$ indeed decreases with increasing W. More realistic calculations show that this dependence can roughly be approximated by

$$\Delta'(W) = \Delta'(0)\left(1 - \frac{W}{W_{\text{sat}}}\right) \tag{9.8}$$

where $\Delta'(0)$ is the linear stability index calculated in the limit of vanishing island width (Eq. (8.30)). We note here that a more refined analysis should take into

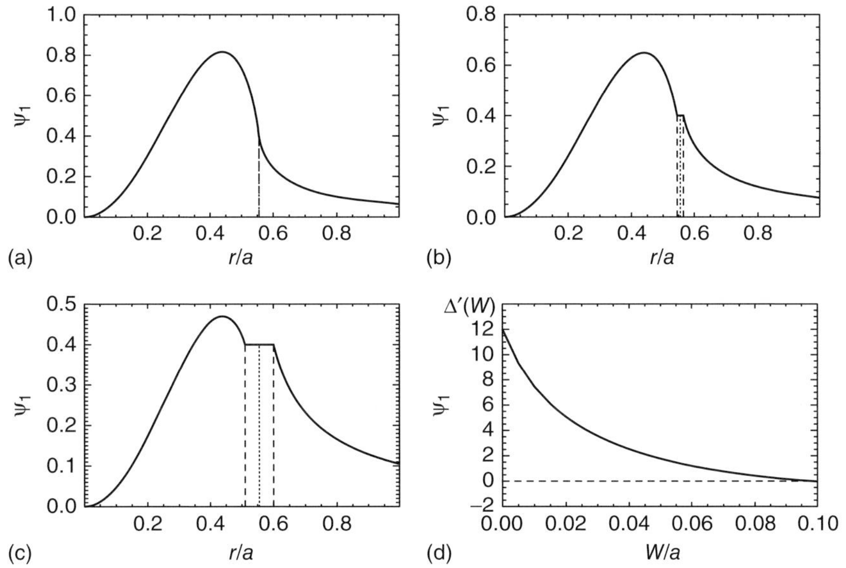

Figure 9.4 The perturbed helical flux function Ψ_1^*, calculated using the tearing mode equation based on the current profile $j \propto (1 - (r/a)^2)^3$, is matched to a region of constant value around the resonant surface representing the island. (a-c) The plots show the cases $W = 0$, $W = 0.05$ and $W = 0.1$. As a result of the change in gradients, the non-linear stability parameter $\Delta'(W)$ is decreasing with increasing W as shown in (d).

account further Fourier harmonics in the perturbed flux, leading to a modification of (9.8). Obviously, the saturated island size W_{sat} is a property specific of the equilibrium current profile. Inserting this relation into the Rutherford equation leads to a non-linear saturation according to

$$W(t) = W_{sat}\left[1 - \exp\left(-t\,\frac{r^2\Delta'(0)}{\tau_R W_{sat}}\right)\right] \tag{9.9}$$

In the saturated state, $\Delta'(W_{sat}) = 0$, that is there is no helical current flowing in the island and the island is rather a new three-dimensional equilibrium state. Such single helicity saturated current driven tearing modes are sometimes observed experimentally, but they do not play a major role in tokamaks. The most important role of current driven islands is rather their contribution to the non-linear processes of the disruptive instability and the sawtooth instability. These will be treated in the following chapters. Before, we treat the important topic of tearing mode rotation and locking in tokamaks.

9.3 Tearing Mode Rotation and Locking

In Chapter 1, it was shown that in ideal MHD, field lines are frozen into the electron fluid and hence flux tubes move together with the plasma with this velocity. As in general, $\mathbf{v}_e \neq 0$ in the laboratory frame, MHD modes in tokamaks are often observed to rotate in this frame. In the following, we describe the rotation of magnetic islands in tokamaks and then treat two important cases where rotation in the laboratory frame is not observed, namely the slowing down and locking of tearing modes due to the electromagnetic interaction with the vessel wall and external helical fields (mode locking) as well as the growth of ab-initio locked modes, which can be considered as a response of the plasma to an external helical field that is stationary in the laboratory frame.

9.3.1 Rotation of Tearing Modes in Tokamaks

As pointed out earlier, tearing modes in tokamaks are usually observed to rotate in the laboratory frame. This is due to the fact that magnetic perturbations are frozen into the electron fluid. For small islands, we have seen above that the pressure is not fully flattened and hence we expect that there is a finite diamagnetic current so that the island will rotate with respect to the fluid frame at roughly the electron diamagnetic velocity. On the other hand, a large island will have vanishing pressure gradient across it and hence, the perpendicular current is zero so that $v_e = v$[4]. We thus expect to find islands rotating at a velocity somewhere between the fluid velocity and the sum of fluid and diamagnetic velocity, depending on the

4) A more detailed discussion involves the ion viscosity but is beyond the scope of this book.

flattening of the pressure gradient. As the plasma rotation is incompressible, that is $\nabla \cdot \mathbf{v} = 0$, the toroidal rotation velocity will vary $\propto R$ on a flux surface and hence the angular frequency $\omega = v_{\text{tor}}/R$ is a constant. The picture described earlier is in general consistent with experimental observations.

Figure 9.5 shows an experimental example of the evolution of a (2,1) tearing mode when the plasma fluid velocity is varied by varying the external torque input due to heating by neutral beam injection (NBI). It can be seen that the frequency follows the rotation at the $q = 2$ surface, giving evidence that for this large island, the motion is indeed close to the fluid velocity.

However, a consistent theoretical description of island rotation in the laboratory frame is complicated by the fact that there is no general quantitative understanding of the fluid velocity in tokamaks. In cases where the plasma is strongly heated by unidirectional NBI, there is a direct momentum input associated with

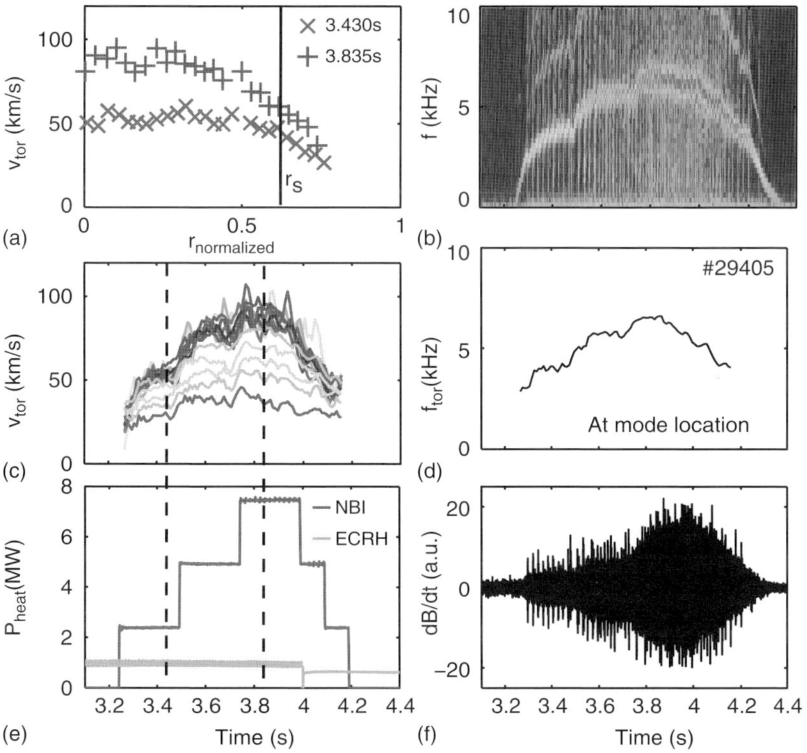

Figure 9.5 Time traces of the evolution frequency of a (2,1) tearing mode in ASDEX Upgrade (spectrogram of magnetic perturbation shown in (b), with the raw signal shown in (f)). (d) The temporal evolution matches the evolution of the fluid rotation at the $q = 2$ surface. (e) The variation of the fluid rotation is achieved by changing the momentum input due to heating by neutral beam injection (NBI). (a) Rotation profiles for different time points are shown together with the temporal evolution at different radii (c). (Please find a color version of this figure on the color plates.)

the injected neutrals that are ionized and transfer momentum and energy while slowing down by collisions. This leads to a torque balanced by viscous damping, resulting in a radial profile of mainly toroidal plasma rotation. Examples are shown on the left of Figure 9.5. However, tokamak plasmas are also observed to rotate in the absence of NBI, indicating that there are other torques acting on the plasma. These are much less understood than the NBI-heated cases and are hence an area of active research. For example, in low β ohmic discharges, plasma rotation is usually in the electron diamagnetic drift direction, with a magnitude of roughly $v_{e,dia}$.

In the following, we discuss the interaction of tearing modes and plasma rotation. In the absence of a good model for plasma fluid rotation, we will assume that there is a 'natural' frequency ω_0 at which the modes will rotate if there is no force acting on them. If a force is acting on a mode, such as the force due to the interaction with the vessel wall or an error field discussed in the following sections, there will be a viscous force that tries to restore the original rotation. In a 0-d model, the torque balance in the absence of direct electromagnetic forces on the mode can be modeled by assuming that the 'natural' toroidal rotation is due to an external force F_{ext}

$$m_p R_0^2 \frac{d\omega}{dt} = R_0 F_{ext} - \frac{m_p R_0^2}{\tau_M} \omega \tag{9.10}$$

where τ_M is the momentum confinement time that characterizes the viscosity and m_p the mass of the plasma involved in the momentum balance. For a single mode, this is the mass of the island and not of the whole plasma. The stationary solution is $\omega_0 = F_{ext}\tau_M/(m_p R_0)$, and hence the normalized torque balance can be written as

$$\frac{d\omega}{dt} = \frac{1}{\tau_M}(\omega_0 - \omega) \tag{9.11}$$

In the following Sections, we extend this equation by the different forces acting on the plasma when interacting with the wall and an external error field.

Finally, we note here that from the measurement of the propagation of MHD modes that have constant phase on field lines with mode numbers $q = m/n$, it is impossible to decide if they rotate poloidally or toroidally as an angular velocity of nv_{tor}/R is equivalent to an angular velocity of mv_{pol}/r[5]. However, we expect the fluid poloidal rotation to be strongly damped as the fluid elements would undergo continuous compression and expansion while rotating.

9.3.2
Locking of Pre-existing Magnetic Islands

In Section 9.3.1, it was shown that tearing modes usually rotate in the laboratory frame with the electron fluid. In this section, we analyse the process of mode

5) In German language, this is called the *Maibaum Theorem*; the American equivalent could be a barber shop pole. For both of them, it is not possible to decide if they are rotated or translated along the axis just from observing the motion of the helical stripes on them.

locking that is the slowing down and stopping of tearing mode rotation. In many tokamaks, it is found that the position of a locked mode of given helicity is always the same with respect to the vacuum vessel, indicating that there is a breaking of the axisymmetry of the external field, for example an error field due to imperfections in shape and position of the external field coils. As the plasma cannot flow through a large island at any significant rate, mode locking also leads to a stopping of the plasma rotation at the resonant surface and an appreciable slowing down of the plasma as a whole due to viscous coupling. Mode locking is usually observed if tearing modes grow to large amplitude. It is especially important as locked modes often precede disruptions, which is thought to be due to the fact that stochastization due to the interaction between tearing modes at different resonant surfaces can only take place if the modes are locked together in phase, which is true by definition for locked modes.

Figure 9.6 shows an experimental example where an initially rotating mode slows down in rotation frequency while it grows until is finally stops rotation. It can be seen that in the first phase, the signals representing the magnetic perturbation are rather harmonic, indicating a gradual slowing down of the plasma, while in the final phase, the signals become strongly anharmonic. This behaviour is due to the combination of two different forces acting on a rotating magnetic perturbation, namely the drag due to the eddy currents induced in the resistive wall and the interaction with the resonant helical component of the static error field due to imperfections mentioned earlier. Both can be understood from a model that represents the mode and the wall current as surface currents with a vacuum in between, as discussed in Section 7.3.2.

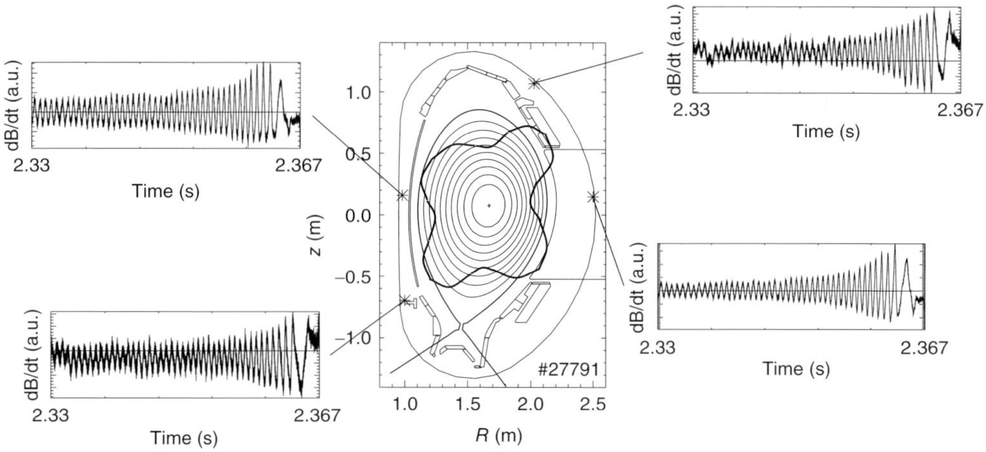

Figure 9.6 Time traces of the magnetic perturbation field due to a (2,1) magnetic island in ASDEX Upgrade, recorded at various positions in the poloidal cross section as indicated. Initially, the island rotates and the time trace is sinusoidal. During the island growth, the frequency drops and the signals become more and more anharmonic, indicating that the mode no longer rotates with constant angular velocity due to the interaction with the external error field.

The drag from the eddy currents in the wall comes from the finite phase average of the $\mathbf{j} \times \mathbf{B}$ force in case of finite conductivity. As we are interested in the force perpendicular to the equilibrium field, we need to evaluate the phase averaged $j_w B_r$, which is conveniently calculated as $\langle \Re(j_w) \Re(B_r) \rangle = 1/2 j_w CC(B_r)$ where $\Re$ means the real part and CC the complex conjugate. Using Eq. (7.18) together with Eq. (7.22) and multiplying by the volume $4\pi^2 r_w d R_0$ on which the force is acting, we arrive at

$$F_w = -4\pi^2 \frac{R_0 m^2}{\mu_0 r_s} \omega \tau_w \Psi^* CC(\Psi^*) \tag{9.12}$$

where the force at the resonant surface has been calculated from actio = reactio for the torques, that is $F_w(r_s) = (r_w/r_s) F_w(r_w)$. Inserting the expression for the helical flux in the wall (Eq. (7.22)), the force on the mode due to interaction with the wall becomes

$$F_w = 4\pi^2 R_0 \frac{(m\overline{\Psi}^*)^2}{\mu_0 r_s} \left(\frac{r_s}{r_w} \right)^{2m} \frac{\omega \tau_w}{1 + (\omega \tau_w)^2} \tag{9.13}$$

In the case of an ideal wall, $\omega \tau_w \to \infty$, there is no drag force because the radial field in the wall is zero. On the other hand, if there is no rotation ($\omega \tau_w \to 0$), there is no induced current and no force either. In between, there is a maximum force when the mode rotates at a frequency that equals the inverse wall time, that is at $\omega \tau_w = 1$.

The force due to the external error field is calculated in a straightforward manner from the vacuum field generated by a static error field current located at r_{ef} producing a helical flux $\overline{\Psi}^*_{\text{ef}}$ at r_s [6]

$$\Psi^*_{\text{ef}} = \overline{\Psi}^*_{\text{ef}} \left(\frac{r}{r_s} \right)^m e^{\mathrm{i}m\theta_0} \text{ for } r < r_{\text{ef}} \tag{9.14}$$

where θ_0 represents the phase of the error field in the laboratory frame. The force due to interaction with the surface current at r_s according to Eq. (7.8) is

$$F_{\text{ef}} = 4\pi^2 R_0 J_s m \overline{\Psi}^*_{\text{ef}} \Re(ie^i(m\theta - \theta_0)) = -4\pi^2 R_0 J_s m \overline{\Psi}^*_{\text{ef}} \sin m(\theta - \theta_0) \tag{9.15}$$

and we see that the force on the mode is due to the component of the error field that is out of phase with the island perturbation. Rotation in the presence of this force will lead to a periodic modulation of the phase velocity as can be seen by analysing the change of ω^2, keeping in mind that $\omega = d\theta/dt$ and $m_{isl} r_s d\omega/dt = F_{\text{ef}}$:

$$\frac{d(\omega^2)}{dt} = 2\omega \frac{d\omega}{dt} = -8\pi^2 \frac{m J_s R_0}{m_{\text{isl}} r_s} \overline{\Psi}^*_{\text{ef}} \omega \sin(m(\theta - \theta_0)) \tag{9.16}$$

where m_{isl} is the mass of the island(s), that is viscous coupling to the rest of the plasma has been neglected. As $\omega \sin(m(\theta - \theta_0)) = -(1/m) d(\cos(m(\theta - \theta_0)))/dt$, the solution is

$$\omega^2 = \omega_0^2 + \omega_{\text{ef}}^2 \cos(m(\theta - \theta_0)) \tag{9.17}$$

6) Note that as the error field is static in the laboratory frame, wall shielding does not play a role in this calculation.

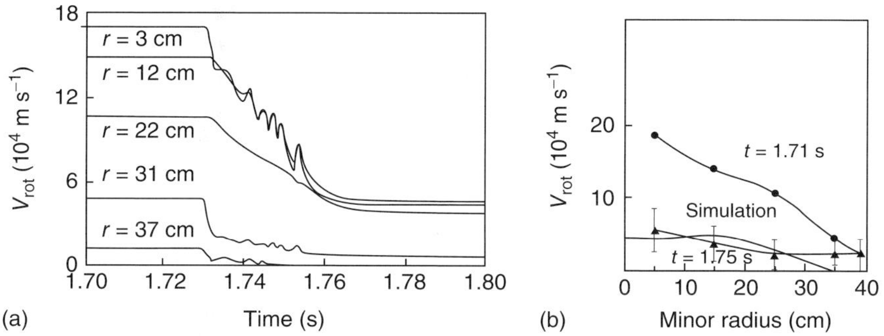

Figure 9.7 Modelling of mode locking by combining the forces due to wall drag and error field with a model of plasma rotation due to viscous drag and NBI torque for a (2,1) tearing mode in ASDEX. Prescribing the temporal evolution of the island width, the slowing down and locking of the plasma can be modelled (a) and the final rotation profile matches well the experimental measurement (b). *Source:* Zohm 1990 [51], reproduced with permission of IOP.

where

$$\omega_{ef}^2 = 8\pi^2 \frac{J_s R_0}{m_{isl} r_s} \overline{\Psi}_{ef}^* \tag{9.18}$$

One can see that in this case, there is no net deceleration, but the angular velocity is modulated around the initial value ω_0 as long as $\omega_0 > \omega_{ef}$. For $\omega_0^2 < \omega_{ef}^2$, the mode is trapped in the error field and does not rotate. A mechanical analogon is a mass point moving in a periodic potential which, depending on its initial velocity, will either move across it with varying velocity or be trapped.

Hence, the mode locking process can be explained as follows: at small island width, that is small $\overline{\Psi}^*$, the mode rotates as described in Chapter 8 due to a balance of torque and viscosity. Usually, the 'natural' rotation frequency is so high that $\omega_0 \gg \omega_{ef}$[7] and $\omega_0 \gg \tau_w^{-1}$ so that the motion is at constant angular velocity. As the island grows, the wall force becomes larger and the rotation is slowed down leading to a further increase of the wall force. Depending on the parameters, this can lead to a 'runaway' situation which leads to an abrupt breaking of mode rotation, such as the one shown in Figure 7.7, albeit for an RWM and hence with different underlying physics. Finally, the mode will enter the regime where ω_0 is of the order of ω_{ef}, which, through $\overline{J}_s$, also increases with island width, and the motion becomes strongly anharmonic until finally, the mode is trapped in the error field. This picture explains in detail the experimental observations, as is shown in Figure 9.7, where the temporal evolution of plasma rotation during mode locking has been modelled using the forces (Eq. (9.13)) and (Eq. (9.15)) coupled to a model of plasma rotation balancing viscous torque and angular momentum input from NBI (Section 9.3.1).

7) In present-day tokamaks, intrinsic error fields are of the order $\overline{B}_r/B_t \leq 10^{-4}$.

In the simulation shown in Figure 9.7, the viscosity has been determined for a state without mode indicated by the time point '$t = 1.71s$' in Figure 9.7(b) and its value was left unchanged during the simulation shown in (a). Also, the temporal evolution of the island width (a (2,1) classical tearing mode in this case) has been prescribed according to the experimental observation of the growth of the mode. A model of this evolution should also consider the effect of the error field on stability, which will be treated in Section 9.3.3.

Experimentally, the rotation profile is observed to slow down over the whole radius, which indicates that more than one mode takes part in the process. This has been incorporated into the simulation shown in Figure 9.7 by assuming interaction between the (2,1) mode and a (1,1) mode in the plasma centre. In the simplified model neglecting the plasma response, such a coupling between tearing modes can be described by the same formalism as the interaction with the error field because the modes are represented by surface currents, just as was done for the error field[8]. A more realistic modelling, however, should also consider the effect of plasma shielding and the change of mode stability due to the presence of an additional field of same helicity discussed in Section 9.3.3.

9.3.3 *Ab-initio* Locked Modes

We now turn to the case of ab-initio locked modes. These are cases where magnetic islands grow in a locked state from the beginning, indicating that the external error field plays a role already in their generation. In fact, calculating the ideal response of a plasma to an external perturbation of the form (Eq. (9.14)), one finds that the instability parameter Δ' becomes positive because of the presence of the error field[9]. This can be seen in a model where the plasma outside the resonant surface is assumed to be current free, that is in the absence of the error field, we have $\Psi^* \propto r^{-m}$. For $\Psi_i^*(r)$, the part inside the resonant surface, the form of the solution is still given by the integration of the tearing mode equation (Eq. (8.29)) with boundary condition $\Psi_i^*(0) = 0$ so that the shape of $\Psi_i^*(r)$ does not change, but of course its amplitude can be adjusted to match the outer solution as the tearing mode equation is linear and homogeneous. The helical flux function can then be written as [52]

$$\Psi^*(r,\theta) = \Psi_i^*(r)e^{\mathrm{i}m\theta} \text{ for } r < r_s \tag{9.19}$$

$$\Psi^*(r,\theta) = (\Psi_i^*(r_s)e^{\mathrm{i}m\theta} - \overline{\Psi}_{\mathrm{ef}}^* e^{\mathrm{i}m\theta_0})\left(\frac{r_s}{r}\right)^m + \overline{\Psi}_{\mathrm{ef}}^*\left(\frac{r}{r_s}\right)^m e^{\mathrm{i}m\theta_0} \text{ for } r_s < r < r_{\mathrm{ef}} \tag{9.20}$$

where, as before, the solution has been constructed such that it is continuous across r_s, but its first derivative is not. It can be seen that for $r = r_s$, the error field is

8) The coupling between modes of different helicity is due to the sidebands as described by Eq. (7.3).
9) The term *error field* includes here also the case of external helical fields that are deliberately generated by external coils, usually referred to as resonant magnetic perturbations (RMPs).

shielded as if the plasma was an ideal conductor[10]. However, while $\overline{\Psi}^*_{ef}$ is given, the amplitude of Ψ^*_i is still a free parameter, with $\Psi^*_i(r_s)$ being the reconnected helical flux. For example, if $\Psi^*_i(r_s) = 0$, the plasma response is just that of ideal MHD, that is complete shielding towards the inside $r < r_s$ by a surface current on the resonant surface. For finite $\Psi^*_i(r_s)$, there will be an island of magnitude $W \propto \sqrt{\Psi^*_i(r_s)}$ replacing the surface current.

From the solution above, we can calculate Δ' in the usual way and obtain

$$\Delta' = \Delta'_0 + \frac{2m}{r_s}\frac{\overline{\Psi}^*_{ef}}{\Psi^*_i(r_s)}\Re(e^{im(\theta_0-\theta)}) = \Delta'_0 + \frac{2m}{r_s}\frac{\overline{\Psi}^*_{ef}}{\Psi^*_i(r_s)}\cos m(\theta-\theta_0) \qquad (9.21)$$

where Δ'_0 is the linear value calculated in the absence of the error field. We see that the influence of the error field on stability is related to the component that is in phase with the island (remember that the force due to the error field was due to the component out of phase) and for $\theta = \theta_0$, the most unstable situation is obtained, meaning that in this case, $\Psi^*_i(r_s)$ will grow and an island will develop, thereby reducing Δ'. This process is also called *error field penetration.*

The growth of the island due to the increase in $\Psi^*_i(r_s)$ is described by the Rutherford equation and Eq. (9.21) may be turned into a dispersion relation introducing the W-dependence of Δ' by Eq. (9.7) and expressing the flux at the resonant surface through W according to Eq. (8.55)

$$\Delta'(W) = \Delta'_{w/o}(W) + \frac{2m}{r_s}\left(\frac{W_{ef}}{W}\right)^2 \cos m(\theta-\theta_0) \qquad (9.22)$$

where W_{ef} is the island width that would be created by the error field alone according to Eq. (8.55) and $\Delta'_{w/o}(W)$ is the nonlinear value in the absence of the error field.

For $\theta = \theta_0$, in a tearing unstable case, ($\Delta'_0 > 0$), the saturated island width will be larger than in the absence of the error field as a positive term is added to Eq. 9.8. In the case of a plasma that is stable to tearing without error field, the saturated island width for $\theta = \theta_0$ can be calculated from $\Delta'(W) = 0$ to be

$$W = \sqrt{\frac{2m}{r_s(-\Delta'_0)}}\,W_{ef} \qquad (9.23)$$

which, for a vacuum field, where $\Delta'_0 = -2m/r_s$, yields $W = W_{ef}$ but, in the presence of the plasma, can show 'error field amplification' that is islands that are larger than W_{ef} if $(-\Delta') < (-\Delta'_0)$.

In the discussion earlier, we have assumed that the island is in phase with the error field, $\theta = \theta_0$. If the phase difference is finite, for example due to an external torque acting on the plasma, the saturated island width will be smaller, and eventually, the situation can even be stable. For a rotating island, the effect of the error field is actually alternating between destabilizing and stabilizing, depending on the phase difference. It can be shown that, as the mode moves faster when moving

10) Contrary to an ideal wall, the plasma can only shield components of same helicity.

through the destabilizing region, there is a net stabilizing effect when integrating over time.

It is obvious that Eq. (9.22) predicts instability for arbitrarily small error field and the growth of an island in phase with the error field, $\theta = \theta_0$, which means that any small error field would create an island. This is usually not observed in tokamak experiments which rather show a threshold amplitude of $\overline{\Psi}^*_{ef}$ before penetration occurs. Figure 9.8 shows an example of an experiment where a dominantly helical (2,1) field was deliberately produced by coils located in the interior of the ASDEX Upgrade vacuum vessel. While the amplitude of the helical field is increased linearly on a timescale slow with respect to the reconnection time, there is little response until at a critical error field, a large island occurs that finally leads to a disruption. While here, the helical field has been produced deliberately, we remind that there is always a residual error field due to an unavoidable lack of perfect symmetry of the coil system. Hence, the understanding of this threshold is very important as it will determine the required precision for the coil system to avoid ab-initio locked modes.

One ingredient to the explanation of the threshold behaviour is the rotation discussed earlier. As pointed out in the previous sections, tokamak plasmas usually

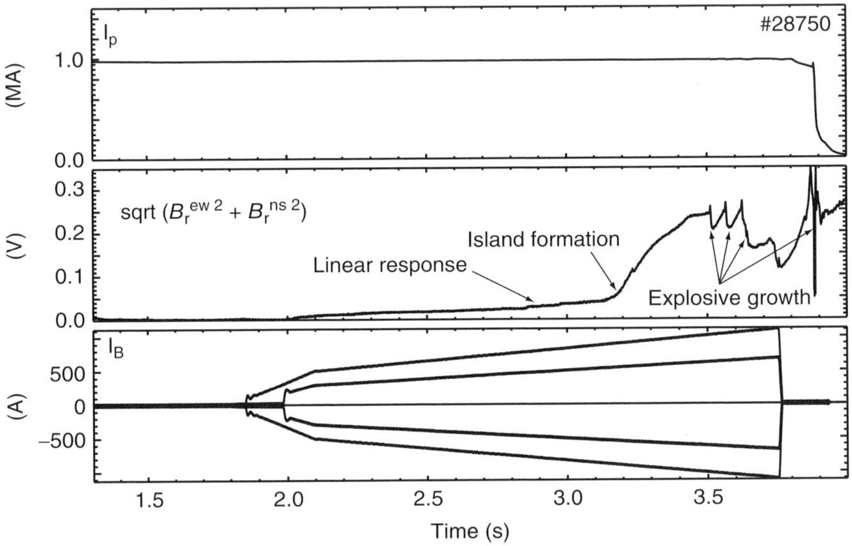

Figure 9.8 Time traces of an experiment where the amplitude of a (2,1) helical field created by external coils is slowly ramped up, as indicated by the coil currents shown in the lowest panel. In the initial phase, the plasma response as seen from the amplitude of the (2,1) radial field created by the plasma (middle panel) is small until at a critical value, a large (2,1) island appears. In the following, this island leads to a disruption, seen on the abrupt termination of the plasma current in the top panel. *Source:* Courtesy of M. Maraschek, IPP.

rotate in the laboratory frame and hence the phase difference between reconnecting flux and error field flux will continuously change. Thus, if the plasma spins much faster than a typical reconnection time, we may expect a vanishing net effect, that is a solution with fast rotation and no island (full shielding). On the other hand, if there is no rotation, the locked position will be close to $\theta = \theta_0$ and an island will occur with width according to Eq. (9.23) (full penetration). This can be viewed as an analogon to the wall locking process that has a rotating solution at $\omega\tau_w \gg 1$ (full shielding current in the wall) and a locked one at $\omega\tau_w \ll 1$ (no shielding current in the wall), with a transition between the two states as the amplitude of the perturbation is increased. Hence, the formalism used in Section 9.3.2 to calculate the torque due to the wall drag can be applied to our problem in a similar manner, assuming that there is a typical reconnection time τ_R in the resistive layer at the resonant surface so that the combination of Ohm's law and Faraday's law for the shielding current (Eq. (7.19)) now reads

$$\frac{m}{r_s}\Psi_i^*(r_s) = -\frac{i}{2\omega\tau_R}\left(\Psi^{*\prime}|_{r_s+} - \Psi^{*\prime}|_{r_s-}\right)$$
$$= -\frac{i}{2\omega\tau_R}\left(\Delta_0'\Psi_i^*(r_s) + \frac{2m}{r_s}\overline{\Psi}^*_{\text{ef}}e^{\text{i}m(\theta-\theta_0)}\right) \tag{9.24}$$

where in the last step, we have made use of Eq. (9.21). We note that (9.24) is essentially the Rutherford equation for mode penetration. The expression for the reconnected flux then becomes

$$\Psi_i^*(r_s) = \frac{\overline{\Psi}^*_{\text{ef}}e^{\text{i}m(\theta-\theta_0)}}{-\frac{\Delta_0' r_s}{2m} + i\omega\tau_R} \tag{9.25}$$

and the net drag force on the resonant surface can be calculated again as the temporally and poloidally averaged $j_s B_r$ force to become

$$F_s = 4\pi^2 R_0 \frac{(m^2\overline{\Psi}^*_{\text{ef}})^2}{\mu_0 r_s}\frac{\omega\tau_R}{\left(\frac{-\Delta_0' r_s}{2m}\right)^2 + (\omega\tau_R)^2} \tag{9.26}$$

which indeed shows the same structure as the wall drag, except that for the vacuum field in the wall, $\Delta_0' = -2m/r_s$. The rotating solution $\omega\tau_R \to \infty$ is the one with $\overline{\Psi}^*_i(r_s) = 0$, that is no reconnected flux, and the locked solution $\omega\tau_R \to 0$ represents the fully reconnected state. As mentioned earlier, balancing this torque with an external torque such as that from NBI, viscous drag and error field torque can lead to a bifurcation in the solution so that the transition between rotating and locked state can be abrupt. An example for this is shown in Figure 9.9, where the total normalized torque

$$\frac{d\omega}{dt} = \frac{1}{\tau_M}(\omega_0 - \omega) + \frac{1}{m_p r_s}F_s \tag{9.27}$$

is shown as a function of ω for increasing $\overline{\Psi}^*_{\text{ef}}$. Here, we have used the total mass of the plasma, assuming that the localized force at the resonant surface slows down the whole plasma by viscous coupling as discussed in Section 9.3.1.

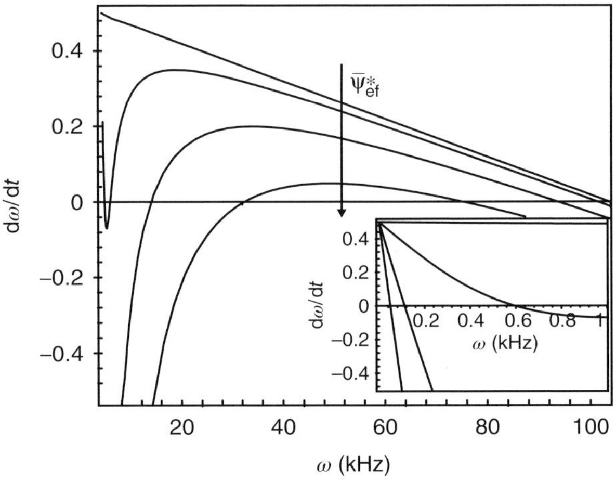

Figure 9.9 Normalized torque $d\omega/dt$ according to Eq. (9.27) as a function of ω for increasing $\overline{\Psi}^*_{\text{ef}}$. In the absence of the error field, there is one stationary state ($d\omega/dt = 0$) at high $\omega = \omega_0$, but with increasing $\overline{\Psi}^*_{\text{ef}}$, two more stationary states appear at lower rotation (see insert for a zoom into the region of small ω) until finally, only the low rotation state exists.

It can be seen that without error field, there is one stable stationary state where the viscous torque balances the NBI input[11]. With increasing $\overline{\Psi}^*_{\text{ef}}$, two more stationary points appear, one of them being unstable ($d\omega/dt$ increases with increasing ω), the other one stable ($d\omega/dt$ decreases with increasing ω). With further increase in $\overline{\Psi}^*_{\text{ef}}$, only the low-frequency stable stationary point remains. Thus, if we start from a state at high ω, small $\overline{\Psi}^*_{\text{ef}}$ and increase the latter, the rotation will decrease accordingly until it rapidly switches to the low rotation state. From the insert in Figure 9.9, one can see that the value of ω is so low that including the force due to the error field will usually lead to a trapping of the mode in this state so that the solution shows indeed the above-mentioned bifurcation character. The values inserted in Eq. (9.27) to produce Figure 9.9 are typical for NBI-heated medium-sized tokamaks, that is $n_e = 0.5 \times 10^{20}\text{m}^{-3}$, $\tau_M = 0.1$ s, $r_s = 0.3$ m, $F_{\text{NBI}} = 0.5$ N and the helical field required to lock the mode is predicted to be of the order $B_{r,\text{ef}}/B_t \approx 10^{-3}$, well in line with the experimental observations.

For low β plasmas, this model of mode locking and penetration gives a good description of the basic physics, although a quantitative prediction of the tolerable error field is difficult as it relies on the prediction of plasma rotation and the knowledge of the reconnection time τ_R. For example, the experimentally observed density dependence of the threshold for error field penetration might be related

11) A stable stationary state is one where $d\omega/dt = 0$ and the slope is negative as a small perturbation to ω will create a restoring torque.

to the diamagnetic motion of islands. Finally, we note that the discussion earlier did not consider the effect of finite β. As discussed in Section 7.4, we expect that close to the no-wall β-limit, the physics of the RWM becomes important and the threshold for mode penetration can be reduced. Such a behaviour has been identified in experiments, but for a correct description, one must include the detailed RWM physics which, as outlined in Section 7.4, is beyond the scope of this book.

10
Disruptions

The so-called disruptive instability or disruption is an instability that leads to an uncontrolled loss of the plasma current in a tokamak, abruptly terminating the discharge. While disruptions are generally linked to the occurrence of magnetic islands, the most pronounced often being the (2,1) mode, the reasons for disruptions can be quite different and hence a rich phenomenology exists. This chapter describes some of the main features of disruptions, with a special emphasis on the consequences of disruptions as they massively impact the way tokamaks are designed and operated.

10.1
Phenomenology of Disruptions

As mentioned earlier, the phenomenology of disruptions is rich, but the basic phases of a disruption are characterized by the following sequence of events:

- In the precursor phase, the process usually starts by the occurrence of several large tearing modes, with the (2,1) mode usually being the most prominent. These modes slow down and lock to the wall (Section 9.3.2). In some cases at high β, the initial MHD modes are found to be ideal kinks rather than tearing modes that deform the plasma on a macroscopic scale.
- A sudden loss of most of the thermal energy on a very fast (<1 ms in present-day tokamaks) timescale. This is attributed to stochastization due to the overlap of the locked islands (Section 9.1).[1] In addition, macroscopic pressure driven modes can lead to a rapid mixing of the outer and inner regions of the plasma. While the density is usually not too strongly affected by this phase, the temperature falls rapidly down to values of typically 10–100 eV. This phase is usually referred to as the thermal quench.
- A subsequent appearance of a negative spike on the loop voltage measured outside the plasma, together with a temporary increase in the plasma current. This phase is interpreted as a rapid flattening of the current profile due to the

1) Experimentally, the vast majority of disruptions occurs only after mode locking, supporting the stochastization hypothesis that requires phase locking.

Magnetohydrodynamic Stability of Tokamaks, First Edition. Hartmut Zohm.

flattening of the temperature profile, which leads to a lowering of ℓ_i and, on a timescale on which flux conservation inside the vacuum vessel holds, therefore to an increase in the plasma current as $\psi \sim \ell_i I_p$.

- A consecutive resistive decay of the plasma current on a timescale consistent with the low temperatures mentioned earlier: for a drop of T_e of factor 100, the resistivity will increase by a factor of 1000, which means that in order to sustain the plasma current, the loop voltage induced by the transformer would have to be increased by that factor, which is far beyond the technical capabilities of present and future tokamaks, resulting in a loss of the plasma current. This phase is usually referred to as the current quench.

An example for a disruption is shown in Figure 10.1.

We note that occasionally, a reheat of the plasma during the current quench is observed such that the discharge can be recovered, especially if the stored energy after the thermal quench is not too low. Such an event is called a *minor disruption*. This indicates that in principle, the plasma can recover from stochastization after the thermal quench, consistent with the fact that the flattened current density should remove the drive for the magnetic islands that caused the disruption.

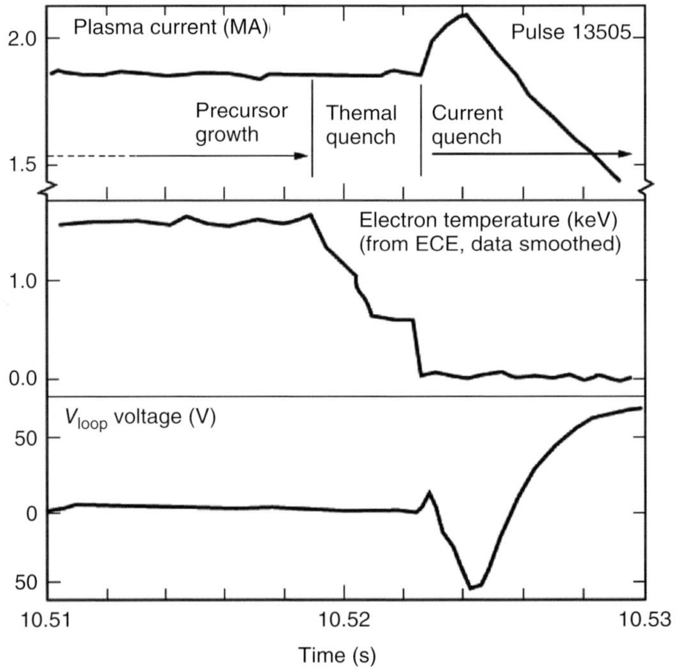

Figure 10.1 Sequence of events during a disruption on the JET tokamak. *Source:* Wesson *et al.* 1989 [53], reproduced with permission from the IAEA.

While there is a number of possible causes leading to a disruption, generally disruptions occur either due to the violation of an operational limit or due to technical failure of one or more tokamak subsystems [54]. The operational limits that may lead to disruptive termination are as follows:

- The β-limit, either in its ideal form (Section 7.5) or the resistive limit due to NTMs (Section 12.2).
- The low-q current limit (Kruskal Shafranov limit, see Section 4.2.1).
- The density limit (Section 10.1.1).

While the first two limits are described in the sections mentioned above the density limit is described in Section 10.1.1. As for the technical failures leading to disruptions, there is a variety of possibilities, but most of them lead to a large impurity influx because of the plasma touching the wall in an uncontrolled way.[2)]

10.1.1 The Density Limit

The term *density limit* is not unambiguous as upper limits of density exist for H-mode operation (usually leading to a backtransition to L-mode) and for L-mode operation. As the latter usually leads to a disruptive termination, we will describe here only the main mechanisms leading to this limit. Phenomenologically, as the plasma density is increased, there is a point at which MHD modes, usually dominated by a large (2,1) tearing mode, become unstable and initiate a disruption as described earlier. As this can occur at very low β, the main destabilizing mechanism must be the current gradient. It is not straightforward to deduce a change of the current profile from an increasing density profile: while usually, the temperature decreases roughly as $T_e \propto 1/n_e$, the current profile will only change when the shape of the temperature profile changes. Obviously, this must be the case at the onset of the tearing modes at the density limit, which indicate a steepening of the current gradient at the $q = 2$ surface. Such a steepening can be due to a preferential cooling of the edge plasma, which is indeed observed due to the onset of a radiation instability as the density increases.

The radiative instability of the edge can be understood as follows: in thermodynamic equilibrium, the radiation emitted by a body will increase with temperature, leading to a stable equilibrium between heating power and radiated power[3)] as a positive (negative) perturbation of the temperature from its equilibrium value will lead to an increase (decrease) in radiation and hence the system will return towards the equilibrium point. For a fully ionized plasma, the radiation loss is due to Bremsstrahlung radiation, which indeed is proportional to $n_e^2\sqrt{T_e}$ and we have a stable situation. However, if in a certain

2) The failure mode of wall components falling into the plasma can be seen as a somewhat special variant of this case.
3) For simplicity, we neglect in this discussion the losses due to convection and conduction.

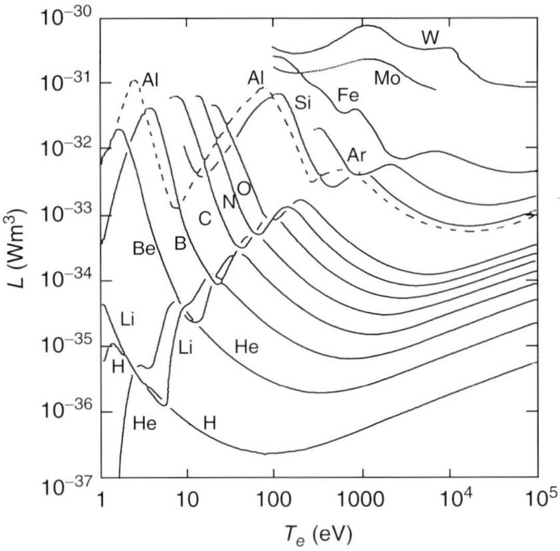

Figure 10.2 Radiated power density as a function of temperature for different elements, normalized to the product of electron and impurity density. For light elements (C,O), pronounced maxima exist in the region 10–100 eV. *Source:* Courtesy of R. Neu, IPP.

temperature range, partially ionized species exist, the radiation will generally be a non-monotonic function of temperature in this range. For light impurities such as oxygen or carbon, that are usually present in tokamak discharges, this range is of the order of 10–100 eV. Figure 10.2 shows the radiation loss function for a number of elements, indicating local maxima in this temperature range.

Now suppose that, at low density, we have a stable operating point at $T_e > 1$ keV. As we increase the density at constant input power, the radiative losses will increase as n_e^2 and the operational point will move to lower temperature (by the MHD force balance, the pressure is still constant on the flux surface). If we assume that the radiative loss is for example dominated by C (Figure 10.2), there will be a point where there is no more stable solution and the plasma will rapidly cool down to around 10 eV where a new stable operational point exists. As this process sets in below a certain temperature, it will first appear at the plasma edge where the temperature is lowest.

Once the plasma has locally cooled to 10 eV, the parallel heat conductivity is so much reduced that the temperature is no longer constant on the cold flux surfaces and a local area of low temperature and high density develops. This highly radiating zone will be toroidally symmetric and sit in the area where the heating power density is lowest, which is typically on the high field side around the X-point of a diverted plasma. This phenomenon, called a *MARFE*,[4]

4) Acronym for multi-faceted axisymmetric radiation from the edge [55].

Figure 10.3 Visible light camera image of a plasma discharge in ASDEX Upgrade exhibiting a MARFE that can clearly be recognized by the toroidally symmetric radiation zone in the lower part of the vacuum vessel. (Please find a color version of this figure on the color plates.)

is well documented experimentally and can even be seen directly in visible light as the cold plasma in the MARFE emits significantly more radiation than the surrounding plasma. An example from the ASDEX Upgrade tokamak is shown in Figure 10.3.

Owing to the very low temperature in this radiative zone, the current in the edge region of the plasma will effectively be quenched[5] and hence the current gradient at the edge resonant surfaces such as the $q = 2$ surface steepens until a tearing mode is triggered which initiates the disruptive instability. We note that such a region of very low temperature due to a MARFE can be observed in Figure 9.3 in the area where the (3,1) mode occurs.

While this picture is generally observed and has been confirmed in a number of experiments, it cannot explain the usually observed scaling of the tokamak density limit with plasma current. Empirically, it was found that the line averaged density achievable in tokamaks with circular corss section before the disruptive density limit could be expressed in a universal curve if the density is normalized as $n_e R/B_t$ with this quantity showing a roughly linear variation with $1/q_{\text{edge}}$, that is the plasma current. However, while this so-called Hugill-limit was established for tokamak plasmas with circular cross section, an analysis including discharges with shaped poloidal cross section found that all data points could be combined assuming a simple current scaling of the maximum achievable line averaged density of the form

5) Note that for a current following the field lines, a local region of high electrical resistance appears in series with the rest of the plasma.

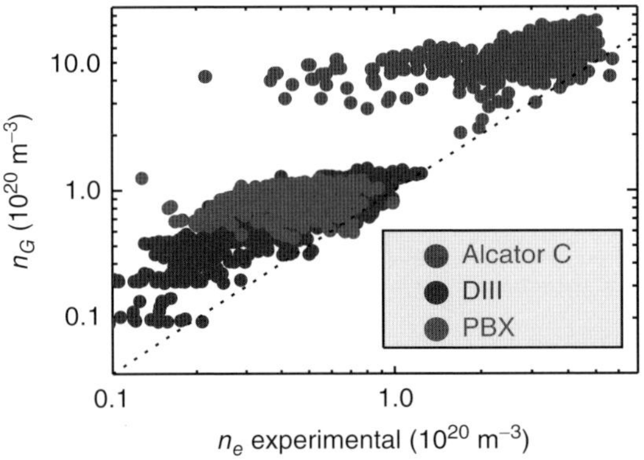

Figure 10.4 Operational points from the Alcator C, DIII, and PDX tokamaks plotted against the empirical density limit n_{GW} described by Eq. (10.1). The data points support the empirical scaling for different cross sections. *Source:* Replotted from Greenwald 1988 [49], reproduced with permission of the IAEA. (Please find a color version of this figure on the color plates.)

$$\overline{n}_{e,\mathrm{max}} \approx n_{\mathrm{GW}} = \frac{I_p}{\pi a^2} \tag{10.1}$$

where n_{GW} is the so-called Greenwald density. This scaling indicates that it is rather the average current density than the total plasma current that determines the density limit. An example is shown in Figure 10.4 where data points from several tokamaks are plotted against the scaling (Eq. (10.1)).

Experimentally, it is found that the density limit depends only weakly on the heating power while the description in terms of a power balance issue in the edge might suggest that increasing the heating power should help to significantly increase the density limit. Hence, another physics element is needed to explain the parametric dependency of the density limit. Such an element could be a significant increase of the edge transport above a certain density, which should depend on the current and might also be related to the experimental observation that with increasing density, there is a backtransition from H- to L-mode, usually occurring at some fraction of n_{GW}. While signs for such an additional transport channel have been found recently, its nature is not yet explored to a degree where it would allow a first-principles explanation.

Finally, we note that the Greenwald limit to the line averaged density can be substantially exceeded in the case of peaked density profiles, as long as the edge density stays below n_{GW}, consistent with the earlier-mentioned description of an edge density limit. While in present-day tokamaks, substantial density peaking at high density can only be obtained using central fuelling by pellets, theory predicts that ITER and DEMO will exhibit peaked density profiles even with peripheral particle source due to the turbulent inward pinch occurring at low collisionalities.[6)]

6) This pinch is clearly seen in present-day experiments at low collisionality, which for present-day devices means low normalized density n/n_{GW}.

10.2 Consequences of Disruptions

The disruptive instability is very important as the consequences of setting free in an uncontrolled manner the stored thermal and poloidal magnetic energy of a tokamak discharge can be quite detrimental to the tokamak hardware, potentially producing large thermal and mechanical loads. For these reasons, disruptions have a large impact on the operation and design of tokamaks, which will be even more important for future large devices running higher absolute values of the plasma current than in present-day devices.

10.2.1 Thermal Loads

Concerning the thermal loading of components during disruptions, we are mainly interested in the temperature increase due to the rapid deposition of energy. While in steady state, typical limits for components are for the tolerable heat flux, the limitation for pulsed deposition on a timescale much faster than the time it takes to reach thermal equilibrium in the component is described by the 'energy impact' η, which is proportional to the rise in temperature: for given deposited energy, the temperature increase in the component is inversely proportional to its thermal capacity, which is the product of its heat capacity and effective volume. For a pulsed load, the heat diffuses into the component such that the affected width is proportional to $\sqrt{t}$, and hence the effective volume increases accordingly.

$$\Delta T_{\text{component}} \propto \eta = \text{const.}\frac{\Delta W}{A_c\sqrt{\Delta t}} \tag{10.2}$$

where ΔW is the energy deposited in time interval $\Delta t, A_c$ the surface area of the component and the constant depends on the specific heat capacity and the heat conductivity of the component. For W and C, which are typically used in first wall components in present-day tokamaks, the allowable heat impact before damaging the wall is of the order of $\eta_{\max} = 0.05\ \text{GJ m}^{-2}\text{s}^{-1/2}$.

In order to determine this number for present-day and future experiments, it is crucial to know where the energy is deposited. Clearly, the two limiting cases are to either assume that the energy is distributed uniformly over the first wall or that it reaches the divertor target plates along field lines in the narrow deposition zone observed in normal operation.[7] In reality, a combination of both will occur. For future large experiments such as ITER, the heat impact between these extreme cases may differ by more than two orders of magnitude, but experimental values for the heat deposited on the divertor plates during disruptions generally indicate a broadening of the deposition zone by a factor of 5–10, reducing the power density by this factor.

7) The width of the effective wetted area is only of the order of centimetre in present-day devices and has been found not to scale with machine size R_0 [57].

Concerning the timescales, the typical timescale Δt_{TQ} for the thermal quench in present-day devices is of the order of 0.5–1 ms, with a tendency to be longer in the larger machines, but owing to our limited understanding together with the scatter of experimental values, a clear rule for extrapolation to next generation devices is still missing. For calculating the heat impact η, it is important to note that the time interval in which the heat is deposited on the divertor is usually longer than Δt_{TC}, attributed to the finite heat conduction in the scrape-off layer.

The situation is clearer for the duration of the current quench Δt_{CQ}, where there is also a large scatter of experimental values, but the lower bound is well described by a constant times the poloidal cross-section area A_p. This is consistent with the assumption that during the current quench, mainly the internal poloidal flux[8] is lost, which is given by $\psi = \mu_0 R_0 \ell_i$, so that the L/R timescale for current decay becomes

$$\Delta t_{\mathrm{CQ}} \propto \frac{L}{R_p} \propto \mu_0 \sigma \ell_i A_p \tag{10.3}$$

where R_p is the electrical resistance of the plasma column. It can be seen that for constant σ and ℓ_i, the timescale is indeed proportional to A_p, and the experimental data can be explained assuming that the temperature, and hence the resistivity, after the thermal quench varies, but has a lower limit of 5 eV, which basically is the recombination limit below which the whole plasma will be neutralized. This leads to the numerical value of $\Delta t_{\mathrm{CQ}} \approx 1.8\,\mathrm{ms}\, A_p[\mathrm{m}^2]$ characterizing the experimentally observed lower bound (assuming a flat current profile, $\ell_i = 0.5$).

Inserting numbers of present-day experiments, one finds that the damage limit might already be exceeded if all thermal energy went to the divertor, but this is generally not seen because a sufficient fraction of the energy will end up on the first wall due to radiation. However, extrapolating to ITER, the limit will certainly be exceeded unless a large fraction can be converted into radiation before it reaches the divertor. This is one of the aims of disruption mitigation, which is treated in Section 10.3.

10.2.2 Mechanical Loads

Concerning the poloidal magnetic field energy, we note that it will finally also end up as thermal energy in the wall, but as $\beta_p \approx O(1)$ and $\Delta t_{\mathrm{CQ}} \gg \Delta t_{\mathrm{TQ}}$, its contribution to η will have to be considered, but will not lead to a large aggravation of the problem of thermal loading. A more important aspect of the release of magnetic energy is the inductive generation of currents in components of the tokamak assembly, leading to large $\mathbf{j} \times \mathbf{B}$ forces that can potentially cause mechanical damage to the device. To begin with, a disruption will induce toroidal eddy currents in the vacuum vessel and other conducting structures trying to resist

8) To be more precise, one would also have to count the flux between plasma and vacuum vessel.

the change of the poloidal flux due to the decaying plasma current. As these are largely parallel to the magnetic field, they will not lead to large forces. However, for vertically elongated plasmas, disruptions usually also lead to a vertical displacement event (Section 4.4) because of the abrupt change of $\beta_p + \ell_i/2$ at the thermal quench leading to a mismatch of the external field applied to control the position.

As can be seen in Figure 10.5, this will drive the plasma column to the top or the bottom of the machine, resulting in a scraping off of flux surfaces and hence to a shrinking of the plasma cross section which, in turn, induces a poloidal

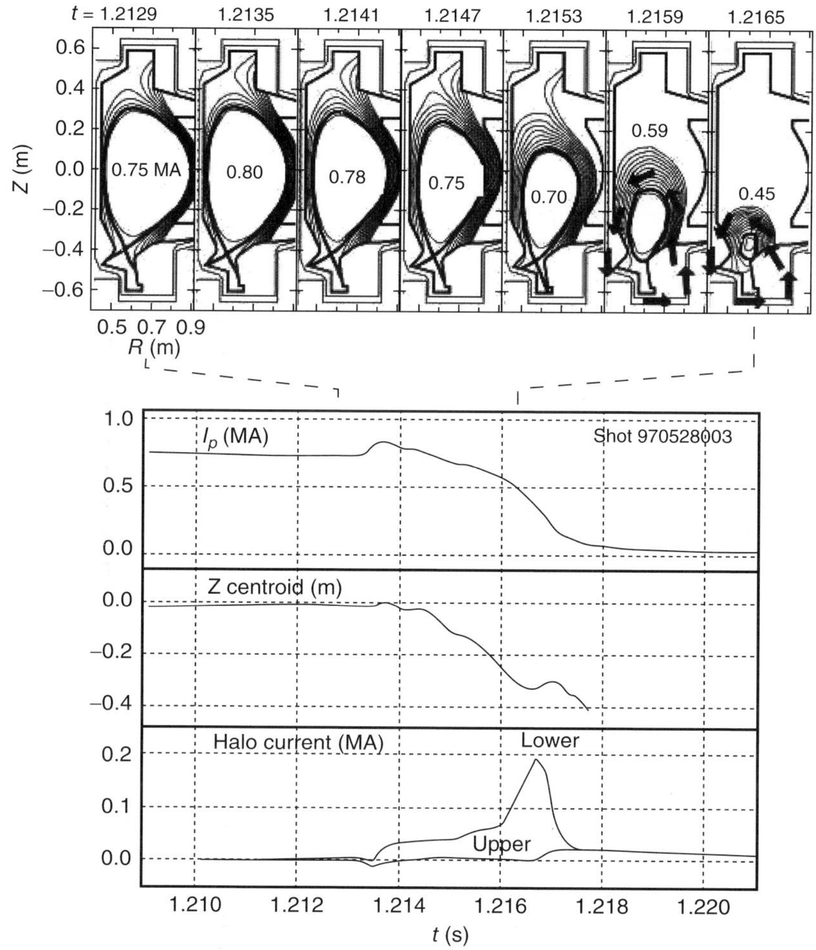

Figure 10.5 Reconstructed series of equilibria during a vertical displacement event on Alcator C-Mod, indicating the generation of mainly poloidal halo currents. Note that for the last time point, while the closed flux surface has nearly been extinguished, there is still substantial plasma current, more than half of the initial value. *Source:* ITER Physics Basis, Chapter 3 [58], reproduced with permission of the IAEA.

loop voltage driving a poloidal current in the 'halo' surrounding the closed flux surfaces. This so-called halo current closes via the vacuum vessel components, mainly in poloidal direction, leading to a large force when crossed with the toroidal field. As can be seen in Figure 10.5, the plasma current can still have substantial magnitude at large displacement. As according to Eq. (4.18), $F \propto I_p \Delta z$, this in turn will generate a large force on the plasma, which is essentially transferred to the vessel by the induced currents as the plasma itself cannot sustain any force because of its very low mass. While the modelling of halo currents is a complex task, experimentally, up to 50% of the total plasma current can be converted into a halo current according to measurements across a number of tokamaks. We also mention here that the shrinking of the plasma cross section at only slightly reduced plasma current leads to a decrease of the edge safety factor that often leads to MHD activity connected to the low-q limit (see also Section 4.2.1). This activity can accelerate the final decay of the plasma current and is also thought to be responsible for toroidal asymmetries in halo forces and thermal loads that have led to the definition of the 'toroidal peaking factor' TPF, which is the ratio of maximum to toroidally averaged halo current. This is usually small at large current but can be as large as 3–5 for small currents, so that the ITER design has adopted the product of halo current normalized by plasma current and multiplied by the TPF to characterize the severeness of the problem. On present-day experiments, the forces generate by halo currents can be as large as several 100s of kN, and in ITER, they may be as large as several MN. It is hence another goal of disruption mitigation to limit the product of plasma current and vertical displacement during a disruption to reduce the forces (Section 10.3).

10.2.3
Runaway Generation

A final detrimental consequence of disruptions is the potential for generation of the so-called runaway electrons: in a plasma, the friction force that a charged particle with velocity $v \gg v_{\text{th}}$ experiences due to Coulomb collisions with the background decreases as $1/v^2$, so that for given particle velocity v, there will be a critical value of the electric field E_c above which the acceleration due to the internal toroidal electric field E_{internal} is no longer balanced by friction and charged particles can 'run away', that is be accelerated without braking force.[9] Owing to the lower mass, this will always first occur for electrons. In the nonrelativistic case, this critical field is given by

$$E_c(v_e) = \frac{e^3 n_e \ln \Lambda}{4\pi\epsilon_0^2 m_e v_e^2} \tag{10.4}$$

9) In a toroidal confinement system, the energy will ultimately be limited by loss due to synchrotron radiation which increases with fourth power of the relativistic electron energy $\gamma m_e c^2$ where $\gamma = 1/\sqrt{1-(v/c)^2}$ is the relativistic mass factor.

where $\ln\Lambda$ is the Coulomb logarithm.[10] It can be seen that the value of E_c decreases with v_e, so that for a Maxwellian distribution, there should always be electrons that run away, but if $v_e \gg v_{\mathrm{th},e}$, their number will be exponentially small. Furthermore, if $v_e \to c$, a relativistic treatment of the problem using the Moeller cross section rather than the Rutherford cross section shows that there is a residual minimum drag leading to a minimum

$$E_{c,\mathrm{rel}} = E_c(v_e = c) = \frac{e^3 n_e \ln\Lambda}{4\pi\epsilon_0^2 m_e c^2} \tag{10.5}$$

for relativistic electrons and hence for $E_{\mathrm{internal}} < E_{c,\mathrm{rel}}$, there will be no runaways at all. However, if the critical v_e approaches $v_{\mathrm{th},e}$, there should be a substantial number of runaway electrons if the electric field is high enough. For the so-called Dreicer field, $E_{c,\mathrm{Dreicer}} = 0.43E_c(v_e = v_{\mathrm{th},e})$, the whole electron population is predicted to run away.

Concerning runaway generation during disruption, we expect a high internal toroidal electrical field during the current quench due to the rapid change of poloidal flux, and a substantial part of the electrons can be converted into runaways and carry the current in the otherwise cold plasma. A simple estimate based on the current quench physics described earlier yields

$$E_{\mathrm{internal}} \approx \frac{\mu_0 \ell_i I_p}{2\pi\Delta t_{\mathrm{CQ}}} \tag{10.6}$$

Comparing with Eq. (10.3) shows that due to $\sigma \propto T_e^{3/2}$, the ratio $E_{\mathrm{internal}}/E_{c,\mathrm{Dreicer}}$ is proportional to $n_e^{-1}T_e^{-1/2}$, that is the situation is most dangerous for low temperature as is the case during the current quench. Using the minimum value of Δt_{CQ} mentioned earlier, E_{internal} can be evaluated to be of the order of 40 V in ITER. On the other hand, the critical field $E_{c,\mathrm{Dreicer}}$ at this temperature is of the order of kilovolts, so that there should be no problem concerning the generation of a large runaway current. In fact, in present-day experiments, during the current quench, a significant runaway population is only observed at low density.

There is, however, an important difference in runaway generation mechanism between present-day devices and future large machines such as ITER. In the latter, it is predicted that the knock-on effect of runaways during occasional close encounters in the collisional deceleration process will generate secondary runaway electrons [59], leading to an exponential growth of the runaway current I_{RA} approximately described by

$$\frac{1}{I_{\mathrm{RA}}}\frac{dI_{\mathrm{RA}}}{dt} \approx \frac{eE_{c,\mathrm{rel}}}{m_e c \ln\Lambda}\left(\frac{E_{\mathrm{internal}}}{E_{c,\mathrm{rel}}} - 1\right) = \frac{1}{\tau_{\mathrm{RA}}} \tag{10.7}$$

It follows that for $E_{\mathrm{internal}} > E_{c,\mathrm{rel}}$, there should be 'avalanche' production and at $E_{\mathrm{internal}} \gg E_{c,\mathrm{rel}}$, the avalanche amplification given by $\exp(\Delta t_{\mathrm{CQ}}/\tau_{\mathrm{RA}})$,

10) For fusion plasmas, $\ln\Lambda$ is of the order of 16.

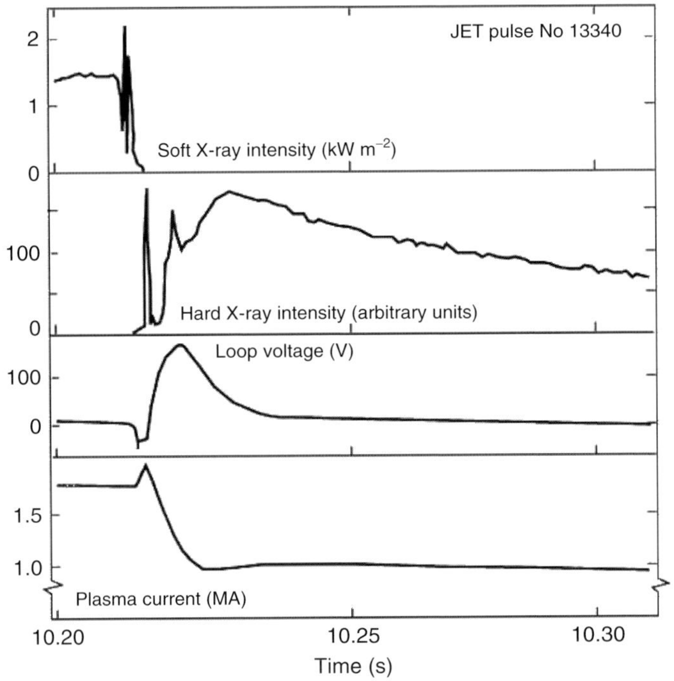

Figure 10.6 Generation of a large runaway tail current during a disruption in JET: while the soft-X radiation indicative of the thermal energy collapses at the thermal quench, the hard-X radiation indicative of relativistic electrons increases and the current decay stops due to the formation of a runaway current. *Source:* Gill 1993 [60], reproduced with permission of the IAEA.

using Eq. (10.6), becomes independent of Δt_{CQ} and can be evaluated to be $\exp(2.4I_p/[\text{MA}])$.

For present-day medium-sized tokamaks, the amplification is hence of the order of 10 and a substantial 'seed run away current' must exist for amplification, while for ITER, the factor is 10^{16}, making it virtually independent of the seed current. For JET, the amplification factor is already 10^4 and in fact, the avalanche process has been identified to be a substantial contributor to the generation of large runaway tail currents during the current quench. Disruptions at high current and low density in JET have shown tails containing more than 50% of I_p in the form of runaway current. An example is shown in Figure 10.6, where the loss of plasma thermal energy leads to the decrease in the soft X-ray radiation, but the buildup of the runaway tail is detected by the increase in the hard X-ray emission. As can be seen, even at $E_{\text{internal}} = 0$, this current will persist for several seconds due to the slow collisional slowing down of the runaways in the relativistic regime. This can be understood from Eq. (10.7) as exponential decay corresponds to the case $E_{\text{internal}} < E_{c,\text{rel}}$, and hence $1/\tau_{\text{RA}}$ will be much smaller than for the exponential growth case.

The main danger of such runaway currents is that if a beam of relativistic electrons of, say, several 10 MeV, locally interacts with matter, it can produce significant damage. Hence, avoidance of runaway electron generation during the current quench is another important goal of disruption mitigation.

10.3
Disruption Avoidance and Mitigation

From the preceding discussion, it is clear that a strategy to protect future large tokamaks against the consequences of disruptions must be found. Given the severity of the problem, a layered approach is proposed in which a hierarchy of protection measures is taken. These layers can be classified as follows:

1. Avoidance of operational limits to β, n_e and q_{95}. The physics of these limits has been discussed in Sections 7.5, 10.1.1 and 4.2.1, respectively.
2. Active control of MHD activity in case an operational limit is touched (Chapter 13) while manoeuvering back to safe operational point.
3. Controlled shutdown if active control is impossible. This will be done by reducing stored thermal and magnetic energy as quickly as possible in a controlled way.
4. Mitigation of the adverse consequences of a disruption in case controlled shutdown is no longer possible. This may include intentional triggering of a disruption and is discussed in this section.

We note that in order to take these actions, an on-line determination of the plasma state must be possible that allows to decide which measure to take, which in itself is a non-trivial task. According to the discussion earlier, the main goals of disruption mitigation in order to protect the machine can be characterized as follows:

- Conversion of the plasma energy into electromagnetic radiation to lower the energy impact on the first wall during the thermal quench by increasing the affected area.
- Limitation of the product $\Delta z I_p$ to minimize mechanical loads during the concomitant VDE.
- Avoidance of generation of runaway current during the current quench, for example by increasing the density such that $E_{\text{internal}} < E_{c,\text{rel}}$.

The present approach to all three goals is to inject a large amount of impurities that would rapidly penetrate the plasma, enhance the radiative losses, shorten the current quench phase such that the current decays already in an early stage of the VDE and increase the density such that no runaways are generated. In order to achieve this, the impurities must be delivered before the thermal quench, that is a disruption prediction system must exist that indicates an upcoming thermal quench with sufficient warning time. This can be as simple as detecting a locked mode but may involve much more sophisticated combinations of real-time diagnostics in future. The impurities should be delivered fast and penetrate deeply; for this reason, the injection of frozen pellets or high pressure (10 s of bar) gas jets from nearby reservoirs have been studied.

Experiments so far have shown that indeed, a large part of the stored energy can be radiated, leading to a substantial reduction of the power conducted to the target plates following the thermal quench. In general, the radiation is not uniformly distributed over the first wall, and studies on how this can be achieved by injecting the impurities from several locations around the torus are ongoing. Concerning the forces, termination of the plasma current can be fast enough to limit the product $\Delta z I_p$ to acceptable values, and it hence looks like the first two goals of disruption mitigation as outlined earlier can be achieved using impurity injection in ITER as well. An example of the decrease in divertor load and forces on the vacuum vessel during an intentionally triggered disruption using massive injection of Ne in ASDEX Upgrade is shown in Figure 10.7.

Concerning the suppression of runaway electron generation, the experiments have not been successful so far to reach the necessary 'Rosenbluth' density n_{RB} given by the condition

$$n_{\mathrm{RB}} = \frac{2n_e + n_b}{2} \geq \frac{4\pi\epsilon_0^2 m_e c^2}{e^3 \ln \Lambda} E_{\mathrm{internal}} \tag{10.8}$$

which can be derived from the condition $E_{\mathrm{internal}} < E_{c,\mathrm{rel}}$, where E_{internal} and $E_{c,\mathrm{rel}}$ are given in the previous section and n_b is the density of bound electrons in partly ionized atoms that contribute to the stopping power for a relativistic electron

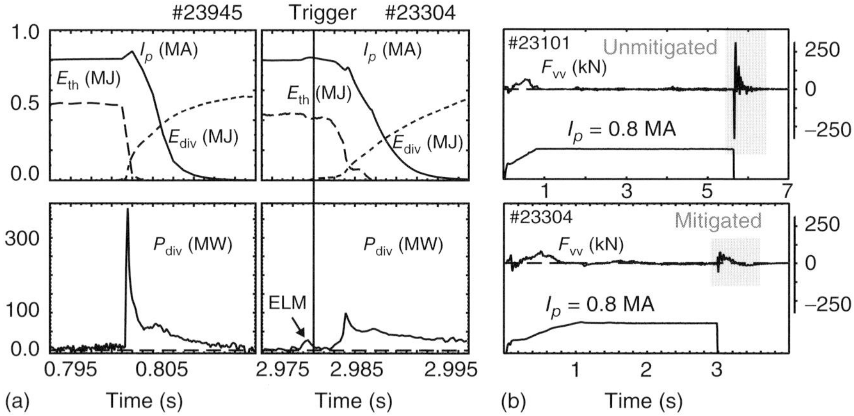

Figure 10.7 Mitigation of thermal and mechanical loads during a disruption initiated by massive gas injection using Ne in ASDEX Upgrade. Two rows in (a): zoom of the thermal quench without (left) and with (right) mitigation – the thermal load on the divertor is greatly reduced due to the symmetrization by radiation. Rows in (b): the force on the vacuum vessel F_{VV} is greatly reduced with mitigation (lower trace) compared to the case without mitigation (upper trace). *Source:* Courtesy of G. Pautasso, IPP.

beam. A main problem on the path towards reaching the Rosenbluth density is the experimentally observed degradation of the fuelling efficiency at high injected amount of gas. It remains to be seen, however, if slowing down of runaway electrons is fully described by classical collisional theory as any additional loss would lead to a reduction of the requirements. Alternatively, it might be possible to deconfine the runaways by application of magnetic perturbation coils during the current quench in order to prevent the buildup of a large population.

11
M=1 Modes beyond Ideal MHD: Sawteeth and Fishbones

In Chapter 4, we have examined the ideal stability of the internal (1,1) kink mode and found that we had to go to fourth order in inverse aspect ratio to determine the stability properties. Consequently, (1,1) stability is subtle and non-ideal effects can play a major role in determining it. This concerns not only finite resistivity but also kinetic effects, that is the interaction of thermal and supra-thermal particles with the mode. Owing to its special geometry, that is the unstable mode and the plasma core have the same topology, there is also the possibility that non-linearly, the mode affects a large part of the plasma centre. This leads to a non-linear cycle, the so-called sawtooth instability which is described in Section 11.1. In addition, there is another non-linear cycle that is due to the interaction of the (1,1) mode with fast particles. This so-called fishbone instability is treated in the last section of this chapter. As already mentioned in Section 6.3.1, both cycles can be described conceptually by the simple model for non-linear cycles presented in the context of ELMs.

11.1
The Sawtooth Instability

In 1974, an analysis of the soft-X-ray emission from the ST tokamak showed a characteristic cyclic instability whose time traces resemble the shape of sawteeth. Soon after, this instability was linked to a reconnection phenomenon at the $q = 1$ surface. Although the basic picture of the sawtooth cycle has not changed since this first explanation, there are still remaining questions concerning the precise nature, especially of the crash. In the following, we take a look at the individual physics elements.

11.1.1
Phenomenology

The sawtooth instability is observed in most conventional tokamak scenarios[1] at reasonably high current, that is when a $q = 1$ surface is present in the plasma.

1) By conventional scenario, we mean here those scenarios that have a peaked current profile dominated by the ohmic contribution (see also Section 7.1).

For example, in ohmic tokamak discharges sawteeth are usually observed up to values of $q_{95} \approx 7 - 10$. Then, a periodic modulation of the central temperature and density is seen, with a time trace resembling a sawtooth. Figure 11.1 schematically shows the temporal evolution of the temperature during a sawtooth cycle. Outside the so-called inversion radius $r_{\rm inv}$, the shape is inverted, indicating a crash of the central temperature profile and, as a consequence, a heat pulse travelling outward through the outer region of the plasma. Obviously, this is a non-linear cycle involving two different timescales, namely the sawtooth period $\tau_{\rm st}$, that is the time between two sawtooth events, in which the profiles of density and temperature are found to continuously peak, thereby leading to the rise of the central values over the sawtooth period, and the fast crash time $\tau_{\rm crash}$, in which the profiles inside the mixing radius $r_{\rm mix} > r_1$, the radius of the $q = 1$ surface, are flattened and the heat pulse is ejected into the region outside the mixing radius[2]. Often, (1,1) MHD mode activity is observed to precede the sawtooth crash. The basic picture of the sawtooth cycle is that under central heating, the temperature profile peaks and the current profile follows due to $\sigma \propto T^{3/2}$. The increase in central current density leads to a decrease in q_0. Once $q_0 < 1$, the (1,1) mode becomes unstable and leads to the crash, mixing the plasma inside $r_{\rm mix}$ and, at the same time, redistributing poloidal flux such that $q_0 \geq 1$. As mentioned earlier, this cycle is conceptually of the nature described in Section 6.3.1, with the gradient of the current density as driver and the (1,1) crash as relaxation instability. Because of the very disparate timescales, the sawtooth-like shape of the signals is very pronounced.

Owing to the redistribution of heat, particles, and poloidal flux, the sawtooth instability plays an important role in determining the quasi-stationary (i.e. sawtooth-cycle averaged) parameters of tokamak discharges. In particular, it limits the peaking of temperature, density, and current density profiles. While the redistribution of energy and particles from the centre of the discharge is decreasing the averaged stored energy and thus has a negative impact on the energy confinement time, the enhanced particle transport can be beneficial to avoid impurity accumulation and, for a reactor, also the accumulation of the He ash. In addition, as will be discussed in more detail in Chapter 12, sawtooth crashes can also act as trigger for the β-limiting neoclassical tearing mode. Hence, active control schemes of the sawtooth instability do not necessarily aim at completely suppressing the sawteeth but may rather target a tailoring to minimize the energy loss while at the same time providing adequate central impurity control. This will be treated in Chapter 13.

11.1.2
Sawtooth Period and Onset Criterion

Following the picture earlier, the sawtooth period must be related to the redistribution of current density by diffusion inside the $q = 1$ surface and hence should,

2) From this discussion, it is clear that the inversion radius itself is not of special importance for the physics of the sawtooth instability but rather for the characterization of the diagnostic signals.

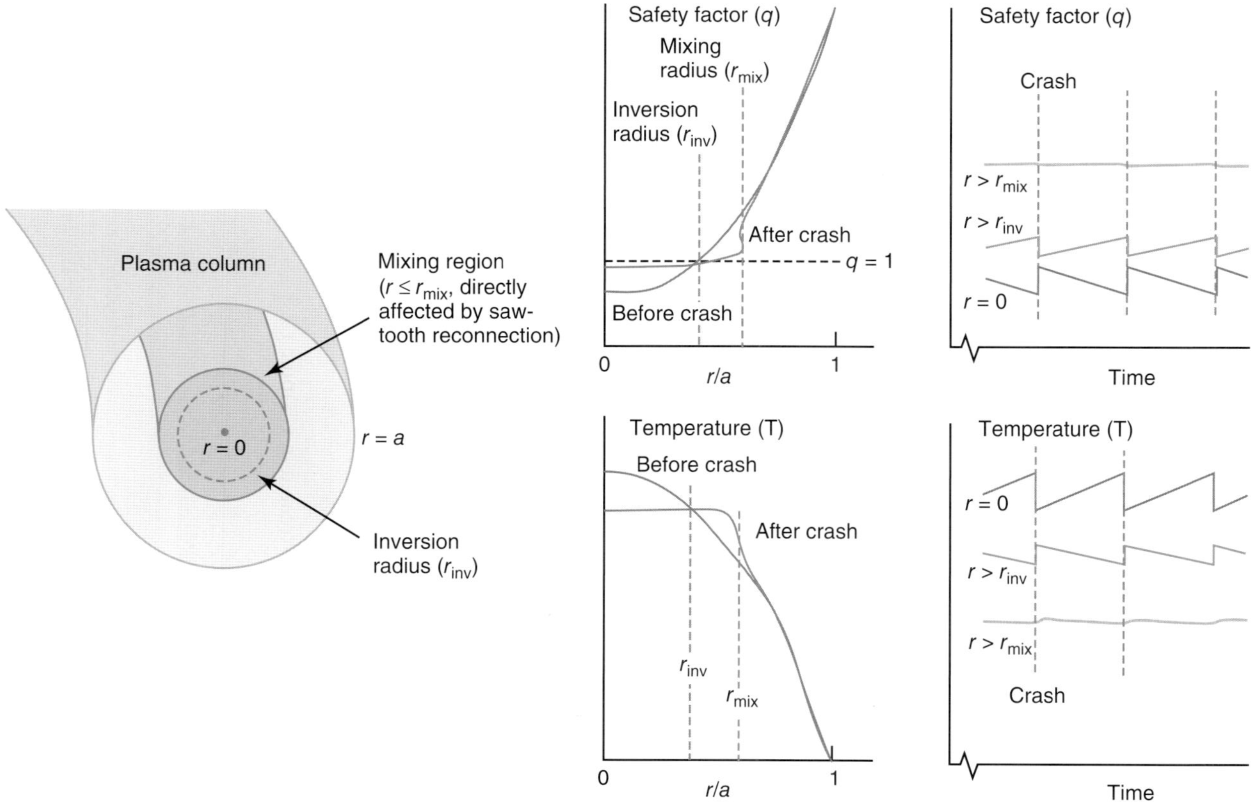

Figure 11.1 Schematic representation of typical time traces and profile evolution during a sawtooth cycle. *Source:* ITER Physics Basis, Chapter 3 [58], reproduced with permission of the IAEA. (Please find a color version of this figure on the color plates.)

according to Eq. (8.3), roughly scale as $\tau_{st} \propto \mu_0 \sigma r_1^2$. While such a scaling is indeed roughly observed in experiments, the precise onset time of the crash can vary appreciably, indicating that the stability criterion is more complex than just $q_0 < 1$. In fact, this is also clear from the observation that the plasma becomes unstable when the $q = 1$ surface is already significantly distant from the magnetic axis, indicating that a finite radial region with $q < 1$ exists, which is affected during the crash.

In Section 4.3, we examined the ideal MHD stability of the (1,1) internal kink mode and found that, including finite β and toroidal effects, there should be a stability window at low enough β_{p1}. It turns out, however, that including finite resistivity in the analysis widens the window for instability, and as the ideal δW, which is negative for the unstable ideal internal kink, approaches zero, a so-called resistive kink becomes unstable [61]. This instability has a radial eigenfunction similar to that of the ideal kink, but the singularity at the rational surface is resolved by resistivity rather than by inertia (as is the case for the ideal mode). A sketch of the different regimes of the (1,1) internal kink is shown in Figure 11.2. Qualitatively, one can see that this mode will have a large positive Δ' as Ψ' is negative inside $q = 1$ and zero outside, indicating instability. Finally, for positive δW, this resistive instability turns into a (1,1) island with an eigenfunction comparable to that of an $m \geq 2$ tearing mode as described in Chapter 9. This solution is also shown in Figure 11.2. In addition, the flow pattern arising due to the internal kink motion is shown, indicating the formation of a very localized layer at the $q = 1$ surface.

Thus, ideal MHD cannot describe adequately the onset of the crash, and the physics of the layer at the $q = 1$ surface arising from compressing the flux surfaces there due to the internal kink motion must be considered when deriving the stability criterion. One important effect is that for a layer width of order of the

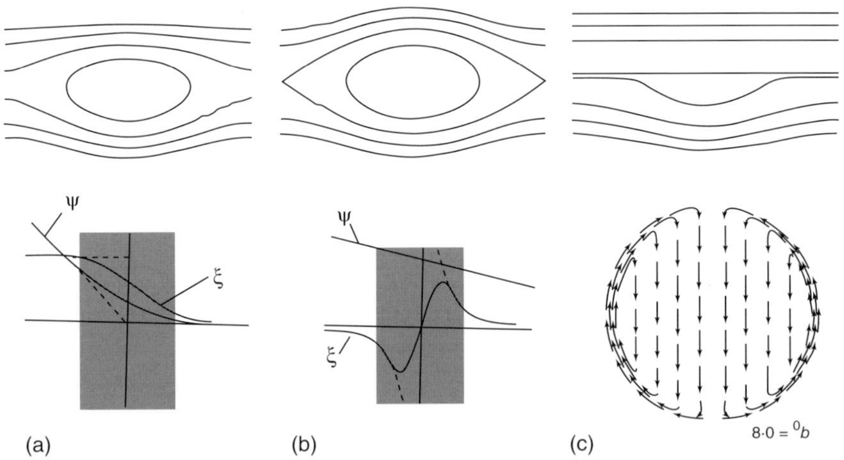

Figure 11.2 Eigenfunctions and flux surface topology for resistive kink (a) and the (1,1) tearing mode (b). The topology and the flow pattern inside the $q = 1$ surface generated by the ideal kink (c) are also shown. Source: courtesy of K. Lackner, IPP.

Larmor radius, the viscosity for the flow pattern is effectively enhanced, damping the growth rate (so-called finite Larmor radius (FLR) effect). Similarly, the diamagnetic drift of ions modifies stability as well because it provides a differential motion between the plasma mass flow and the motion of the magnetic structure of the mode, which is frozen into the electrons. This means that effectively, growth rates that are below the ion diamagnetic drift frequency will be suppressed[3]. However, most important, kinetic effects due to resonances between particle orbits and the mode pattern have a pronounced effect on stability and hence must be considered for the description of (1,1) stability. Similar to the kinetic effects influencing RWM stability discussed in Section 7.4, this concerns the particles trapped in the magnetic mirror between low- and high-field side of the tokamak, performing the so-called banana orbit. While a rigorous analysis of kinetic effects is well beyond the scope of the one-fluid MHD model treated in this book, we will outline here the physics elements that have to be considered in deriving a model that adequately decribes the experimental findings.

First of all, it was found that kinetic effects due to the thermal trapped particles have a stabilizing effect on the $m = 1$ mode (Kruskal–Oberman effect). Then, as discussed for the RWM, kinetic effects due to fast (i.e.$v_i \gg v_{th,i}$) trapped ions can have either a stabilizing or a destabilizing effect: the banana orbits of these ions perform a precession motion in toroidal direction due to the finite poloidal field. If this precession is much faster than the rotation of the (1,1) mode, the preceding banana orbits can be seen as a flux-conserving ring around the $q = 1$ surface and a growing mode that displaces the particles has to do work against them[4]. Hence, for

$$\omega_{prec} = W_{ion}/(eR_0 r_1 B_t) \gg \omega_{mode} \tag{11.1}$$

fast particles have a stabilizing effect. This effect is clearly documented in present-day experiments where ion cyclotron resonance heating (ICRH) can create ions that are fast enough to significantly enlarge τ_{ST}.

An example is shown in Figure 11.3, where the so-called Monster sawteeth are created in JET using ICRH, their sawtooth period being more than an order of magnitude longer than without ICRH. For high enough power, sawteeth can be suppressed for the duration of the whole ICRH pulse. In addition, it can be seen that after ICRH is turned off, there is a time lag of 60–80 ms before the sawteeth reappear. This time is consistent with the slowing down time of the fast ions that are responsible for the suppression. In future reactor grade experiments, it is expected that the fusion born α-particles will play an important role in stabilizing sawteeth.

Conversely, if $\omega_{prec} \approx \omega_{mode}$, the precession of the orbits can be in resonance with the (1,1) mode and actually excite it. This destabilizing effect leads to the so-called fishbone instability treated in the Section 11.2.

3) This effect is often incorporated into stability criteria by replacing γ^2 by $\gamma(\gamma - \omega_i^*)$.
4) This argument is equivalent to invariance of the magnetic flux through the banana, also known as the *third adiabatic invariant*.

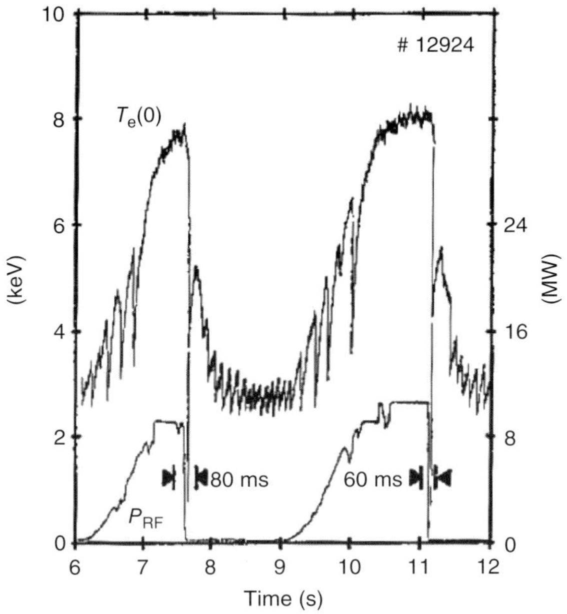

Figure 11.3 Suppression of sawteeth by fast ion created by ion cyclotron resonance heating on the JET tokamak For the highest ICRH power, sawteeth are absent for the full duration of the ICRH phase. *Source:* Porcelli 1991 [62], reproduced with permission by IOP.

Considering all these effects, the onset criterion for the (1,1) mode can be written as

$$\delta W_{\text{MHD}} + \delta W_{\text{kin}} + \delta W_{\text{fast}} > \delta W_{\text{crit}} \tag{11.2}$$

where the first term is the ideal MHD contribution, the second the stabilizing effect of the thermal trapped particles, the third the effect of fast ions and the term on the right-hand side includes all the non-ideal effects connected to the physics of the layer at the $q = 1$ surface. For the latter, one has to distinguish several different regimes that lead to a different physical description in each of the regimes. In the case of a collisionless plasma, as will be adequate for ITER, the criterion can be re-expressed as a condition for s_1 the normalized magnetic shear at the $q = 1$ surface [63]:

$$s_1 > s_{1,\text{crit}} \propto \beta^{7/12} \frac{r_1}{L_n} \left(\frac{r_1}{L_p} \right)^{1/6} S^{1/6} \rho_i^{*1/2} \tag{11.3}$$

where L_n and L_p are the density and pressure decay lengths, respectively, $S = \tau_{\text{res}}/\tau_A$ is the Lundquist number already introduced by Eq. (8.5) and ρ_i^* is the ion Larmor radius, normalized to the minor radius of the plasma. This criterion usually predicts well the onset of sawteeth, showing that this part can be explained by linear MHD theory with the adequate non-ideal corrections. However, for a complete predictive model, one also needs to describe the crash in detail as it sets the boundary condition for the slow evolution of the profiles up to the next crash. This will be the subject of Section 11.1.3.

11.1.3
Models for the Sawtooth Crash

Soon after the discovery of the sawtooth cycle, a model was put forward by Kadomtsev that explained the complete expulsion of the flux inside the $q = 1$ surface on a timescale much faster than the timescale for current re-distribution inside the $q = 1$ surface. This model has some similarity to the Sweet–Parker reconnection model developed in Astrophysics and describes a 'driven' reconnection due to the kinetic energy associated with the unstable internal kink. The flow pattern generated by this instability is shown in the lower right of Figure 11.2. It is important to realize that this is different from the Rutherford reconnection described in Section 8.4 as there, the timescale is determined by the resistive dissipation of perturbed helical flux Ψ_1^*, whereas in the Kadomtsev model, we consider the equilibrium helical flux Ψ_0^* to be convected through the X-point forming at the $q = 1$ surface and the timescale is determined by the flow driven by the unstable (1,1) internal mode. Hence, we expect a different timescale than for the Rutherford reconnection, which will be shown below to be a hybrid between Alfvén and resistive timescales for the sawtooth crash.

The geometry used in the Kadomtsev model is shown in Figure 11.4. Figure 11.4a shows the initial helical flux function Ψ_i^*, which, according to its definition (Eq. (7.9)), has an extremum at the $q = 1$ surface as the helical field $B_\theta^* = d\Psi^*/dr$ changes sign there. The Kadomtsev reconnection model constructs

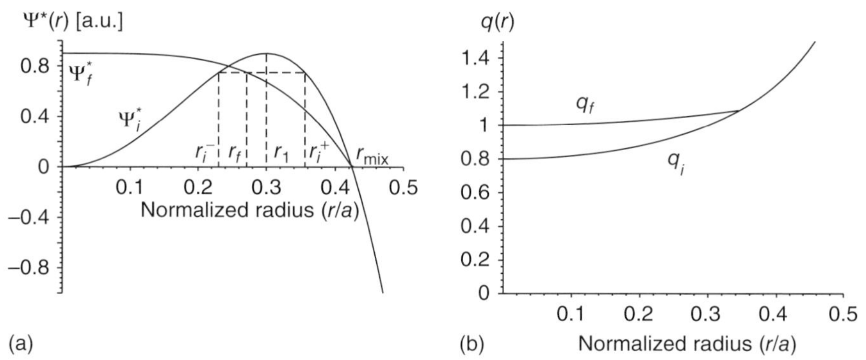

Figure 11.4 Geometry used in the Kadomtsev model for the sawtooth crash. (a) The initial helical flux Ψ_i^* is rearranged to give the final flux Ψ_f^* by joining volume elements with same Ψ_i^* located at r_i^+ and r_i^- to become a new element with $\Psi_f^* = \Psi_i^*$ at $r = r_f$. (b) The initial and final q-profile.

the final helical flux function Ψ_f^*, also shown in Figure 11.4a, through the following process:

- Starting at r_1, flux surfaces with same value Ψ_i^* located at radii r_i^+ outside r_1 and at r_i^- inside r_1 merge to form a new flux surface at radius r_f. As the flux is assumed to be convected through the X-point of the (1,1) mode, the value of Ψ_f^* at r_f is the same as the initial value of Ψ_i^* at r_i^+ and r_i^-. As the volume of the flux surfaces at r_i^+ and r_i^- will in general not be the same, the rule for the combination of the different elements has to be treated differentially, leading to

$$\frac{d\Psi_i^*}{dr}|_{r_i^-} dr = \frac{d\Psi_i^*}{dr}|_{r_i^+} dr = \frac{d\Psi_f^*}{dr}|_{r_f} dr \tag{11.4}$$

  as the rule for conservation of helical flux.
- As only the helical flux is rearranged, the toroidal flux is conserved, which, assuming constant B_t, means that the poloidal area of the plasma elements must be conserved. Noting that as r_i^+ moves outwards, r_i^- moves inwards, this condition leads to

$$r_i^+ \, dr - r_i^- \, dr = r_f \, dr \tag{11.5}$$

  for area conservation[5)]

Applying this rule to the function Ψ_i^* shown in Figure 11.4, one sees that the $q = 1$ surface, where $r_+ = r_-$ will move according to Eq. (11.5), to $r = 0$, that is the initial $q = 1$ surface becomes the final plasma axis. This rule can be applied through decreasing Ψ_i^* values until $r_- = 0$. At this point, all initial flux inside r_+ has been reconnected through the $q = 1$ surface and there is no more r_- to match with r_+. We therefore identify the mixing radius from the condition $\Psi_i(r_{\text{mix}}) = \Psi_i(0) = \Psi_f(r_{\text{mix}})$. Outside r_{mix}, the helical flux function is unperturbed and we have completely mixed the plasma elements inside r_{mix}, explaining the flat kinetic profiles after the crash. We also note that the maximum of Ψ_f^* occurs in the plasma centre, indicating that there, $q = 1$, that is we have also expelled all flux surfaces with $q < 1$. Looking at the corresponding q-profiles on the right-hand side of Figure 11.4, one sees that this is the case. We also note that while Ψ_f^* connects continuously to the (unchanged) Ψ_i^* outside r_{mix} at r_{mix}, the first derivative is not continuous, indicating the existence of a current sheet after the reconnection. This can also be seen from the jump of $q'(r)$ at r_{mix} and is due to the idealized assumption of the flux being purely convected through the X-point while in reality, there will also be reconnection, leading to a resistive decay of the current sheet.

As was pointed out earlier, owing to the convective nature of this reconnection process, its timescale is different from that for Rutherford reconnection. A simple estimate for the timescale can be obtained by the following analysis: consider Ohm's law for the flow v_{in} into the layer at the $q = 1$ surface:

$$j_{\|} = \sigma E_{\|} = \sigma v_{\text{in}} B_\theta^* \tag{11.6}$$

5) In circular geometry, this is readily evaluated to give $r_f^2 = r_i^{+^2} - r_i^{-^2}$.

where, as before, the sheet current $j_{\|}$ can be related to the change of B_θ^* across the layer of width δ by $B_\theta^* \approx \mu_0 j_{\|}\delta$. This can be combined to give an equation for the speed at which the plasma flows into the X-point:

$$v_{\text{in}} \approx \frac{1}{\mu_0 \sigma \delta} \tag{11.7}$$

On the other hand, the outflow of plasma into the newly formed flux surfaces can be thought to be driven by the magnetic pressure of the plasma core moving into the layer, which determines the outflow speed via the Bernoulli equation to be

$$\rho v_{\text{out}}^2 \approx \frac{B_\theta^*}{2\mu_0} \quad \rightarrow \quad v_{\text{out}} \approx v_A \tag{11.8}$$

that is the Alfvèn speed evaluated using the helical field. Geometrically, the flow through the X-point resembles that through a nozzle as the influx is happening through the narrow region δ, whereas the outflux into the newly formed flux surface is along a distance of order r_1, so that continuity requires $v_{\text{in}}/\delta \approx v_{\text{out}}/r_1$, and the limitation of the outflux to the Alfvèn speed leads to a relatively slow influx determining the timescale of the process. This is quite similar to the analysis in Section 8.1. We estimate δ using Eq. (11.7) and replacing v_{in} by the relation just derived so that

$$\delta = \sqrt{\frac{r_1}{\mu_0 \sigma} \frac{1}{v_A}} = \sqrt{\frac{\tau_A}{\tau_{\text{res}}}} r_1 \tag{11.9}$$

Finally, the estimate for the crash time is obtained assuming that it is the time it takes the core to move the distance r_1 at speed v_{in}:

$$\tau_{\text{crash}} \approx \frac{r_1}{v_{\text{in}}} = \mu_0 \sigma \delta r_1 = \sqrt{\tau_A \tau_{\text{res}}} = \sqrt{S}\tau_A \tag{11.10}$$

where S is the Lundquist number defined by Eq. (8.5). As mentioned earlier, this timescale is also known as the *Sweet–Parker timescale.* It is significantly smaller than τ_{res} and was able to explain the experimentally observed crash times at the time the model was derived. However, with more experiments adding a broader set of experimental observations, it became clear that the model in its simplest form, as outlined earlier, could not explain the variety of experimental observations. Specifically, the variation of the crash time and the assumption of complete reconnection were found to be more complex than initially predicted and continue to be subject to active research.

Concerning the crash time, it was found on the larger tokamaks JET and TFTR that the sawtooth crash time did not increase with S as predicted by Eq. (11.10) but showed a much weaker increase. In fact, the crash time seems to saturate with S at values around 100μs, even if S is substantially increased for cases at high temperature in large devices. A variety of modifications have been proposed to the original model in order to account for this experimental observation. For the timescale of the crash, it is obviously necessary to include additional terms in Ohm's law to account for the fast reconnection. However, no complete picture has emerged yet. Recently, numerical simulations using two-fluid MHD have been able to address a realistic range of magnetic Reynolds number S and they indicate that the crash time can saturate with increasing S. This is due to the formation of an X-point geometry instead of a thin current layer which occurs in one-fluid MHD. This finding is hence a very promising candidate to explain the fast crash and the absence of a clear scaling with S in hot plasmas.

Another issue with the Kadomtsev model is the assumption of complete reconnection. Measurements of the temporal evolution of $q(0)$ on the TEXTOR tokamak indicated that it can significantly stay below 1 throughout the whole sawtooth cycle, in contrast to the assumption of complete reconnection. This measurement is consistent with the experimental observation of (1,1) modes being present also directly after the crash, again hinting at what has been dubbed 'incomplete reconnection', that is the reconnection process stops before the whole region with $q \leq 1$ has been expelled. An experimental example from ASDEX Upgrade for such a sawtooth crash with a 'postcursor' oscillation is shown in Figure 11.5. On the other hand, there are also measurements of q_0 during a sawtooth cycle that indicate complete reconnection with $q \geq 1$ after the crash, and 'postcursor' oscillations are not always observed. This suggests that the Kadomtsev model, when an appropriate Ohm's law is used, captures the basic mechanism of the sawtooth crash, but there is a physics element missing that determines the amount of reconnected flux. Such an element could be stochastization due to the interaction of several harmonics of the (1,1) during the crash, which can increase the transport and hence remove the drive during the crash phase, but this has so far not been identified unanimously from neither theory nor experiment.

11.2 The Fishbone Instability

It was described earlier that fast particles trapped on banana orbits can also destabilize the (1,1) internal kink mode if their precession frequency ω_{prec} as given by Eq. (11.1) is of the order of the typical (1,1) mode frequency. In present-day experiments, heating by neutral beam injection (NBI) can lead to a substantial population of fast ions with energies up to 100 keV. Inserting this value into Eq. (11.1), the precession frequencies will be of the order of 10 s of kilohertz, which is of the order of the 'natural' (1,1) mode frequency in NBI-heated plasmas. Hence, a direct resonant interaction between particles and mode is possible if the NBI geometry is such that it creates a large enough gradient of trapped fast particles at the resonant surface, providing the free energy for the mode–particle interaction.

This picture is confirmed in detail by experimental observations on a number of tokamaks, starting from the PDX device in Princeton, where the instability was discovered. In particular, it is found experimentally that fishbones indeed only occur when NBI with sufficient radial injection is used, leading to a large population of trapped fast particles, and that a threshold for the gradient of fast particles at the $q = 1$ surface exists for the instability to occur. The signature of the instability is a growth and decay of a (1,1) mode on the timescale of some milliseconds.

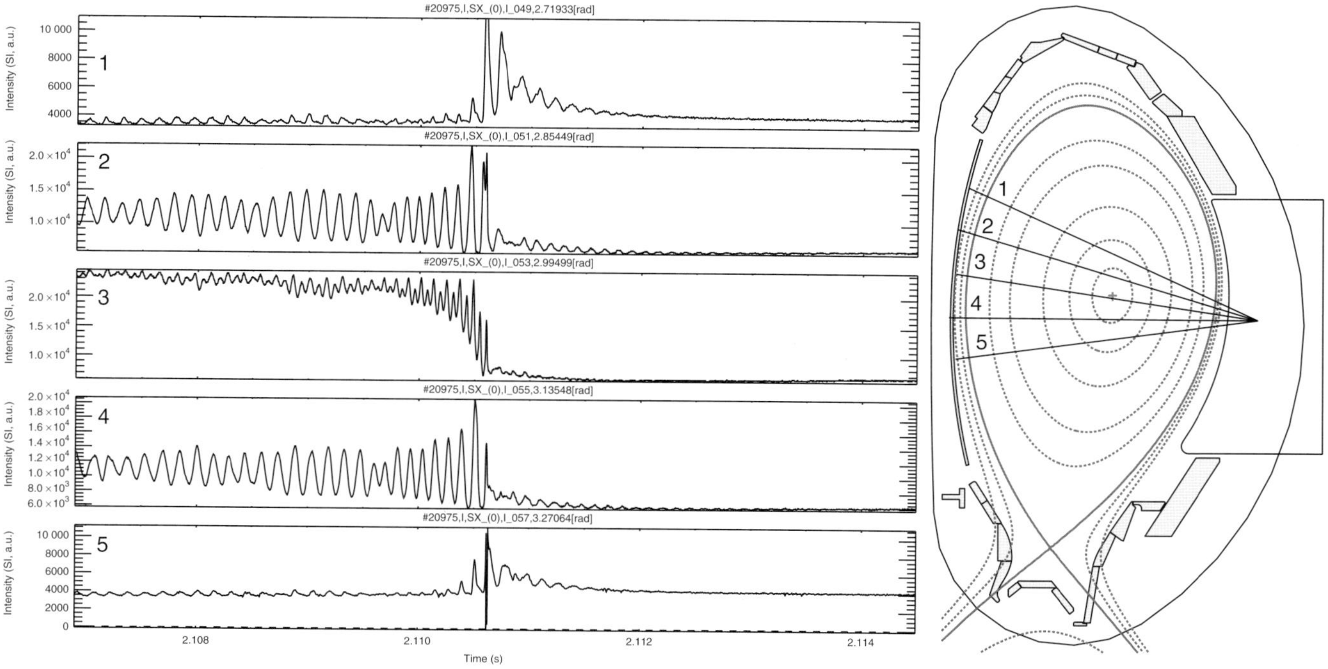

Figure 11.5 Observation of a (1,1) 'postcursor' oscillation following a sawtooth crash on the ASDEX Upgrade tokamak, indicating incomplete reconnection in contrast to the simple Kadomtsev model. Line-integrated signals of the soft-X-ray diagnostic are shown, with the viewing geometry indicated on the right. Here, lines 1 and 5 are outside r_{inv}, lines 2 and 4 roughly coincide with r_{inv} and line 3 views through the centre. Note the suppressed zero of the signals. *Source:* Courtesy of V. Igochine, IPP. (Please find a color version of this figure on the color plates.)

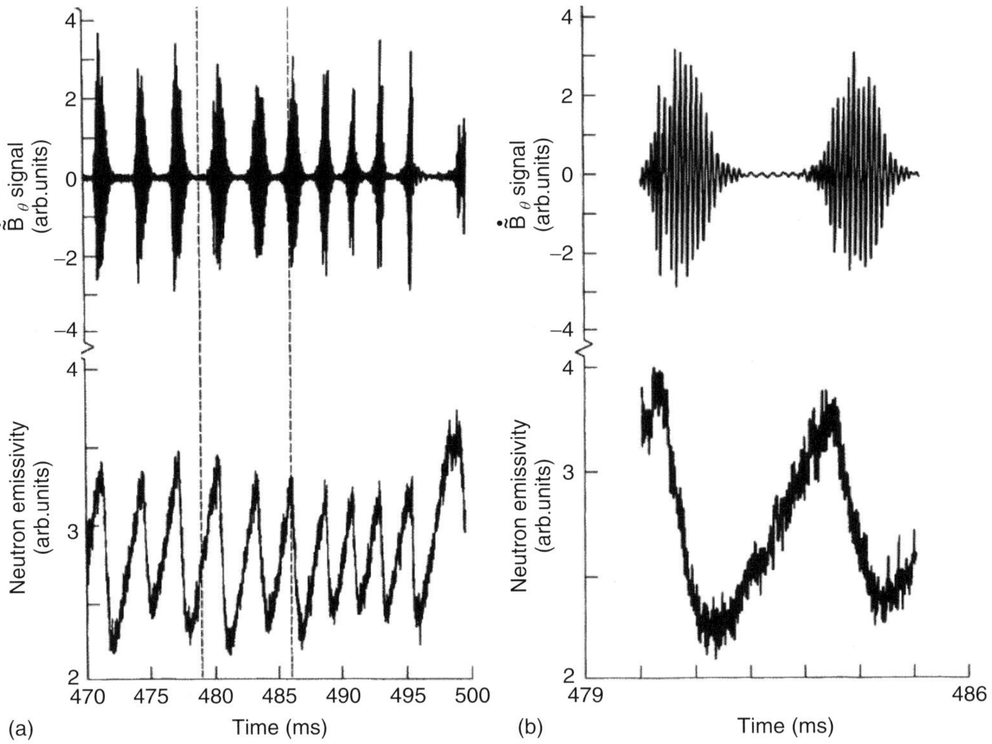

Figure 11.6 Characteristic time traces of the fishbone instability as seen on magnetic pick-up coils (upper trace) and its effect on the neutron rate (lower traces) on the PDX tokamak. (b) An expansion of the time interval from (a). *Source:* Mc Guire *et al.* 1983 [64], reproduced with permission of the APS.

Fourier analysis of the mode frequency shows a characteristic drop in frequency during the burst (so-called chirping), which is interpreted as a change in the resonance condition $\omega_{\text{prec}} = \omega_{\text{mode}}$, that is as the mode grows, it redistributes the fast particles across the $q = 1$ surface, thus removing its drive. The drop of frequency indicates that this process starts with the highest particle energy and then proceeds down to the lower particle energies. An example for fishbones occurring in an NBI-heated tokamak discharge in the PDX tokamak is shown in Figure 11.6 (a). The repetitive character of the instability can be seen, with the characteristic shape of the signal of the magnetic pick-up coils giving the name to the instability. The lower traces in Figure 11.6 show the modulation of the neutron rate arising from the instability. As in present-day NBI-heated experiments, the neutron rate is usually dominated by reactions of the fast beam ions with the target plasma, it is a sensitive indicator for the fast ion content, in this case proving the expulsion of fast beam ions from the plasma core. We note that in the time-expanded traces on the right of Figure 11.6, it is also evident that, as described above, the mode frequency decreases during a fishbone burst due to the change of the resonance condition as the energy of the affected fast ions decreases.

Finally, we mention here that also this non-linear cycle can be described by the simple model presented in Section 6.3.1, this time with the fast particle pressure gradient across the $q = 1$ surface as the driver. Contrary to the ELM and sawtooth cycle, the timescales of mode growth and restoring of driving force are not too disparate and hence the typical time traces is not a sharp sawtooth but rather a modulation[6].

6) The fast particles pressure gradient is restored by the collisions of the fast particles with the background electrons and ions, leading to typical timescales of milliseconds to 10 s of milliseconds.

12
Tearing Modes in Finite β-Tokamaks

We now include the effects of finite β and toroidicity in the analysis of tearing modes. As for the ideal case, both effects should be included together in order to arrive at a consistent description. As noted in Chapter 8, the Rutherford equation can be modified to include any non-inductive current that will appear, properly averaged over the island flux surfaces, as additional term on the right-hand side of Eq. (8.63). Hence, in this chapter, we discuss the contribution of non-inductive helical currents arising from finite pressure and toroidicity, leading to what is known as the *modified Rutherford equation* [65]. They give rise to a new branch of the tearing mode instability that is essentially due to finite pressure, the neoclassical tearing mode (NTM).

12.1
The Modified Rutherford Equation

Including finite β and toroidicity, the following non-inductive currents will have to be included in the derivation of the modified Rutherford equation:

- The bootstrap current due to the pressure gradient of trapped particles that gives rise to a mostly toroidal current carried by the circulating electrons.
- The Pfirsch–Schlüter current that guarantees $\nabla \cdot \mathbf{j} = 0$ in a toroidal plasma leading to a force free current. In tearing mode theory, this term is usually referred to as the *Glasser–Greene–Johnson* term [66].

While these two terms govern the behaviour at large island width, a number of additional terms that describe the physics at small island width exist and two of them are discussed in Section 12.2.

For the bootstrap current (see also Section 7.1), we have

$$j_{bs} \propto \sqrt{\frac{r}{R_0}}\frac{\nabla p}{B_\theta} \approx -\sqrt{\frac{r}{R_0}}\frac{1}{L_p}\frac{p}{B_\theta} \tag{12.1}$$

where we have defined the pressure scale length as $L_p = -p/(dp/dr)$ assuming a monotonically decreasing pressure profile so that $L_p > 0$.

Magnetohydrodynamic Stability of Tokamaks, First Edition. Hartmut Zohm.

Using the same definition also for the magnetic shear length, that is $dq/dr = q/L_q$, the Glasser–Greene–Johnson contribution can be shown to read [67]

$$j_{GGJ} \propto \frac{r}{R_0^2}\left(1-\frac{1}{q^2}\right)\frac{L_q}{L_p}\frac{p}{B_\theta} \tag{12.2}$$

Both terms can be combined and inserted into the Rutherford equation which becomes

$$\frac{\tau_R}{r_{\text{res}}}\frac{dW}{dt} = r_{\text{res}}\Delta'(W) + c_{\text{sat}}\frac{r_{\text{res}}^{3/2}}{R_0^{1/2}}\frac{L_q}{L_p}f_{\text{GGJ}}\beta_p\frac{1}{W} \tag{12.3}$$

where the coefficient $c_{\text{sat}} = 4.63$ can be obtained in the cylindrical constant ψ approximation by appropriate averaging over the island and is related to the saturated island width at finite β (see below). Furthermore,

$$f_{\text{GGJ}} = 1 - c_{\text{GGJ}}\frac{r^{1/2}}{R_0^{3/2}}\left(1-\frac{1}{q^2}\right)L_q \tag{12.4}$$

is the correction due to the Glasser–Greene–Johnson term. One can see that for positive shear, the bootstrap contribution will be destabilizing while the Glasser–Greene–Johnson term is always stabilizing. For small aspect ratio, this can be an important effect as the coefficient c_{GGJ} is of order unity. However, for usual parameters, the bootstrap contribution will dominate and finite pressure effects can be destabilizing. This leads to the occurrence of the so-called NTM discussed in Section 12.2. As mentioned earlier, several additions to Eq. (12.3) have been proposed to remove the singular behaviour at small island width and some of them are discussed in the following section.

12.2 The Neoclassical Tearing Mode (NTM)

As discussed in Section 12.1, including the bootstrap current into the tearing mode equation leads to an unstable situation. This can be understood as follows: suppose an initial perturbation leading to a small island on a rational surface. Owing to the flattening of the kinetic profiles in the island as discussed in Chapter 9, the bootstrap current density will be reduced in the island, effectively creating a helical hole in the bootstrap current at the resonant surface. This helical deficit current will, for positive shear, increase the width of the magnetic island, which, in turn, increases the magnitude of the deficit current[1]. This unstable situation is the basic mechanism of the NTM. In some sense, this can be regarded as a pressure driven tearing mode as the bootstrap current deficit arises from the finite pressure gradient at the resonant surface. This interpretation is in line with

1) We remind of the finding from Chapter 8 that for positive shear, the current creating an island is antiparallel to the plasma current in the O-point. For negative shear, the situation is reversed and hence stable.

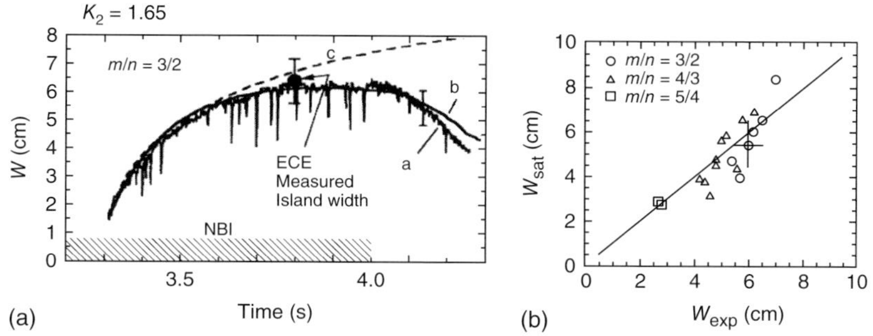

Figure 12.1 Typical time trace for the evolution of the magnetic island width W of an NTM in TFTR (a, experimental data labelled 'a'), shown together with the evolution modelled using the modified Rutherford equation (curves labelled 'b' and 'c', where 'c' considers the measured time dependence of β_p). (b) Comparison of saturated island width for different cases with model prediction (Eq. (12.5)). *Source:* Chang *et al.* 1995 [68], reproduced with permission of APS.

the observation that NTMs often grow to a saturated state which, according to Eq. (12.3), with finite pressure drive, is only possible if Δ' is negative, that is the plasma is stable to classical, current gradient driven tearing modes. The saturated island width W_{sat} is then given by $dW/dt = 0$ as

$$W_{sat} = c_{sat} f_{GGJ} \sqrt{\frac{r_{res}}{R_0} \frac{L_q}{L_p} \frac{\beta_p}{(-\Delta')}} \tag{12.5}$$

NTMs are observed frequently in tokamak experiments at sufficient β_N and can occur well below the ideal Troyon limit. The confinement reduction in the presence of NTMs is usually consistent with the belt model, that is it can approximately be described by Eq. (9.5), and the fact that the saturated amplitude grows with β_N means that the occurrence of NTMs results in a 'practical' β-limit at which the confinement degrades such that achieving higher β_N can require substantially higher heating power than in the case without NTM. The occurrence of large (2,1) NTMs can lead to mode locking and disruptions, representing a 'hard' limit in this case. Figure 12.1 shows typical time traces for a (3,2) NTM (Figure 12.1a) and results from an analysis of saturated amplitude W_{sat} against the island size predicted by Eq. (12.5) in Figure 12.1(b). Good agreement can be seen, although usually, c_{sat} is used as a fit parameter (held constant for the whole set of data points shown in Figure 12.1) as the numerical value given above has been calculated in an idealized case and also Δ' cannot be determined experimentally with sufficient accuracy. The correlation of W_{sat} with β_p is usually taken as indicator of the neoclassical character of a tearing mode[2)].

While the saturated state of NTMs is well described by Eq. (12.3), this equation obviously predicts instability for arbitrarily small island width due to the $1/W$ dependence of the bootstrap drive. Experimentally, however, NTMs only occur

2) NTMs can also be distinguished from classical ones by their different initial growth behaviour $W(t) \sim \sqrt{t}$ for an NTM as opposed to $W(t) \sim t$ for a classical tearing mode.

above a critical β_p, while below they are stable. It is hence clear that the Rutherford equation has to be further modified concerning its behaviour at small island width. One obvious modification is to account for the incomplete flattening of the pressure profile when the island is so small that the radial heat flux can compete with the parallel one around the island that was discussed in Section 9.1. In this case, the bootstrap current within the island will no longer be reduced and the neoclassical drive will vanish as $W \ll W_c$, where W_c is given by Eq. (9.4).

Another term that has been conjectured to be important at small island width is due to the above-mentioned neoclassical polarization current. Assuming that the island moves with respect to the plasma rest frame at speed $v_{\mathrm{NTM}} = \omega_{\mathrm{NTM}} R$, there will be a time-dependent radial electric field E_r affecting the banana orbits. This field would mainly lead to a drift in the direction of $\mathbf{E} \times \mathbf{B}$. However, owing to the conservation of canonical angular momentum, such a drift is not possible and hence the banana orbits will experience a net toroidal velocity E_r/B_θ that compensates the poloidal component. As a result, there will be a net current resulting from the electron return current compensating the drift of the particles in the time-varying E-field. The exact calculation of this current is quite complex, but in large aspect ratio circular geometry, it can be written as [69]

$$j_{\|} \propto \left(\frac{r}{R_0}\right)^{3/2} \frac{L_q}{L_p^2} \frac{p}{B_\theta} \frac{\omega_{\mathrm{NTM}} \left(\omega_{\mathrm{NTM}} - \omega_i^* \left(1 + \frac{L_n}{L_{T_i}}\right)\right)}{\omega_i^{*2}} \left(\frac{\rho_{\theta i}}{W}\right)^2 \tag{12.6}$$

where ω_i^* is the ion diamagnetic drift frequency. It can be seen that depending on the direction of mode rotation in the flusid frame, the term can be positive and give rise to a stabilizing contribution. Owing to its strong W dependence at small island size, it is also a candidate to explain the threshold behaviour. Furthermore, the form of the polarization current given in Eq. (12.6) has been derived in the 'collisionless' limit, that is $\nu_{ii}/(\omega_e^* r/R_0) \ll 1$, whereas in the opposite limit, it will be larger by a factor of $(R_0/r)^{3/2}$, giving rise to a much larger contribution. From this term, one expects hence a strong collisionality dependence of the onset β of NTMs. However, a quantitative prediction of the magnitude of the effect is at present not possible and we mention here that several other contributions to the modified Rutherford equation have been proposed to describe the behaviour at small island width, such as the reduction of bootstrap current once the island size becomes comparable to the banana size [70].

Including both effects into the modified Rutherford equation (Eq. (12.3)) leads to

$$\frac{\tau_R}{r_{\mathrm{res}}} \frac{dW}{dt} = r_{\mathrm{res}} \Delta'(W) + r_{\mathrm{res}} \beta_p \frac{L_q}{L_p} \times \left(c_{\mathrm{sat}} f_{\mathrm{GGJ}} \sqrt{\frac{r_{\mathrm{res}}}{R_0}} \frac{W}{W_0^2 + W^2} - c_{\mathrm{pol}} \left(\frac{r_{\mathrm{res}}}{R_0}\right)^{3/2} \frac{L_q}{L_p} \frac{\rho_{\theta i}^2}{W^3} \right) \tag{12.7}$$

where we have assumed that $\omega_{\mathrm{NTM}} \approx \omega_e^*$ and $W_0 = 1.8 W_c$.

This equation leads to interesting stability properties that are visualized in Figure 12.2.

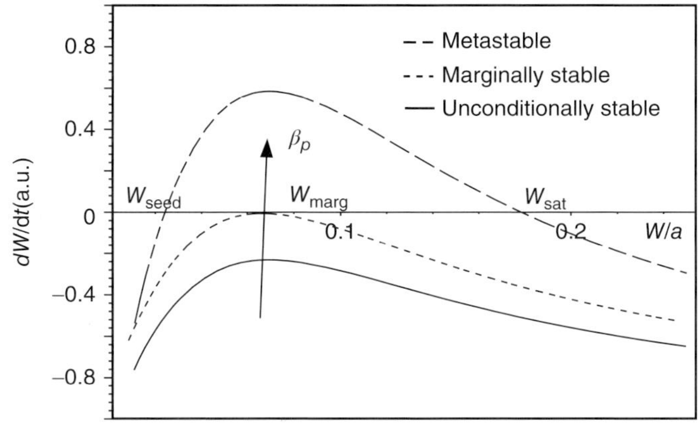

Figure 12.2 Schematic stability diagram for NTMs, indicating the existence of unconditional stability (no stationary point), marginal stability (one stationary point), and metastability (two stationary points of which only W_{sat} is stable). For simplicity, only W_0 has been included as small island term for this plot.

First, owing to the presence of the small island effects, dW/dt becomes negative for small islands, that is the NTM is linearly stable. Second, there is a critical value, the marginal $\beta_{p,\text{marg}}$, below which dW/dt is always negative, that is unconditionally stable. For $\beta > \beta_{p,\text{marg}}$, two values of W exist at which $dW/dt = 0$. The larger one is the saturated island width W_{sat} discussed earlier. This is a stable stationary point as for $W > W_{sat}$, $dW/dt < 0$ and the island will return to W_{sat}. However, for the smaller value, $dW/dt > 0$ if W is slightly larger, and hence, this point is an unstable stationary point. Physically, this means that once an island of this size is created, it will grow to the saturated island size, but islands below that size will decay. This point is hence called the *seed island width* W_{seed}. For $\beta_p > \beta_{p,\text{marg}}$, the NTM is metastable, that is can be excited by a finite seed perturbation. Finally, assuming that an NTM is triggered at finite W_{seed} when $\beta_{p,\text{trigger}} > \beta_{p,\text{marg}}$, it will exist in the range $\beta_{p,\text{marg}} < \beta_p < \beta_{p,\text{trigger}}$, that is there is a hysteresis in β_p for the existence of NTMs.

All the features described earlier are consistent with the experimental observations of NTMs in various tokamaks, making the Modified Rutherford Equation a very successful model for NTMs, with the above-mentioned caveat that the uncertainty in the small island physics limits its predictive capability for describing the seeding process and, as will be shown in Chapter 13, also for the description of its suppression by external current drive. Figure 12.3 shows an experimental example of triggering NTMs, in this case by a sawtooth crash. During the sawtooth crash (see also Chapter 11), a magnetic perturbation can be observed. It can be seen on the enlarged time traces that the predominantly odd ($n = 1$) perturbation also has an even ($n = 2$) component that grows before the sawtooth crash together with the ($n = 1$) component. This ($n = 2$) component will generate the seed for a (3,2) island at the $q = 1.5$ surface. Obviously, in the first time frame shown, the seed is not large enough to trigger an NTM and after the sawtooth crash, the

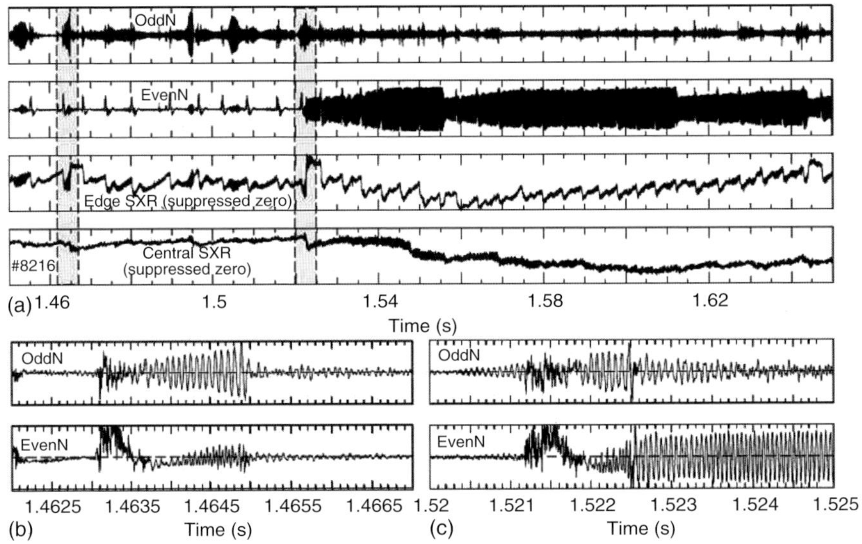

Figure 12.3 Seeding of a (3,2) NTM by the ($n = 2$) component of a sawtooth crash on ASDEX Upgrade. (a) The upper traces show measurements of the magnetic perturbation with odd and even toroidal mode number n together with measurements of the soft-X radiation from plasma edge and centre. (b and c) The two expanded panels show the time frames of two sawtooth crashs creating $n = 1$ and $n = 2$ seed perturbations. In (c), this triggers a (3,2) NTM. The more frequent modulation seen on edge soft-X radiation and even n magnetic perturbation is due to the occurrence of ELMs in this discharge. *Source:* Zohm *et al.* 1997 [71], reproduced with permission of IOP.

$n = 2$ component decays together with the $n = 1$ component. In the second time frame shown, however, the $n = 2$ perturbation remains while the $n = 1$ perturbation decays after the sawtooth crash and a (3,2) NTM persists.

A systematic study of the triggering process shows that, in line with the physics picture developed earlier, the onset value $\beta_{p,\mathrm{onset}}$ exhibits a large scatter, as $\beta_{p,\mathrm{onset}} > \beta_{p,\mathrm{marg}}$ is a necessary but not sufficient condition for NTMs to occur. At higher β_p, W_{seed} will be smaller, and hence, we expect $\beta_{p,\mathrm{onset}}$ to depend on the strength of the seed perturbation. Indeed, one finds experimentally that large sawtooth crashes can trigger NTMs at relatively low β_p, whereas in sawtooth-free plasmas, triggering of NTMs by fishbones or ELMs usually occurs at higher β_p. On the other hand, in experiments where the heating power is gradually reduced to find the value of β_p at which an NTM decays, there is little scatter as this value will always be very close to $\beta_{p,\mathrm{marg}}$.

12.3 Onset Criteria for NTMs

The discussion earlier highlights a problem when trying to predict the onset of NTMs: while the value of $\beta_{p,\mathrm{marg}}$ can be calculated from the modified Rutherford equation, the exact onset will depend on details of the triggering process, which is generally not easy to predict. One possible assumption to overcome this problem is to assume that in similar plasma regimes, $\beta_{p,\mathrm{onset}}/\beta_{p,\mathrm{marg}}$ = const.. For usual H-mode operation, this assumption is reasonably justified by experimental findings, as shown in Figure 12.4. However, one must keep in mind that the occurrence of a single 'off-normal' large MHD event can trigger NTMs at lower β_p than deduced by the assumption earlier.

In principle, the value of $\beta_{p,\mathrm{marg}}$ can be calculated in a straightforward way from the modified Rutherford equation. However, it does involve the uncertainty about the small island terms discussed earlier and it is not clear if one of them is dominating or several of them have to be considered. As can be seen in Figure 12.4, the value of $\beta_{p,\mathrm{marg}}$ usually exhibits a linear scaling with ion poloidal gyroradius $\rho_{\theta i}$ [73]. Such a scaling can indeed be derived for the polarization current model, where, neglecting W_0 for the moment, the criterion for marginal stability gives

$$\beta_{p,\mathrm{marg}} = \frac{3\sqrt{3}}{2}\frac{\sqrt{c_{\mathrm{pol}}}(-\Delta')\rho_{\theta i}}{(c_{\mathrm{sat}}f_{\mathrm{GGJ}})^{3/2}}\sqrt{\frac{L_p}{L_q}} \tag{12.8}$$

On the other hand, neglecting the polarization current and keeping only W_0 results in

$$\beta_{p,\mathrm{marg}} = 2W_0\sqrt{\frac{R_0}{r_{\mathrm{res}}}\frac{L_q}{L_p}\frac{(-\Delta')}{c_{\mathrm{sat}}f_{\mathrm{GGJ}}}} \tag{12.9}$$

that is in order to derive a scaling, we have to make an assumption about the parameter dependence of perpendicular and parallel heat transport over the island. If we assume that the perpendicular transport is anomalous, as is the case without island, we can use a gyro-Bohm scaling, $\chi_\perp \sim T^{3/2}/B^2$. For the parallel transport, one can use the Spitzer parallel heat conductivity $\chi_\parallel \sim T^{5/2}/n$ as long

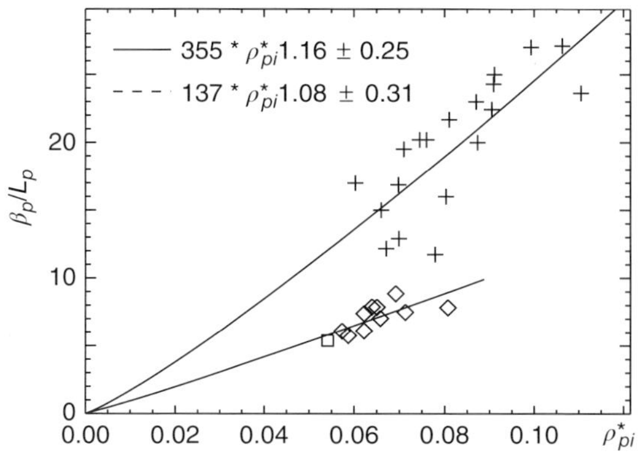

Figure 12.4 Experimental results for $\beta_{p,\mathrm{onset}}$ and $\beta_{p,\mathrm{marg}}$ for (3,2) NTMs in H-mode discharges in ASDEX Upgrade. As the local drive comes from the pressure gradient, the β values have been normalized by the pressure scale length L_p. *Source:* Maraschek *et al.* 2003 [72], reproduced with permission of IOP.

as the plasma is collisional, that is the mean free path is short compared to the connection length, but in the opposite limit, heat transport will be limited by the free flow of electrons along field lines and the scaling will rather be $\chi_{\|} \sim T^{1/2}$. Using the latter, we obtain

$$W_0 \propto \left(\frac{\chi_\perp}{\chi_\|}\right)^{1/4} \propto \left(\frac{T}{B^2}\right)^{1/4} \propto \sqrt{\rho_{\theta i}} \tag{12.10}$$

that is the scaling obtained from this expression is weaker than that observed experimentally.

On the other hand, the observation of $\beta_{p,\text{onset}}/\beta_{p,\text{marg}} \approx$ const. implies a decrease in the necessary seed island size with increasing β_p if there is no β dependence of the seeding mechanism itself. Such dependence can be derived neglecting the polarization current but keeping W_0, which, for $W_0 \ll W_{\text{sat}}$, leads to

$$W_{\text{seed}} \approx \frac{W_0^2}{W_{\text{sat}}} = W_0^2 \frac{(-\Delta')}{\beta_p c_{\text{sat}} f_{\text{GGJ}}} \sqrt{\frac{R_0}{r_{\text{res}}} \frac{L_p}{L_q}} = W_0 \frac{1}{2} \frac{\beta_{p,\text{marg}}}{\beta_p} \tag{12.11}$$

Conversely, the expression for W_{seed} using only the polarization current leads to

$$W_{\text{seed}} = \rho_{\theta i} \sqrt{\frac{r_{\text{res}}}{R_0} \frac{L_q}{L_p} \frac{c_{\text{pol}}}{c_{\text{sat}} f_{\text{GGJ}}}} \tag{12.12}$$

which does not depend on β_p, implying that an NTM is triggered as soon as $\beta_p > \beta_{p,\text{marg}}$. This is in contrast to the wide hysteresis between the onset and the marginal β often observed in the experiment as well as the observation that larger seed islands trigger NTMs at lower β_p.

It was mentioned earlier that experimentally, a 'threshold' behaviour in collisionality is found such that NTMs are preferentially triggered at low collisionality, with no explicit collisionality dependence once it is low enough. It was already outlined earlier that the polarization term exhibits such a switch between collisional and collisionless case. Similarly, for the W_0-term, such a behaviour was found above assuming that the perpendicular transport is gyro-Bohm like and the parallel transport is heat flux limited, whereas in the collisional case, $W_0 \sim \nu*^{1/4}$. It follows that we cannot easily determine a single small island mechanism that would explain all experimental observations. Rather, it is likely that several terms play a role in determining the β_p dependence.

Furthermore, studies of NTM onset in plasmas where the torque input was varied point towards a lower onset β with low rotation. This is shown in Figure 12.5. While this figure shows a large scatter in the value of β at which NTMs occur, consistent with the idea of a metastability window, it is clear that the maximum value that can be reached before an NTM is triggered substantially decreases with decreasing rotation and exhibits a minimum around zero rotation. Here, the rotation has been normalized to the Alfvén speed at the resonant surface. Clearly, the rotation dependence of NTM onset cannot be described by the theory outlined earlier which did not consider rotation and might either be due to a direct dependence of Δ' on rotation (or its shear) or there could be an effect of

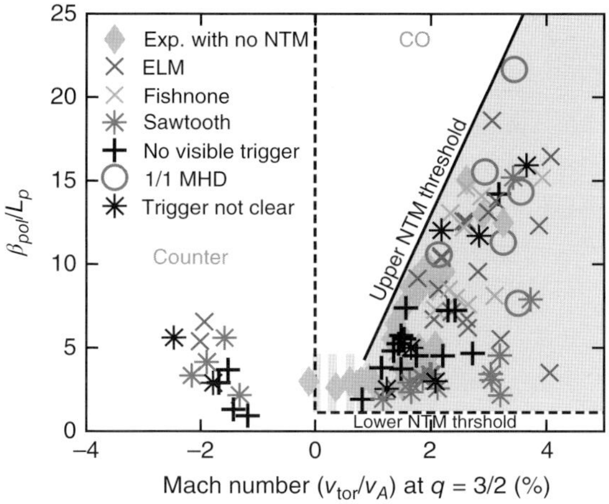

Figure 12.5 Onset β for (3,2) NTMs as function of the normalized rotation of the resonant surface in ASDEX Upgrade. The maximum achievable β has a minimum around zero rotation. The points are labelled according to the triggering event. *Source:* Fietz *et al.* 2013 [74], reproduced with permission of IOP. (Please find a color version of this figure on the color plates.)

differential rotation on the triggering mechanism, although in Figure 12.5, such a mechanism is not obvious (the different symbols refer to different trigger mechanisms as explained in the box).

We note that the scaling outlined earlier leads to a pessimistic prediction for future reactor-grade fusion devices such as ITER and DEMO: while the saturated island size scales with machine size, the onset condition scales with $\rho_{\theta i}$, which decreases relative to the machine size in future devices. In addition, the low normalized rotation that is expected in these devices can also lower the maximum achievable β without NTMs, as shown in Figure 12.5. It is hence likely that NTMs are metastable at very low β_p in these devices, and active means for their suppression should be foreseen, at least for the (2,1) mode due to the danger of disruptive discharge termination when this mode is present. Active control is treated in Chapter 13. Alternatively, one may explore the possibility of operating in a regime of 'acceptable' NTMs. Such a regime is presented in Section 12.4.

12.4 Frequently Interrupted Regime (FIR) NTMs

In several tokamaks, it has been observed that NTMs can either grow to saturated island size or exhibit a regime in which the growth of the NTM is frequently interrupted, leading to sudden (of the order of 1 ms) decrease of the amplitude and then a consecutive growth until the next interruption of the growth. That way, the

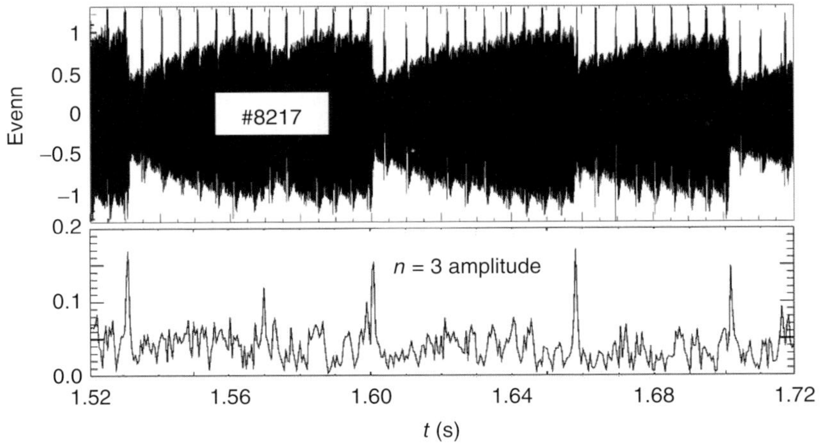

Figure 12.6 Magnetic perturbation signals characterizing the dynamics of the FIR (frequently interrupted regime) NTM regime in ASDEX Upgrade. The growth of the (3,2) NTM (signal of a combination of pick-up coils representing the even toroidal mode number shown in the upper panel) is regularly interrupted due to interaction with an ideal (4,3) mode (envelope of the signal representing the odd toroidal mode number shown in the lower panel). The regular spikes on the signals are due to the occurrence of ELMs in this discharge. *Source:* Gude *et al.* [75], reproduced with permission of the IAEA.

NTM never reaches its full saturated size and its average amplitude can be considerably lower than W_{sat}, thereby also leading to a smaller decrease of confinement relative to the state without NTM.

An example is shown in Figure 12.6. It can be seen that each of the drops in (3,2) NTM amplitude is correlated with a burst of $n = 3$ activity. More generally, an analysis of the MHD modes involved in the fast drop of the (m, n) NTM amplitude in the FIR shows that the amplitude drops of the (m, n) NTM are related to the sudden growth of an $(m + 1, n + 1)$ mode, that is on a resonant surface close to, but just inside the NTM resonant surface. From the growth rate (below 1 ms^{-1}), it is clear that this mode must be an ideal mode. Considering the local flattening of the current profile induced by the NTM together with the strong pressure gradient observed just adjacent to the island, this mode is thought to be an infernal mode as discussed in Section 7.2, and linear ideal MHD stability analysis supports this idea. The sudden drop in amplitude coincides with a rapid flattening of the pressure gradient at the NTM resonant surface. A possible explanation for this sudden drop is the onset of stochastization due to the coupling of the ideal $(m + 1, n + 1)$ infernal mode and the (m, n) NTM. This hypothesis is supported by the observation that the drop only occurs when the modes rotate in a phase-locked position, that is $\omega_{\mathrm{rot},infernal}/(n + 1) = \omega_{\mathrm{rot},NTM}/n$, which has been identified as a necessary condition for stochastization in Section 9.1.

As the infernal mode is pressure driven, the FIR occurs at higher β_N than the conventional regime in which saturated NTMs are usually seen. Figure 12.7 shows a comparison of the occurrence of FIR NTMs and saturated NTMs in ASDEX

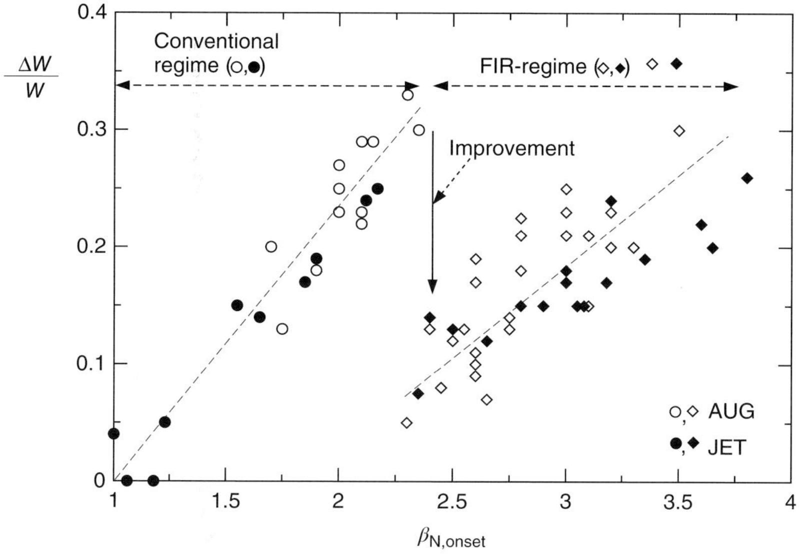

Figure 12.7 Occurrence of the FIR of NTMs in ASDEX Upgrade and JET as function of β_N. In the FIR, the confinement degradation due to the mode, $\Delta W/W = (W_{w/o\ NTM} - W_{w/\ NTM})/W_{w/o\ NTM}$ drops to a low value just at the transition between the two regimes. *Source:* Courtesy of M. Maraschek (IPP).

Upgrade and JET. The transition between the conventional NTM and the FIR NTM regimes can be seen from the abrupt change in the loss of confinement induced by the NTM, which is plotted on the ordinate of Figure 12.7. For both devices, the transition between the two states occurs around $\beta_N = 2.3$, confirming that this is a relatively robust feature, favouring H-mode operation in the range above $\beta_N = 2.3$ for conventional H-modes. Experiments aiming at triggering the FIR by decreasing the local shear at the $(m+1, n+1)$ surface have also shown some success so that the transition into this regime may also be actively controlled in future machines, for example by ECCD as is discussed in Chapter 13.

13
Control of Resistive MHD Instabilities by External Current Drive

In the previous chapters, we have described the stability properties of tokamak discharges with respect to the occurrence of MHD instabilities. In principle, these instabilities can be avoided by properly choosing the operational point, but the tendency to maximize fusion performance usually pushes it close to the stability limits described earlier. It was mentioned before that in general, MHD instabilities due to an ideal MHD limit occur on the Alfvén timescale, making it difficult to apply active control techniques, and hence, conducting wall elements are introduced to slow down the growth rate such that the instability can be controlled. A prominent example of such a scheme is the RWM control by active feedback coils described in Section 7.4. On the other hand, resistive MHD instabilities occur on the much longer resistive timescale, making them directly accessible to active feedback control. In this chapter, we discuss how resistive MHD stability can be actively controlled by external current drive, a method that is especially interesting as it can be applied locally, that is in the vicinity of the resonant surface of the mode of interest, minimizing both the impact on the global discharge characteristics and the power requirements. Two main schemes can be distinguished, namely the local change of the equilibrium current distribution, which results in a change of the magnetic shear at the resonant surface, and the generation of a helical current within a magnetic island. The latter is due to the equilibration of the externally driven current on the island flux surfaces and can hence be generated by applying a current drive source that is distributed uniformly over the helical angle, but its effect can be enhanced by restricting the current drive to a certain range of helical angle, usually around the island's O-point.

As will become apparent in the remainder of the chapter, the ability to generate a localized current with good control of the radial location is crucial to the success of the control schemes for resistive MHD instabilities. In this chapter, we focus on electron cyclotron current drive (ECCD), which possesses these properties and has so far been used most successfully for the control of resistive MHD instabilities using external current drive. We note however that the stabilization schemes described will work equally well with other methods that can generate this localized additional current.

Magnetohydrodynamic Stability of Tokamaks, First Edition. Hartmut Zohm.

13.1 Basic Properties of Localized Electron Cyclotron Current Drive (ECCD)

ECCD generates current by asymmetrically distorting the electron distribution function due to resonant interaction of an injected wave with electrons gyrating at the electron cyclotron frequency $\omega_{ce} = eB/m_e$. For typical magnetic field values in present-day fusion experiments of the order of several Tesla, the wave frequency has to be of order of $\nu_{ECCD} = \omega_{ECCD}/(2\pi) \approx 100 - 200$ GHz, that is in the microwave range with typical wavelength of order millimetre. Hence, the wave can be injected as a Gaussian beam of typical width of several centimetres, usually generated by gyrotron vacuum tubes of source power of the order of 1 MW. If the beam is injected via a movable mirror, as shown in Figure 13.1, it can be steered across the flux surfaces to control the deposition location, that is the region where it is absorbed by the gyrating electrons. This occurs where the resonance condition

$$\omega_{ECCD} = n\omega_{ce} - k_{\|}v_{\|} \tag{13.1}$$

is met for an electron. Here, n is an integer denoting the harmonic at which absorption occurs [1)]. Owing to the variation $B(R) \propto 1/R$, the major radius of the deposition is well defined by the intersection of the microwave beam with the surface on which the resonance condition holds. Although the wave–particle interaction takes place in a poloidally and toroidally localized region, the equilibration on flux surfaces is usually fast (for typical plasma parameters below 1 ms) and the deposition can be thought of as a flux function $P_{ECCD}(\rho)$. Beam steering as shown in Figure 13.1 can even be used in feedback control schemes to control the location, for example to deposit in magnetic islands as shown in Figure 13.1.

From Eq. (13.1), it can be seen that the resonance frequency is Doppler shifted due to the motion of the electrons parallel to the magnetic field. For wave injection from the low field side, as depicted in Figure 13.1, this means that electrons traveling antiparallel to the wave will absorb power already in a region where the magnetic field is lower than would be determined by $\omega_{ce} = \omega_{ECCD}$ (indicated by the blue line in Figure 13.1). Thus, if the beam has a component in the toroidal direction, a preferential heating of electrons traveling in direction $-k_\phi$ will occur. While this heating is only increasing the perpendicular energy of the electrons, it decreases their collision frequency and thus, by the longer slowing down of the hotter electrons with respect to the bulk, generates a net toroidal current. For typical parameters in present-day tokamaks, the Doppler broadening is small and hence the width d of the driven current profile is of the order of typically only few centimeters. Similar to the equilibration of the heating source on the flux surface, also the driven current will become a flux function so that in the following, we will

1) Usually, the plasma is optically thick for electron cyclotron radiation at the first harmonic for injection of the ordinary mode (electric field of the wave parallel to the equilibrium magnetic field) and at the second harmonic for X-mode (E-field perpendicular to equilibrium B).

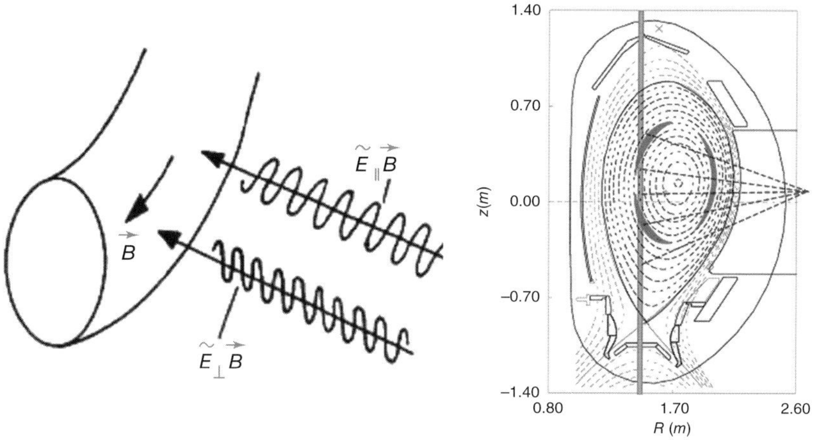

Figure 13.1 Sketch of generating localized current at a location controlled by the injection geometry of a microwave beam at the ECRH frequency. The different beam trajectories correspond to a motion of the injection mirror. (Please find a color version of this figure on the color plates.)

assume that the additional current generated by ECCD at deposition location r_{dep} that is assumed to have a Gaussian distribution

$$j_{\text{ECCD}}(r) = \frac{I_{\text{ECCD}}}{2(\pi)^{3/2} d r_{\text{dep}}} e^{-\frac{\left(r-r_{\text{dep}}\right)^2}{d^2}} \tag{13.2}$$

will lead to an ECCD current density $j_{\text{ECCD}}(\rho)$ where ρ is a flux surface label.

We note here that the local heating can also generate additional current by changing locally the conductivity. As the current drive efficiency increases with electron temperature, this effect will be less important in future reactor grade plasmas but may well play an important role in present- day experiments. The ability of ECCD to generate localized current at a well-controlled radial position makes it particularly well suited for MHD control. This will be discussed in the following sections.

13.2
Criteria for Control of Resistive Instabilities

In this section, we review the criteria for control of resistive MHD instabilities by localized ECCD. The impact on stability can be due to either a change in equilibrium current density or the generation of a helical current inside a magnetic island. Both will be discussed in Sections 13.2.1 and 13.2.2.

13.2.1
Control by Changing the Equilibrium Current Density

In the previous chapters, it was discussed how the stability of the resistive (1,1) mode as well as that of $m \geq 2$ modes depends on the equilibrium current profile.

In both cases, the drive for the instability comes from the gradient of equilibrium current density, and hence, stability can be changed by altering this gradient. For the (1,1) mode, it was outlined that it becomes unstable (and hence, a sawtooth crash will occur), if the shear at the resonant surface exceeds a critical value, that is $s_1 > s_{1,\mathrm{crit}}$ according to Eq. (11.3). In addition, in Section 12.4, the FIR regime of NTMs was discussed, which is due to a combination of small shear and high-pressure gradient driving an ideal mode next to the NTM under consideration. Hence, also here, an additional change in shear by local current drive can be beneficial.

Assuming very localized current drive (as is the case for ECCD) at the resonant surface, the poloidal field will jump by an amount $\delta B_\theta \approx \mu_0\, j_{\mathrm{ECCD}} d \sim \mu_0\, I_{\mathrm{ECCD}}/(r_{\mathrm{dep}})$ (see Eq. (7.7)) and the resulting jump of q is

$$\delta q = \delta\left(\frac{rB_\mathrm{t}}{RB_\theta}\right) \approx -q\frac{\delta B_\theta}{B_\theta} \approx -q\frac{I_{\mathrm{ECCD}}}{I(r_{\mathrm{dep}})} \tag{13.3}$$

where $I(r_{\mathrm{dep}})$ is the total plasma current flowing inside the surface on which ECCD is deposited. This finally leads to a change of shear by

$$\delta s = \delta\left(\frac{r}{q}\frac{dq}{dr}\right) \approx -\frac{r_{\mathrm{dep}}}{d}\frac{I_{\mathrm{ECCD}}}{I(r_{\mathrm{dep}})} \tag{13.4}$$

clearly indicating that the local change of shear can be quite large if d is small enough, that is the ECCD current is narrow.

Figure 13.2 illustrates this. For the examples shown, a driven current of only 5% of the total current leads to a change of shear by roughly 100% for $d/a = 0.025$. Inside the deposition radius, the shear is decreased while just outside, it increases. Note that the q-profile itself changes little; owing to the assumption of conservation of the total current and a constant residual current profile made in the simulation shown in Figure 13.2, the current inside the resonant surface is slightly reduced, leading to an increase in $q(0)$. We also note that the analysis shown in Figure 13.2 is under stationary conditions, that is for timescales long compared to that for resistive diffusion τ_R of the equilibrium current profile inside the resonant surface and hence will only occur on this relatively slow timescale. For timescales $\tau \ll \tau_\mathrm{R}$, the effect will be smaller by roughly a factor τ/τ_R.

Concerning $m \geq 2$ tearing modes, the linear stability is described by the tearing mode equation (Eq. (8.29)), in which the density of the equilibrium current density enters explicitly. Hence, when integrating the tearing mode equation in order to evaluate Δ', an additional term Δ'_{ECCD} due to the added current density j_{ECCD} occurs. For the above-described case of a Gaussian distributed external current, the integration can be carried out analytically. For a classical tearing mode, where Δ'_0, the value evaluated without the additional ECCD current, is positive, the condition $\Delta' = \Delta'_0 + \Delta'_{\mathrm{ECCD}} < 0$ yields [76]

$$\frac{I_{\mathrm{ECCD}}}{I(r_{\mathrm{dep}})} \geq \left.\frac{r_{\mathrm{dep}}}{q}\frac{dq}{dr}\right|_{r_{\mathrm{dep}}} \frac{d^2}{2r_{\mathrm{dep}}^2}\, r_{\mathrm{dep}}\Delta'_0 \tag{13.5}$$

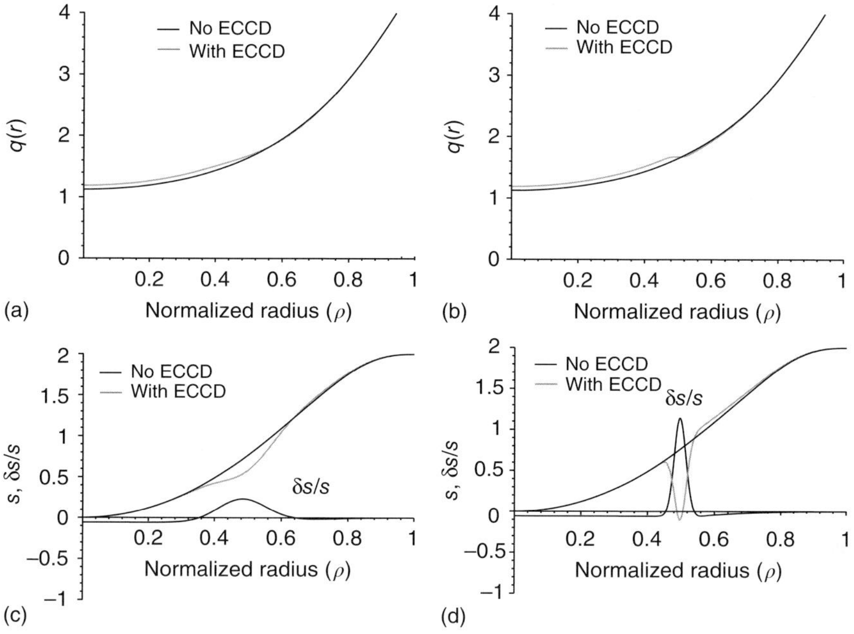

Figure 13.2 Change of shear by localized current drive for different deposition widths. In both cases, the driven current is 5% of the total current, but for (a and c), the deposition width is $d/a = 0.1$ while for (b and d), it is $d/a = 0.025$. While the change of the q-profile (upper figures) is small, the change in shear, as shown by $\delta s/s = (s_{\text{with ECCD}} - s_{\text{w/o ECCD}})/s_{\text{w/o ECCD}}$, can be substantial.

where we have assumed that the current is perfectly localized at the resonant surface, $r_{\text{dep}} = r$. We note here that if this is not the case, the additional current by ECCD will be less efficient in stabilizing the mode and can even have a destabilizing effect if the deposition is off the resonant surface by more than d. For NTMs, where $\Delta_0' < 0$, we note that $\beta_{p,\text{marg}}$ depends linearly on $(-\Delta')$ according to Eqs (12.9) and (12.8), and hence, the region of unconditional stability against NTMs will increase accordingly applying the additional current drive. The condition (Eq. (13.5)) would, for example, increase the marginal β_{p} by a factor of 2. As is the case for the critical shear criterion for sawtooth stabilization, it is not just the total external current that matters, but rather the current density (I_{ECCD}/d), or even its derivative (I_{ECCD}/d^2), that is as outlined in the introduction of this chapter, the highly localized nature of ECCD makes it especially suited for these applications.

We note here that the above-mentioned derivation is for an axisymmetric case, that is in the absence of magnetic islands, relating to a 'pre-emptive' control scheme where the linear mode stability is altered to prevent the mode from becoming unstable. In the case of a pre-existing mode, the situation changes as the driven current will equilibrate on the island flux surfaces. For a 'large' island, that is $W \geq d$, the re-distribution of the driven current will lead to a width larger than d, and the formulae given above are strictly no longer valid. On the other

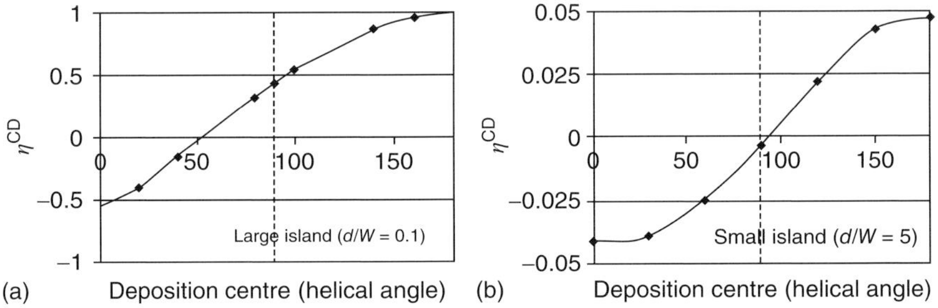

Figure 13.3 Efficiency for generating a helical current by depositing a poloidally localized ECCD beam at helical angle ζ. (a) Small deposition width $d < W$; (b) large deposition width, $d > W$. *Source:* Zohm *et al.* 2007 [77], reproduced with permission of IOP.

hand, the equilibration leads to the generation of a helical current that can be stabilizing. This is discussed in Section 13.2.2.

13.2.2
Control by Generating Helical Currents

In Section 13.2.1, the effect of an additional localized current on the linear mode stability was analysed. Such control schemes can be applied in a pre-emptive manner, that is changing linear stability such that the mode is always stable and does not occur. However, for resistive MHD instabilities involving the formation of islands due to helical currents, another scheme exists that relies on the generation of a helical current within the islands once they have formed. To illustrate how a helical current with helicity of the mode can be created, we assume that we deposit a current filament localized in helical angle and with a radial localization $d < W$ at the rational surface. By equilibration on the island flux surfaces, this will lead to a current distribution $j(\Psi)$, where $\Psi(r, \zeta) = const.$ describes the perturbed flux surfaces. Integrating this current density in the radial direction will lead to a surface current distribution along the helical angle $J(\zeta)$, which, in general, will exhibit a Fourier component of the helicity of the mode. Thus, the equilibration of the current on the island flux surfaces modifies the initial distribution and generates a helical component that can stabilize the mode.

In Figure 13.3, we show the normalized magnitude of this helical component as function of the helical angle where the poloidally localized source is deposited. The efficiency function shown in Figure 13.3, usually denoted as η_{CD}, has been normalized such that it is one for a δ-function in the island's O-point, that is at $\zeta = 180^{\circ}$ [78]:

$$\eta_{CD} = \frac{\int dr \oint d\zeta \cos(m\zeta)\langle j_{ECCD}\rangle}{\int dr \oint d\zeta \langle j_{ECCD}\rangle} \tag{13.6}$$

where $\langle j_{ECCD}\rangle$ is the ECCD current density, averaged over the island flux surfaces according to Eq. (8.60). From Figure 13.3, one can see that ECCD in the X-point will have a destabilizing effect, while it will be stabilizing for injection in the

O-point[2]. An important observation is that for $d < W$, as shown in Figure 13.3a, the function η_{CD} is not perfectly antisymmetric with respect to $\zeta = 90^\circ$. This means that for the case of a mode that is rotating with respect to the laboratory frame, injection of a constant ECCD power, which corresponds to a uniform spread over ζ, will generate a net helical current. Integrating η_{CD} over ζ in the case $d < W$ yields a value of approximately 0.4. If the ECCD power was pulsed to only inject when the phase with respect to the helical angle is stabilizing, a similar value can be achieved as the total (time averaged) ECCD power is reduced accordingly, and in conclusion, modulation of the ECCD power in phase with a rotating island is not necessary in the case $d < W$.

Conversely, if $d > W$, as shown in Figure 13.3b, the curve becomes antisymmetric with respect to $\zeta = 90^\circ$, meaning that without modulation, η_{CD} will approach zero. In this case, we just add a uniform current density that is not changed by the flux surface averaging process and hence just represents a constant 'offset' in current density that does not contribute a helical component. Note that with broad deposition, the current driven within the island is reduced by a factor W/d, leading to a strong reduction of the absolute value of η_{CD} with respect to the case $d < W$ also for the case of modulated CD.

To describe the effect of the helical current generated within the island on stability, we include it in the modified Rutherford equation as outlined in Chapter 8. For the ECCD term, the total current is proportional to $j_{ECCD}d$ and we use the function η_{CD} as defined earlier, which includes the appropriate averaging, to write

$$\frac{\tau_R}{r_{res}}\frac{dW}{dt} = r_{res}\Delta' + \frac{r_{res}L_q}{B_\theta}\left(c_{bs}f_{GGJ}\frac{\mu_0\, j_{bs}}{W} - c_{CD}\eta_{CD}\frac{\mu_0\, j_{ECCD}d}{W^2}\right) \tag{13.7}$$

where for the sake of simplicity, we have neglected the small island terms discussed in Chapter 12. Here, we have explicitly kept the bootstrap current in the neoclassical drive term rather than replacing it expressing it through β_p as was done when discussing NTM stability in Section 12.2 as the following discussion of stability is best done in terms of the non-inductive currents appearing in the modified Rutherford equation.

This equation can be rewritten using the saturated island width in the absence of ECCD, that is Eq. (12.5) which, using the form for the neoclassical drive from above, reads $W_{sat} = c_{bs}f_{GGJ}L_q\mu_0\, j_{bs}/(\Delta' B_\theta)$, to give a requirement for complete stabilization, $dW/dt \le 0$ for arbitrary W:

$$-\frac{W}{W_{sat}} + 1 - \frac{c_{CD}}{c_{bs}\, f_{GGJ}}\frac{j_{ECCD}}{j_{bs}}\frac{d}{W}\eta_{CD} \le 0 \tag{13.8}$$

In the case of $d < W$, it was outlined earlier that η_{CD} becomes a constant of order *0.37* (for continuous CD) to *0.4* (for modulated CD) and hence Eq. (13.8) becomes a quadratic equation for W. Postulating that no real root exists leads to the inequality

$$dj_{ECCD} \ge \frac{W_{sat}\, j_{bs}}{4\eta_{CD}}\frac{c_{bs}\, f_{GGJ}}{c_{CD}} \tag{13.9}$$

2) These statements refer to current drive in the direction of the plasma current (co-ECCD).

which, due to the j_{bs} dependence of W_{sat}, is quadratic in β. In this case, it is the total current rather than the current density that has to be maximized.

For $d \geq W$, we have to distinguish between modulated and continuous CD. In the case of modulated CD, the efficiency drops with the fraction of current driven within the island, that is $\eta_{CD} = \eta_{mod} W/d$, and Eq. (13.8) becomes linear in W. Unconditional stability ($W \leq 0$) is given by

$$\frac{j_{ECCD}}{j_{bs}} \geq \frac{c_{bs} f_{GGJ}}{c_{CD} \eta_{mod}} \tag{13.10}$$

which states that the local current density should be of the order of the bootstrap current density, consistent with the physics picture that the ECCD current replaces the missing bootstrap current in the island. In fact, for ITER, the criterion $j_{ECCD}/j_{bs} > 1.2$ has been adopted assuming broad deposition profiles.

Finally, for unmodulated CD and $d > W$, the reduction of the efficiency is quadratic in W/d, that is $\eta_{CD} \approx \eta_{cont}(W/d)^2$ and unconditional stability can no longer be obtained. Evaluating Eq. (13.8) for this case yields the condition

$$W \geq \frac{W_{sat}}{1 + \frac{c_{CD}\eta_{cont}}{c_{bs} f_{GGJ}} \frac{W_{sat}}{d} \frac{j_{ECCD}}{j_{bs}}} \tag{13.11}$$

which states that unconditional stability ($W \leq 0$) cannot be obtained any longer, but the island width can be reduced with respect to W_{sat}. For sufficient j_{ECCD}, the second term in the denominator will be larger than 1 and it can be seen that the island width can be reduced to a fraction of the deposition width d.

In the following, we will show how experiments on ECCD control of resistive MHD instabilities conducted in various tokamaks validate the physics elements outlined in this section.

13.3 Sawtooth Control

In Chapter 11, it was shown that the onset of a sawtooth crash can be described by a criterion on s_1, the shear at the $q = 1$ surface, which evolves on the timescale for current redistribution inside $q = 1$. On the other hand, the crash itself was shown to be very fast on the scale of 100 s of μs even in large experiments. Hence, the control strategy using ECCD is to change the linear stability by affecting s_1 rather than to influence the crash itself. As sawteeth may be needed in tokamak scenarios to redistribute He ash and other impurities, 'control' does not necessarily mean suppression, but could also aim at a tailoring of the amplitude to maximize the beneficial effects while minimizing the negative impact on energy confinement and the possible triggering of NTMs by sawtooth crashes.

Experiments clearly show the predicted effect of a change of sawtooth period τ_{st} when driving localized current near $q = 1$. Figure 13.4 shows a scan of the deposition of ECCD in ASDEX Upgrade. Here, the sawtooth period τ_{st} in the presence of ECCD is plotted as a function of deposition radius. τ_{st} has been normalized

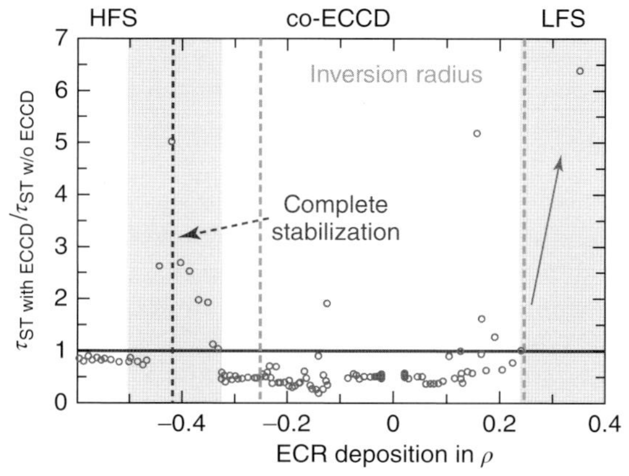

Figure 13.4 Effect of localized ECCD deposited at ρ_{dep} on the sawtooth period τ_{st} in ASDEX Upgrade, normalized to a similar case without ECCD. Negative values of ρ refer to deposition of the high field side, the shaded region is an estimate of the location of the $q = 1$ surface. *Source:* Mück *et al.* 2004 [79], reproduced with permission of IOP. (Please find a color version of this figure on the color plates.)

to the value in reference discharges without ECCD. Consistent with the expectation, there is no change in sawtooth period when current is driven outside $q = 1$ (in the figure, negative values of ρ_{dep} correspond to deposition on the high field side). On the other hand, clear changes are seen when the deposition is around $q = 1$, with a lengthening when deposition is just outside (which will decrease the shear, see Figure 13.2) and a shortening when ECCD is deposited inside $q = 1$. For deposition just on the resonant surface, complete suppression can be observed in some cases. In these experiments, the current driven by ECCD is of the order of 5% of the current inside the deposition radius and the heating power from ECCD is of the order of 20%, indicating that the changes are indeed local and not global and the method has the potential for control application in future large devices.

Detailed comparisons of similar experiments on the TCV tokamak to the Porcelli model outlined in Chapter 11 show good agreement with the theoretical prediction [80]. In addition, the sawtooth amplitude, measured in terms of loss of stored energy, decreases with sawtooth period under these conditions, which is expected in the regime where the sawtooth period is of the order of the energy confinement time, that is the profiles do not attain their equilibrium values during the cycle. Owing to the sensitivity on the deposition location, it will be beneficial to develop feedback control algorithms for the deposition location as well, which is a goal that this method has in common with tearing mode stabilization and is briefly discussed in Section 13.4.

While the experiments discussed earlier provide a clear validation of the theoretical picture outlined before, they are only partly addressing an important issue for future reactor grade devices, namely the control of sawteeth in the presence of a large population of fast particles. In this area, changes of the sawtooth period by applying ion heating schemes such as neutral beam injection and ion cyclotron resonance heating have also been observed, but the interpretation of these results is more difficult as the methods change the fast ion distribution substantially in addition to a possible effect of current drive. On the other hand, these heating systems can be used to prepare, in present-day experiments, conditions under which the sawtooth period becomes very long due to the stabilizing effect of the fast ions generated by the heating systems, as was shown in the case of 'Monster' sawteeth in Figure 11.3. Application of ECCD in this situation has also shown a substantial decrease in the sawtooth period, as is shown in an example from Tores Supra in Figure 13.5. Here, ICRH (lowest panel) is used to create a population of fast particles that leads to a substantial increase in τ_{st} (third panel from above). In the discharge shown in the second panel, ECCD is added and the deposition is swept across the $q = 1$ surface by beam steering using a movable mirror. It can be seen that as soon as the deposition is close to $q = 1$, the sawtooth period decreases with respect to the reference discharge without ECCD shown in the first panel. On the basis of these experiments and concomitant modelling, including the effect of the fast particles, this method is foreseen for sawtooth control in ITER.

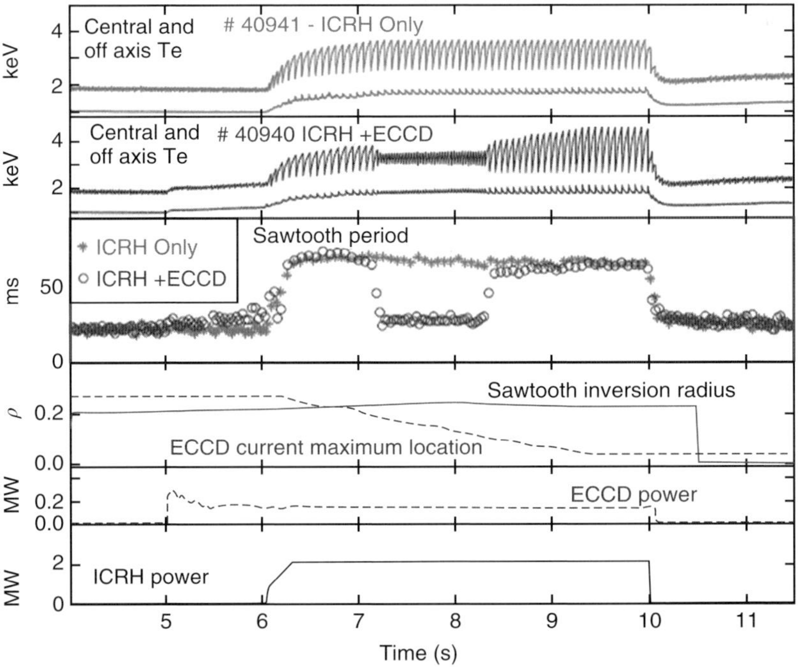

Figure 13.5 Demonstration of change of sawtooth period by application of ECCD on the Tore Supra tokamak in a discharge where fast particles created by ICRH initially lead to large sawteeth with long period τ_{st}. *Source:* Lennholm *et al.* 2009 [81], reproduced with permission of the APS. (Please find a color version of this figure on the color plates.)

13.4
Tearing Mode Control

Control of tearing modes can either be done in a 'pre-emptive' manner by altering the equilibrium current profile, similar to the strategy for sawtooth control described in Section 13.3, or by generating a helical current in an already existing island. The latter has been demonstrated experimentally in a number of tokamaks to be in good agreement with the theoretical expectations outlined earlier. One of the theoretical findings was that the deposition needs to be precisely within the island and a deposition outside will not have an effect.

This can be seen in Figure 13.6 where a slow sweep of the deposition of an ECCD beam across a pre-existing magnetic island has been conducted in ASDEX Upgrade by slowly changing the toroidal field. A decrease in mode amplitude is seen in the time interval where ECCD is deposited within the island. The narrow range in which the island width is influenced (about 4 cm in this experiment, which roughly corresponds to the island width) also proves the localization of the ECCD-generated current, which was evaluated to be $d \approx 2.5$ cm in this experiment. Furthermore, the current was injected continuously into the island region, verifying the generation of a helical current component by equilibration on the island flux surfaces as well. The experiment shown in Figure 13.6 was done deliberately at low ECCD power ($P_{\mathrm{ECCD}}/P_{\mathrm{tot}} = 5\%$). Increasing the power, pre-existing NTMs can be fully stabilized, with the power requirements of the order of $P_{\mathrm{ECCD}}/P_{\mathrm{tot}} 10-20\%$, in agreement with modeling based on the modified Rutherford equation discussed earlier. As for the sawtooth control discussed earlier, this also means that the method has the potential to be used as a control tool in future large devices.

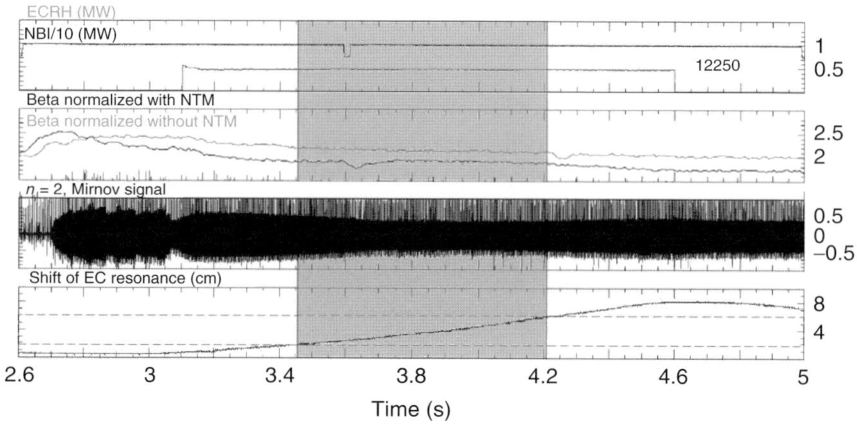

Figure 13.6 Variation of NTM amplitude as measured by a magnetic pick-up coil ('Mirnov signal', third panel) in an experiment where localized ECCD is swept across the resonance surface by changing the ECCD resonance condition through a ramp of B_t (fourth panel). *Source:* Zohm *et al.* 2001 [82], reproduced with permission of AIP.

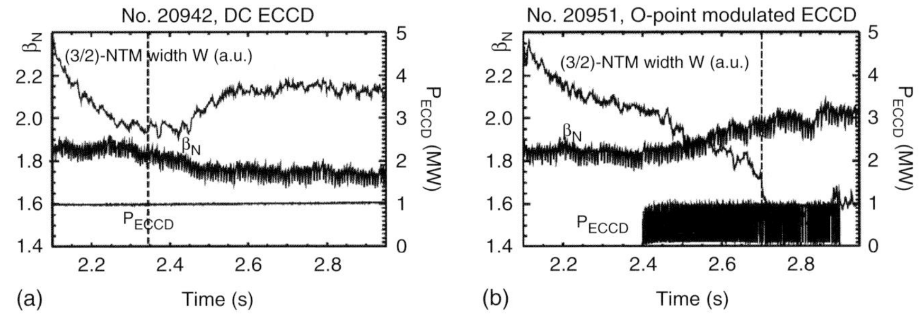

Figure 13.7 Comparison experiments with continuous (a) and island phase-modulated (b) ECCD using broad deposition ($d > W$). Only the modulated ECCD leads to full suppression of the NTM. *Source:* Maraschek *et al.* 2007 [83], reproduced with permission of the APS.

While the experiments shown earlier were carried out using continuous ECCD in a situation where $d \leq W$, we remind that for future devices, this criterion may no longer be fulfilled and modulation of the injected current in phase with the island's O-point could be required. Experiments employing a variation of the ratio W/d have been conducted by changing the toroidal injection angle of the ECCD beam[3)] and have verified the reduced efficiency of ECCD with large deposition width $d > W$. In this case, modulated ECCD shows indeed a substantially higher efficiency for mode stabilization, as shown in Figure 13.7.

Here, Figure 13.7a shows a situation where the applied DC ECCD power is not sufficient to completely suppress a pre-existing (3,2) NTM, although the amplitude is temporarily reduced while ECCD is swept across the island. In a comparison experiment in which the ECCD power is modulated in phase with the island's O-point, shown in Figure 13.7b, complete suppression is achieved and β_N rises after the mode amplitude has dropped to zero.

While the experiments discussed earlier mainly rely on the generation of helical current in a pre-existing island, also the beneficial effect of pre-emptive ECCD in a situation where no island exists was demonstrated on several tokamaks [84]. Owing to the sensitivity of the effect on the deposition location (Section 13.1), it is desirable to feedback control the deposition. Figure 13.8 shows an example from the DIII-D tokamak where the plasma radial position (and hence the position of the ECCD resonance) was feedback controlled in order to match the $q = 2$ surface obtained from real-time equilibrium reconstruction.

Here, an initial rise in β_N prompted by an increase in total heating power leads to the occurrence of a (2,1) NTM that limits β_N. Then, ECCD is applied to remove the mode, mainly by the helical current generation discussed earlier. However, in the following phase, ECCD is still applied to the resonant surface, and β_N is increased close to the ideal no-wall limit without the (2,1) NTM reappearing. This is attributed to the pre-emptive current drive at the $q = 2$ surface, which, in the absence of an island, is predicted to have a large effect on Δ' according to Eq. (13.5).

3) For increasing toroidal launch angle, the Doppler broadening of the resonance leads to a substantial increase in deposition width.

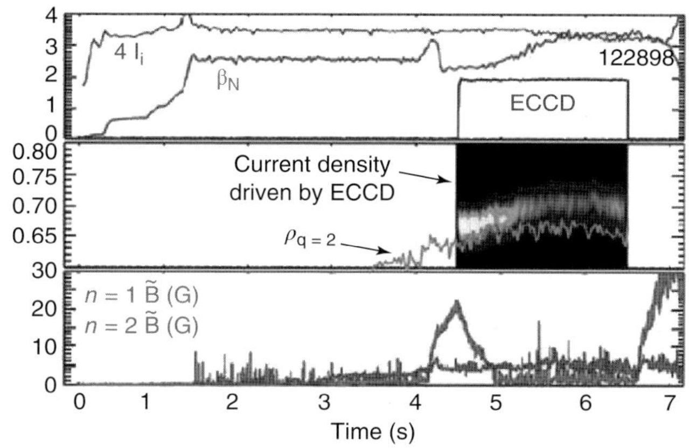

Figure 13.8 Suppression of a pre-existing (2,1) NTM by ECCD. Following complete suppression, β_N (shown in the upper panel) can be increased above the original onset level and the mode, shown by the $n = 1$ amplitude in the lower panel, does not reappear as long as ECCD is on. The deposition is feedback controlled to track the radial motion of the $q = 2$ surface (see middle panel) due to the increasing β. *Source:* Adapted from Prater *et al.* 2007 [85], reproduced with permission of the IAEA. (Please find a color version of this figure on the color plates.)

Consequently, as soon as pre-emptive ECCD is turned off, the (2,1) NTM is triggered again, leading to disruptive termination of the discharge.

Thus, localized ECCD is a viable method for NTM control, provided the deposition location can be controlled with sufficient precision and the deposition width is narrow enough. While in present-day experiments, this can be achieved by changing the plasma position or the toroidal field, it will have to be accomplished by steering of the beam as indicated in Figure 13.1 in future devices where B_t and the plasma position are fixed by other constraints. It is also possible to combine sawtooth control and NTM control as large sawteeth potentially trigger NTMs. In the TCV tokamak, the sawtooth control scheme discussed earlier has been used together with pre-emptive NTM control in a manner that ECCD at the $q = 2$ surface was only applied just before the sawtooth crash was triggered, minimizing the power requirements for NTM control [86]. In this case, the sawtooth crash did not trigger the NTM, whereas in the absence of pre-emptive ECCD, NTMs occurred.

Another important aspect is that control of (2,1) (N)TMs will also decrease the risk of disruptions, which have been shown in Chapter 10 to often be linked to the occurrence of this mode. In fact, classical tearing modes occurring close to the density limit or triggered by impurity influx that leads to a peaking of the current profile have also been suppressed by ECCD in various experiments. As close to the density limit, ECCD usually has a low efficiency (the CD efficiency roughly scales with T_e/n_e, that is $1/n_e^2$ at constant β), this also gives evidence that local heating of islands can lead to stabilization by increasing the electrical conductivity there. In future devices, where temperatures are in general higher, it is however expected that the direct current drive by ECCD will always dominate.

References

1. Freidberg, J.P. (1987) *Ideal Magnetohydrodynamics*, Plenum Press, New York.
2. Balescu, R. (1988) *Transport Processes in Plasmas, 1. Classical Transport Theory*, Elsevier, Amsterdam.
3. Biskamp, D. (1993) *Nonlinear Magnetohydrodynamics*, Cambridge University Press, Cambridge.
4. Wesson, J.A. (2011) *Tokamaks, Oxford University Press*, 3rd edition.
5. Solov'ev, L.S. (1976) Hydromagnetic stability of closed plasma configurations. *Rev. Plasma Phys.*, **VI**, 239–331.
6. Lackner, K. (1976) Computation of ideal MHD equilibria. *Comput. Phys. Commun.*, **12**, 33–44.
7. Helander, P. (2014) Theory of plasma confinement in non-axisymmetric magnetic fields. Rep. Prog. Phys. 77, 087001.
8. Bernstein, I.B., Frieman, E.A., Kruskal, M.D., and Kulsrud, R.M. (1958) An energy principle for hydromagnetic stability problems. *Proc. R. Soc. London, Ser. A*, **244** (1236), 17–40.
9. Greene, J.M. and Johnson, J.L. (1968) Interchange instabilities in ideal hydromagnetic theory. *Plasmas Phys.*, **10**, 729–745.
10. Connor, J.W., Hastie, R.J. and Taylor, J.B. (1979) High mode number stability of an axisymmetric tokamak plasma. *Proc. R. Soc. London, Ser. A*, **365** (1720), 1–17.
11. Zanca, P., Marrelli, L., Paccagnella, R., Soppelsa, A., Baruzzo, M., Bolzonella, T., Marchiori, G., Martin, P. and Piovesan, P. (2012) Feedback control model of the m = 2, n = 1 resistive wall mode in acircular plasma. *Plasma Phys. Controlled Fusion*, **54**, 094004.
12. Wesson, J.A. (1978) Hydromagnetic stability of tokamaks. *Nucl. Fusion*, **18** (1), 87–131.
13. Laval, G., Pellat, R. and Soule, J.S. (1974) Hydromagnetic stability of a current carrying pinch with noncircularcross section. *Phys. Fluids*, **17** (4), 835–845.
14. Robinson, D.C. (1971) High β diffuse pinch configurations. *Plasma Phys.*, **13**, 439–462.
15. Bussac, M.N., Pellat, R., Edery, D. and Soule, J.L. (1975) Internal kink modes in toroidal plasmas with circular cross section. *Phys. Rev. Lett.*, **35** (24), 1638–1641.
16. Mercier, C. (1960) A necessary condition for hydromagnetic stability of plasma withaxial symmetry. *Nucl. Fusion*, **1**, 47.
17. Wagner, F., Becker, G., Behringer, K., Campbell, D.J., Eberhagen, K., Engelhardt, W., Fussmann, G., Gehre, O., Gernhardt, J., Gierke, V., Haas, G., Huang, G., Karger, M., Keilhacker, F., Klüber, M., Kornherr, O., Lackner, M., Lisitano, K., Lister, G., MMayer, G.G., Meisel, H., Müller, D., Murmann, E.R., Niedermeyer, H.D., Poschenrieder, H., Rapp, G., H Röhr, H., Schneider, F., Siller, G., Speth, E., Stäbler, A., Venus, G., Vollmer, O. and Yu, Z. (1982) Regime of improved confinement and high beta in neutral-beam-heateddivertor discharges of the asdex tokamak. *Phys. Rev. Lett.*, **49** (19), 1408–1412.
18. Pütterich, T., Wolfrum, E., Dux, R., Maggi, C. and ASDEX UpgradeTeam

(2009) Evidence for strong reversed shear of toroidal rotation in the edge transport barrier in the ASDEX Upgrade. *Phys. Rev. Lett.*, **102**, 025001.

19. Ozeki, T., Chu, M.S., Lao, L.L., Taylor, T.S., Chance, M.S., Konoshita, S., Burrell, K.H. and Stambaugh, R.D. (1990) Plasma shaping, edge ballooning stability and ELM behaciour inDIII-D. *Nucl. Fusion*, **30** (8), 1425–1432.
20. Suttrop, W., Kaufmann, M., de Blank, H.J., Brüsehaber, B., Lackner, K., Mertens, V., Murmann, H., Neuhauser, J., Ryter, F., Salzmann, H., Schweinzer, J., Zohm, H. and theASDEX Upgrade Team (1997) Identification of plasma-edge related operational regime boundariesand the effect of edge instability on confinement in ASDEX Upgrade. *Plasma Phys. Controlled Fusion*, **39**, 2051–2066.
21. Connor, J.W., Hastie, R.J., Wilson, H.R. and Miller, R.L. (1998) Magnetohydrodynamic stability of tokamak edge plasmas. *Phys. Plasmas*, **5** (7), 2687–2700.
22. Snyder, P.B., Groebner, R.J., Leonard, A.W., Osborne, T.H. and Wilson, H.R. (2009) Development and validation of a predictive model for the pedestalheight. *Phys. Plasmas*, **16**, 056118.
23. Constantinescu, D., Dumbrajs, O., Igochine, V., Lackner, K., Meyer-Spasche, R., Zohm, H. and ASDEX Upgrade Team (2011) A low-dimenional model system for quasi-periodic plasmaperturbations. *Phys. Plasmas*, **18**, 062307.
24. Kirk, A., Wilson, H.R., Akers, R., Conway, N.J., Counsell, G.J., Cowley, S.J., Dowling, J., Dudson, B., Field, A.R., Lott, F., Lloyd, B., Martin, R., Meyer, H., Price, M., Taylor, D., Walsh, M. and the MAST Team (2005) Structure of ELMS in MAST and the implications for energydeposition. *Plasma Phys. Controlled Fusion*, **47**, 315–333.
25. Burrell, K.H., Austin, M., Brennan, D., de Boo, J., Doyle, E.J., Fenzi, C., Fuchs, P., Gohil, P., Greenfield, C.M., Groebner, R.J., Lao, L.L., Luce, T.C., Makowski, M.A., Mc Kee, G.R., Moyer, R.A., Petty, C.C., Porkoab, M., Rettig, C.L., Rhodes, T.L., Rost, J.C., Stallard, B.W., Strait, E.J., Synakowski, E.J., Wade, M.R., Watkins, J.G. and PWest, W. (2001) Quiescent double barrier high-confinement mode plasmas in theDIII-D tokamak. *Phys. Plasmas*, **8** (5), 2153–2162.
26. Lang, P.T., Loarte, A., Saibene, G., Baylor, L.R., Becoulet, M., Cavinato, M., Clement-Lorenzo, S., Daly, E., Evans, T.E., Fenstermacher, M.E., Gribov, Y., DHorton, L., Lowry, C., Martin, Y., Neubauer, O., Oyama, N., Schaffer, M.J., Stork, D., Suttrop, W., Thomas, P., Tran, M., Wilson, H.R., Kavin, A. and Schmitz, O. (2013) ELM control strategies and tools: status and potential for ITER. *Nucl. Fusion*, **53**, 043004.
27. Lang, P.T., Conway, G.C., Eich, T., Fattorini, L., Gruber, O., Günter, S., Horton, L.D., Kalvin, S., Kallenbach, A., Kaufmann, M., Kocsis, G., Lorenz, A., Manso, M.E., Maraschek, M., Mertens, V., Neuhauser, J., Nunes, I., Schneider, W., Suttrop, W., Urano, H. and ASDEX Upgrade Team (2004) ELM pace making and mitigation by pellet injection in ASDEXUpgrade. *Nucl. Fusion*, **44** (5), 665–677.
28. Evans, T.E., Fenstermacher, M., Moyer, R.A., Osborne, T.H., Watkins, J.G., Gohil, P., Joseph, I., Schaffer, M.J., Baylor, L.R., Becoulet, M., Boedo, J.A., Burrell, K.H., SdeGrassie, J., Finken, K.H., Jernigan, T., Jakubowski, M.W., Lasnier, C.J., Lehnen, M., WLeonard, A., Lonnroth, J., Noardon, E., Parail, V., Schmitz, O., Unterberg, B. and PWest, W. (2008) RMP ELM suppression in DIII-D plasmas with ITER similar shapes and collisionalities. *Nucl. Fusion*, **48**, 024002.
29. Progress in the ITER Physics Basis (2007) Chapter 3: MHD stability, operational limits and disruptions. *Nucl. Fusion*, **47** (6), S128–S202.
30. Günter, S., Schade, S., Maraschek, M., Pinches, S.D., Strumberger, E., Wolf, R., Yu, Q. and ASDEX Upgrade Team (2000) MHD Phenomena in reversed shear discharges on ASDEX Upgrade. *Nucl. Fusion*, **40** (8), 1541–1548.
31. Manickam, J., Chance, M.S., Jardin, S.C., Kessel, C.D., Monticello, D., Pomphrey, N., Reiman, A., Wang, C. and Zakharov, L.E. (1994) The prospects

for magnetohydrodynamic stability in advanced tokamakregimes. *Phys. Plasmas*, **1** (9), 1601.

32. Strait, E.J., Bialek, J., Bogatu, N., Chance, M., Chu, M.S., Edgell, D., Garofalo, A.M., Jackson, G.L., Jensen, T.H., Johnson, L.C., Kim, J.S., La Haye, R.J., Navratil, G., Okabayashi, M., Reimerdes, H., Scoville, J.T., Turnbull, A.D., Walker, M.L. and theDIII-D Team (2003) Resistive wall stabilization of high-beta plasmas in DIII-D. *Nucl. Fusion*, **43**, 430–440.
33. Shaing, K.C. (2001) Symmetry-breaking induced transport in the vicinity of a magnetic island. *Phys. Rev. Lett.*, **87**, 245003.
34. Bondeson, A. and Ward, D.J. (1994) Stabilization of external modes in tokamaks by resistive walls and plasma rotation. *Phys. Rev. Lett.*, **72** (17), 2709–2712.
35. Berkery, J.W., Sabbagh, S.A., Betti, R., Hu, S., Bell, R.E., Gerhardt, S.P., Manickam, J. and Tritz, K. (2010) Resistive wall mode instability at intermediate plasma rotation. *Phys. Rev. Lett*, **104**, 035003.
36. Sabbagh, S.A., Berkery, J.W., Bell, R.E., Bialek, J., Gerhardt, S.P., Menard, J.E., Betti, R., Gates, D.A., Hu, B., Katsuro-Hopkins, O.N., LeBlanc, B.P., Levinton, F.M., Manickam, J., Tritz, K. and Yuh, H. (2010) Advances in global MHD mode stabilization research on NSTX. *Nucl. Fusion*, **50**, 025020.
37. Troyon, F., Gruber, R., Saurenman, H., Semenzato, S. and Succi, S. (1984) MHD limits to plasma confinement. *Plasma Phys. Controlled Fusion*, **26** (1A), 209.
38. Strait, E.J. (1994) Stability of high beta tokamak plasmas. *Phys. Plasmas*, **1**, 1415–1431.
39. Sabbagh, S.A., Bell, R.E., Bell, M.G., Bialek, J., Glasser, A.H., LeBlanc, B., Menard, J.E., Paoletti, F., Stutman, D., Fredrickson, E., Garofalo, A.M., Gates, D., Kaye, S.M., LLao, L., Maingi, R., Mueller, D., Navratil, G., Ono, M., Peng, M., Synakowski, E., Zhu, W. and the NSTX Research Team (2002) Beta-limiting instabilities and global mode stabilization in theNational Spherical Torus Experiment. *Phys. Plasma*, **9** (5), 2085–2092.
40. Wesson, J.A. and Sykes, A. (1985) Tokamak beta limit. *Nucl. Fusion*, **25** (1), 85–88.
41. Taylor, T.S. (1994) Experimental Achievement of Toroidal Beta Beyond that Predicted by "Troyon" Scaling, General Atomics, Lab Report, GA–A21821.
42. Lao, L.L., Taylor, T.S., Chu, M.S., Chan, V.S., Ferron, J.R. and Strait, E.J. (1991) Effects of current profile on the ideal ballooning mode. *Phys. Fluids B*, **4** (1), 232–240.
43. Ortolani, S. and Schnack, D.D. (1993) *Magnetohydrodynamics of Plasma Relaxation*, World Scientific.
44. Rutherford, R.H. (1973) Nonlinear growth of the tearing mode. *Phys. Fluids*, **16** (11), 1903–1908.
45. Fitzpatrick, R. (1995) Helical temperature perturbations associated with tearing modes intokamak plasmas. *Phys. Plasmas*, **2** (3), 825–838.
46. Meskat, J.P., Zohm, H., Gantenbein, G., Günter, S., Maraschek, M., Suttrop, W., Yu, Q. and ASDEX Upgrade Team (2001) Analysis of the structure of neoclassical tearing modes in ASDEX-Upgrade. *Plasma Phys. Controlled Fusion*, **43**, 1325–1332.
47. Chang, Z. and Callen, J.D. (1990) Global energy confinement degradation due to macroscopic phenomena intokamaks. *Nucl. Fusion*, **30** (2), 219–233.
48. Yu, Q. (2006) Numerical modelling of diffusive heat transport across magneticislands and local stochastic field. *Phys. Plasmas*, **13** (6), 062310.
49. Rechester, A.B. and Rosenbluth, M.N. (1978) Electron heat transport in a tokamak with destroyed magnetic surfaces. *Phys. Rev. Lett.*, **40**, 38–41.
50. Suttrop, W., Büchl, K., Fuchs, J.C., Kaufmann, M., Lackner, K., Maraschek, M., Mertens, V., Neu, M., Schittenhelm, R., Sokoll, M., Zohm, H. and the ASDEX UpgradeTeam (1997) Tearing mode formation and radiative edge cooling prior to density limit disruptions in ASDEX Upgrade. *Nucl. Fusion*, **37**, 119–125.
51. Zohm, H., Kallenbach, A., Bruhns, H., Fussmann, G. and Klüber, O. (1990) Plasma angular momentum loss by MHD mode locking. *Europhys. Lett.*, **11** (8), 745–750.

52. Fitzpatrick, R. (1993) Interaction of tearing modes with external structures in cylindricalgeometry. *Nucl. Fusion*, **33** (7), 1049–1084.
53. Wesson, J.A., Gill, R.D., Hugon, M., Schüller, F.C., Snipes, J.A., Ward, D.J., Bartlett, D.V., Campbell, D.J., Duperrex, P.A., Edwards, A.W., Granetz, R.S., OGottardi, N.A., Hender, T.C., Lazzaro, E., Lomas, P.J., Lopes Cardoso, N., Mast, K.F., FNave, M.F., Salmon, N.A., Smeulders, P., Thomas, P.R., Tubbing, B.J.D., Turner, M.F. and Weller, A. (1989) Disruptions in JET. *Nucl. Fusion*, **29**, 641–666.
54. de Vries, P., Johnson, M.F., Alper, B., Buratti, P., Hender, T.C., Koslowski, H.R., Riccardo, V. and JET-EFDA Contributors (2011) Survey of disruption causes at JET. *Nucl. Fusion*, **51**, 053018.
55. Lipschultz, B., La Bombard, B., Marmar, E.S., Pickrell, M.M., Terry, J.L., Watterson, R. and Wolfe, S.M. (1984) MARFE: an edge plasma phenomenon. *Nucl. Fusion*, **24** (8), 977–988.
56. Greenwald, M., Terry, J.L., Wolfe, S.M., Ejima, S., Bell, M.G., Kaye, S.M. and Neilson, G.H. (1988) A new look at density limits in to kamaks. *Nucl. Fusion*, **28** (12), 2199–2207.
57. Eich, T., Sieglin, B., Scarabosio, A., Fundamenski, W., Goldston, R.J., Hendrix, J., Herrmann, A. and ASDEX Upgrade Team (2011) Inter-ELM power decay length for JET and ASDEX Upgrade: measurementand comparison with heuristic drift-based model. *Phys. Rev. Lett.*, **107**, 215001.
58. ITER Physics Basis (1999) Chapter 3: MHD stability, operational limits and disruptions. *Nucl. Fusion*, **39** (12), 2251–2390.
59. Rosenbluth, M.N. and Putvinski, S.V. (1997) Theory for avalanche of runaway electrons in tokamaks. *Nucl. Fusion*, **37**, 1355–1362.
60. Gill, R.D. (1988) Generation and loss of runaway electrons following disruptions in JET. *Nucl. Fusion*, **33**, 1613–1625.
61. Ara, G., Basu, B. and Coppi, B. (1978) Magnetic reconnection and m=1 oscillations in current carrying plasmas. *Ann. Phys.*, **112**, 443–476.
62. Porcelli, F. (1991) Fast particle stabilisation. *Plasma Phys. Controlled Fusion*, **33**, 1601–1620.
63. Porcelli, F., Boucher, D. and Rosenbluth, M.N. (1996) Model for the sawtooth period and amplitude. *Plasma Phys. Controlled Fusion*, **38**, 2163–2186.
64. McGuire, K., Goldston, R., Bell, M., Bitter, M., Bol, K., Brau, K., Buchenauer, D., Crowley, T., Davis, S., Dylla, F., Eubank, H., Fishman, H., Fonck, R., Grek, B., Grimm, R., Hawryluk, R., Hsuan, H., Hulse, H., Izzo, R., Kaita, R., Kaye, S., Kugel, H., Johnson, D., Manickam, J., Manos, D., Mansfield, D., Mazzucato, E., McCann, R., McCune, D., Monticello, D., Motley, R., Mueller, D., Oasa, K., Okabayashi, M., Owens, K., Park, W., Reusch, M., Sauthoff, N., Schmidt, G., Sesnic, S., Strachan, J., Surko, C., Slusher, R., Takahashi, H., Tenney, F., Thomas, P., Towner, H., Valley, J. and White, R. (1983) Study of high-beta magnetohydrodynamic modes and fast-ion losses inPDX. *Phys. Rev. Lett.*, **50** (12), 891–895.
65. Sauter, O., La Haye, R.J., Chang, Z., Gates, D.A., Kamada, Y., Zohm, H. *et al.* (1997) Beta limits in long-pulse tokamak discharges. *Phys. Plsamas*, **4** (5), 1654–1664.
66. Glasser, A.H., Greene, J.M. and Johnson, J.L. (1975) Resistive instabilities in general toroidal plasma configurations. *Phys. Fluids*, **18** (7), 875–888.
67. Lütjens, H., Luciani, J.-F. and Garbet, X. (2001) Curvature effects on the dynamics of tearing modes in tokamaks. *Phys. Plasmas*, **8** (10), 4267–4270.
68. Chang, Z., Callen, J.D., Fredrickson, E.D., Budny, R.V., Hegna, C.C., McGuire, K.M., Zarnstorff, M.C. and TFTR group (1995) Observation of nonlinear neoclassical pressure-gradient-driven tearing modes in TFTR. *Phys. Rev. Lett.*, **74** (23), 4663–4666.
69. Wilson, H.R., Connor, J.W., Hastie, R.J. and Hegna, C.C. (1996) Threshold for neoclassical magnetic islands in a low collisionfrequency tokamak. *Phys. Plasmas*, **3** (1), 248–265.
70. Poli, E., Peeters, A.G., Bergmann, A., Günter, S. and Pinches, S.D. (2002)

Reduction of the ion drive and ρ_θ^* scaling of theneoclassical tearing mode. *Phys. Rev. Lett.*, **88** (7), 075001-0750-1.
71. Zohm, H., Gates, D.A., Wilson, H.R., Gantenbein, G., Gruber, O., Günter, S., Maraschek, M., Morris, W., Sokoll, M., Wagner, D., ASDEX Upgrade Team, and COMPASS-D Team (1997) Neoclassical mhd in asdex upgrade and compass-d. *Plasma Phys. Control. Fusion*, **39** (12B), B237–B246.
72. Maraschek, M., Sauter, O., Günter, S., Zohm, H. and ASDEX Upgrade Team (2003) Scaling of the marginal β_p of neoclassical tearing modesduring power ramp-down experiments in ASDEX Upgrade. *Plasma Phys. Controlled Fusion*, **45** (7), 1369–1384.
73. Günter, S., Gude, A., Maraschek, M., Sesnic, S., Zohm, H. and ASDEX UpgradeTeam (1998) β scaling for the onset of neoclassical tearing modes at ASDEXUpgrade. *Nucl. Fusion*, **38** (10), 1431–1434.
74. Fietz, S., Maraschek, M., Zohm, H., Reich, M., Barrera, L., Mc Dermott, R.M. and theASDEX Upgrade Team (2013) Influence of rotation on the (m, n) = (3, 2) neoclassical tearing-mode threshold in the ASDEX Upgrade. *Plasma Phys. Controlled Fusion*, **55**, 085010.
75. Gude, A., Günter, S., Maraschek, M., Zohm, H. and ASDEX Upgrade Team (2002) Temporal evolution of neoclassical tearing modes and its effect onconfinement reduction in ASDEX Upgrade. *Nucl. Fusion*, **42** (7), 833–840.
76. Westerhof, E. (1990) Tearing mode stabilisation by local current desnity perturbations. *Nucl. Fusion*, **30**, 1143–1147.
77. Zohm, H., Gantenbein, G., Leuterer, F., Maraschek, M., Poli, E., Urso, L. and ASDEXUpgrade Team (2007) Control of NTMs by ECCD on ASDEX Upgrade in view of ITER application. *Plasma Phys. Controlled Fusion*, **49**, B341–B347.
78. Hegna, C.C. and Callen, J.D. (1997) On the stabilization of neoclassical magnetohydrodynamic tearingusing localized current drive or heating. *Phys. Plasmas*, **4** (8), 2940–2946.
79. Mück, A., Goodman, T.P., Maraschek, M., Pereverzev, G., Ryter, F., Zohm, H. and ASDEX Upgrade Team (2005) Sawtooth control experiments on ASDEX Upgrade. *Plasma Phys. Controlled Fusion*, **47** (10), 1633–1655.
80. Angioni, C., Goodman, T.P., Henderson, M.A. and Sauter, O. (2004) Effects of localized electron heating and current drive on thesawtooth period. *Nucl. Fusion*, **43**, 455–468.
81. Lennholm, M., Eriksson, L.G., Turco, F., Bouquey, F., Darbos, C., Dumont, R., Giruzzi, G., Jung, M., Lambert, R., Magne, R., Molina, D., Moreau, P., Rimini, F., Segui, J.L., Song, S. and Traisnek, E. (2009) Demonstration of effective control of fast-ion-stabilized sawteethby electron-cyclotron current drive. *Phys. Rev. Lett.*, **102**, 115004.
82. Zohm, H., Gantenbein, G., Gude, A., Günter, S., Leuterer, F., Maraschek, M., Meskat, J., Suttrop, W., Yu, Q., ASDEX Upgrade Team and ECRH Group (AUG) (2001) Neoclassical tearing modes and their stabilization by electroncyclotron current drive in ASDEX Upgrade. *Phys. Plasmas*, **8**, 2009–2016.
83. Maraschek, M., Gantenbein, G., Yu, Q., Zohm, H., Günter, S., Leuterer, F. and ASDEXUpgrade Team (2007) enhancement of the stabilization efficiency of a neoclassical magnetic island by modulated electron cyclotron current drive in the ASDEXUpgrade tokamak. *Phys. Rev. Lett.*, **98**, 025005.
84. Nagasaki, K., Isayama, A., Hayashi, N., Ozeki, T., Takechi, M., Oyama, N., Ide, S., Yamamoto, S. and the TJ-60 team (2005) Stabilization of neoclassical tearing mode by eccd and its evolution-simulation on jt-60u tokamak. *Nucl. Fusion*, **45**, 1608–1617.
85. Prater, R., La Haye, R.J., Luce, T.C., Petty, C.C., Strait, E.J., Ferron, J.R., Humphreys, D.A., Isayama, A., Lohr, J., Nagasaki, K., Politzer, P.A., Wade, M.R. and Welander, A.S. (2007) Stabilization and prevention of the 2/1 neoclassical tearing modefor improved performance in DIII-D. *Nucl. Fusion*, **47**, 371–377.
86. Felici, F., Goodman, T.P., Sauter, O., Canal, G., Coda, S., Duval, B.P., Rossel, J.X. and the TCV Team (2012) Integrated real-time control of MHD instabilities using multi-beamECRH/ECCD systems on TCV. *Nucl. Fusion*, **52**, 074001.

Index

Magnetohydrodynamic Stability of Tokamaks, First Edition. Hartmut Zohm.

Color Plates

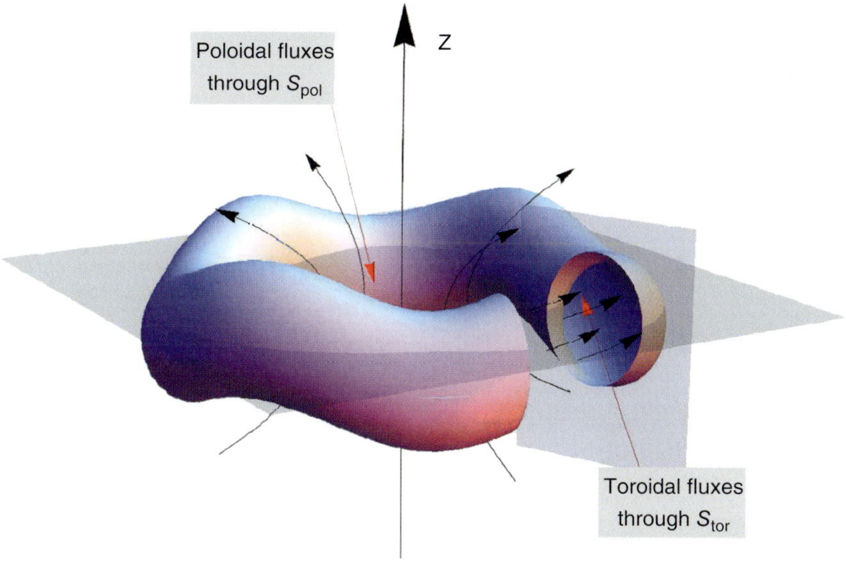

Figure 2.5 General toroidal geometry: two topologically different curves can be used to define flux integrals that can serve as flux surface labels. (This figure also appears on page 22.)

Magnetohydrodynamic Stability of Tokamaks, First Edition. Hartmut Zohm.

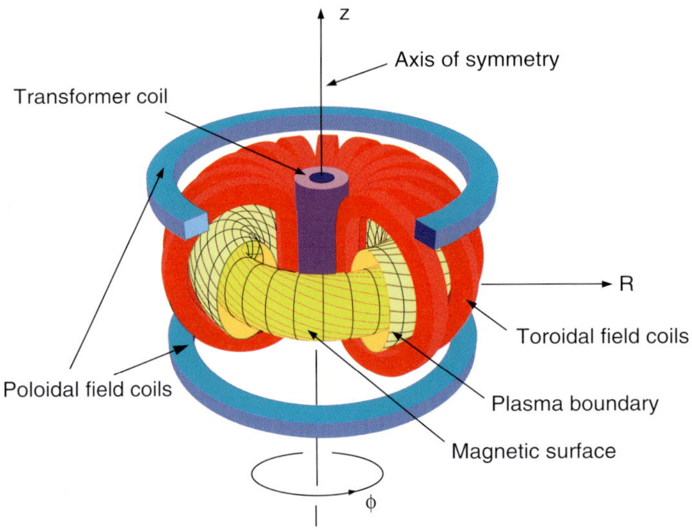

Figure 2.6 Schematic layout of a tokamak coil system as well as the co-ordinates used to derive the equilibrium equation (Eq. 2.33). (This figure also appears on page 24.)

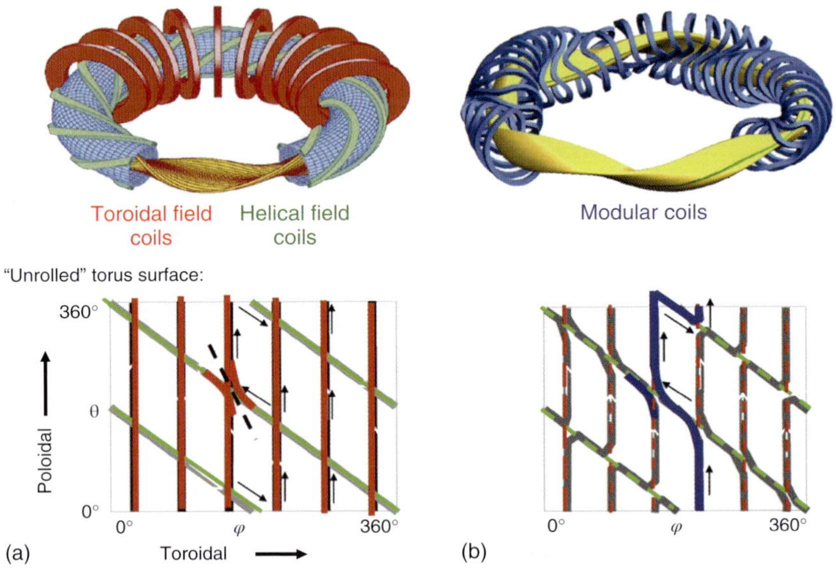

Figure 2.13 Coil configuration of a classical $\ell = 2$ stellarator (a) consisting of a set of planar toroidal field coils and a set of helical windings. In the lower panels, it is shown how this coil set can be replaced by a set of 3D-shaped modular coils that produce a stellarator field without helical windings (b, configuration of the W7-X experiment). (This figure also appears on page 39.)

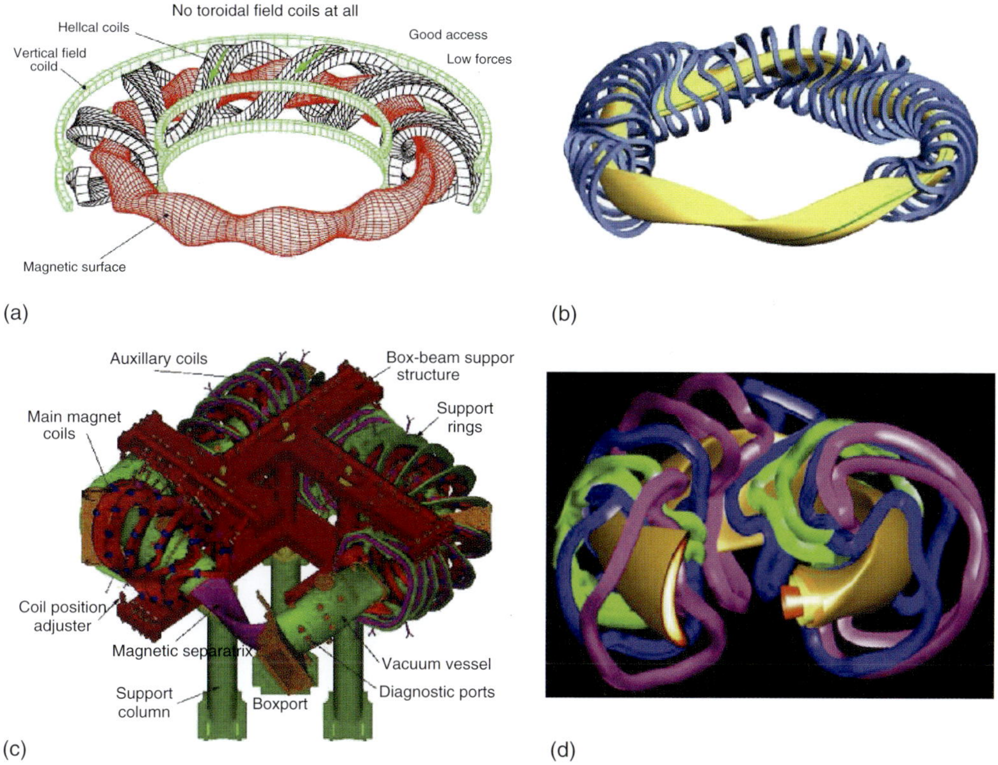

Figure 2.14 Different realizations of the stellarator configuration. (a) LHD (heliotron), (b) W7-X (quasi-isodynamic stellarator), (c) HSX (quasi-helically symmetric stellarator) and (d) NCSX (quasi-axisymmetric stellarator). (This figure also appears on page 40.)

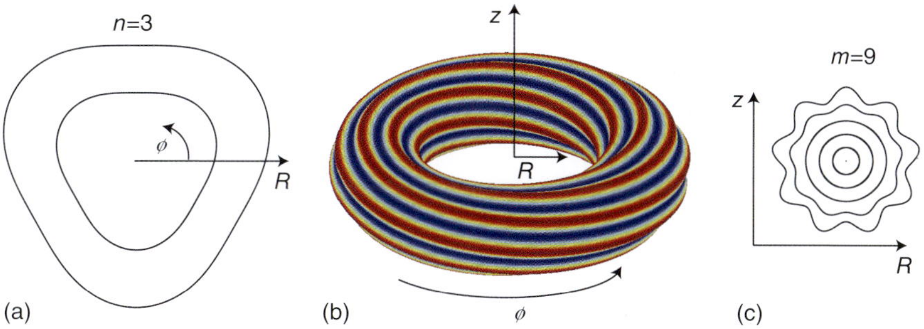

Figure 3.3 Schematic representation of an $m = 9$, $n = 3$ mode following the field lines on the $q = 3$ surface. (a) Horizontal cut through the 3D plot in (b) and (c) the poloidal cross section (in both plots, the amplitude of the perturbation is exaggerated compared to the middle one). (This figure also appears on page 54.)

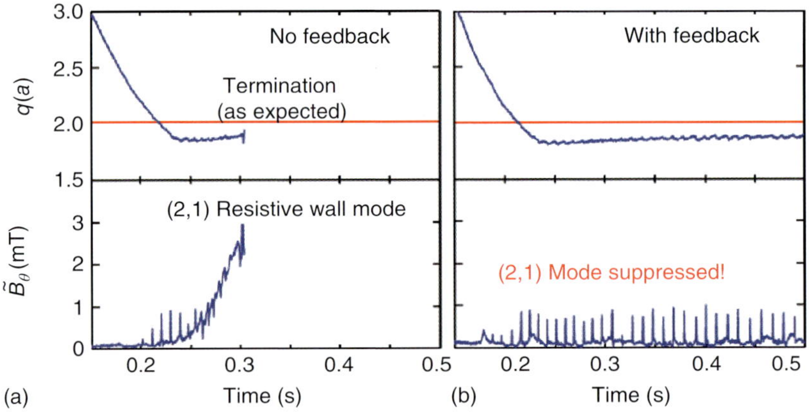

Figure 4.1 Tokamak discharge with circular cross-section run in the RFX reversed field pinch (in tokamak mode). As the current is increased, a disruption occurs at $q_a = 2$ due to the occurrence of an external (2,1) kink (a), whereas with active magnetic feedback control, the mode can be avoided (b). (This figure also appears on page 58.)

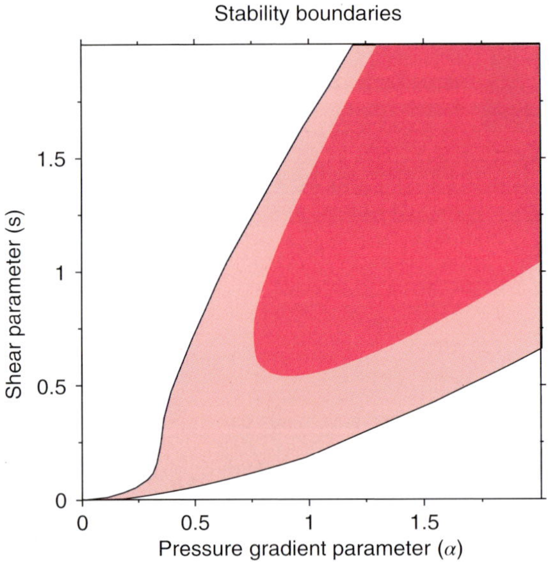

Figure 5.5 Ballooning stability diagram in a plane of normalized pressure α versus normalized shear *s* for a large aspect ratio toroidal equilibrium with circular flux surfaces. For concentric circles, the first and second stability regions are not connected (light shaded instability region) while fully including the Shafranov shift opens up a path to the region of second stability (dark shaded region). *Source:* Courtesy of H.-P. Zehrfeld, IPP. (This figure also appears on page 80.)

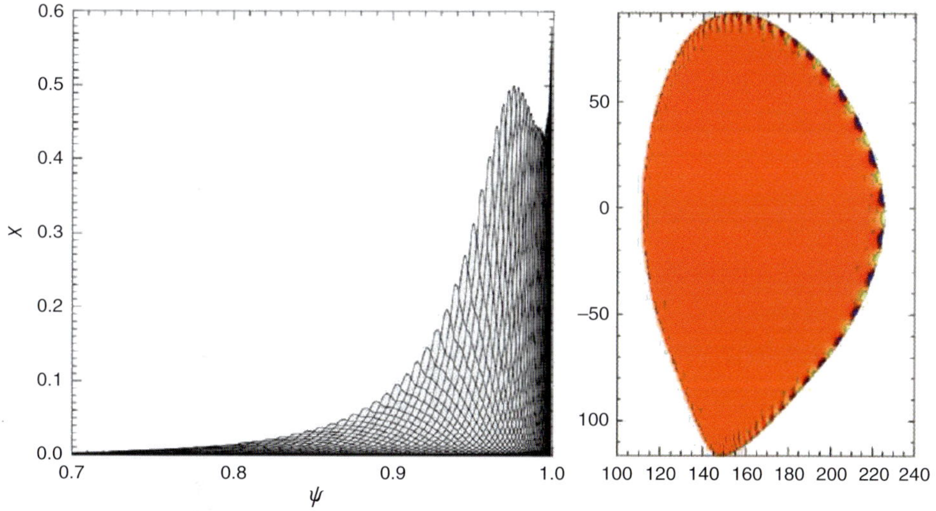

Figure 5.6 Finite *n* ballooning Eigenfunction in the edge of a tokamak plasma (JET-like parameters). For given *n*, there is a spectrum of poloidal wave numbers for which localized Eigenfunctions are calculated. These are combined under an envelope of larger width than the individual modes to form the full Eigenfunction. The ballooning effect is evident from the spatial representation on the right side. *Source:* Courtesy of S. Saarelma, CCFE. (This figure also appears on page 81.)

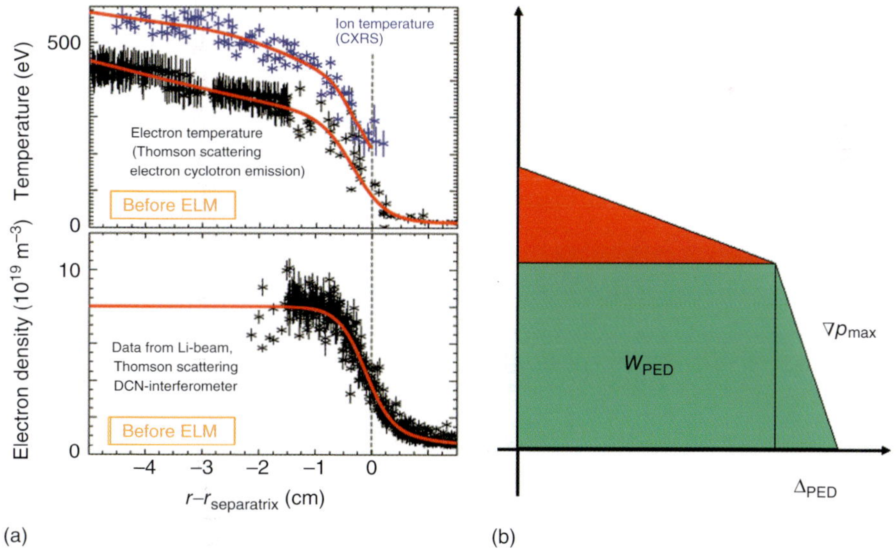

Figure 6.1 Edge kinetic profiles measured during an H-mode discharge on the ASDEX Upgrade tokamak (a) and sketch of the pressure pedestal (b). *Source (a):* Pütterich *et al.* 2009 [18], reproduced with permission of AIP. (This figure also appears on page 84.)

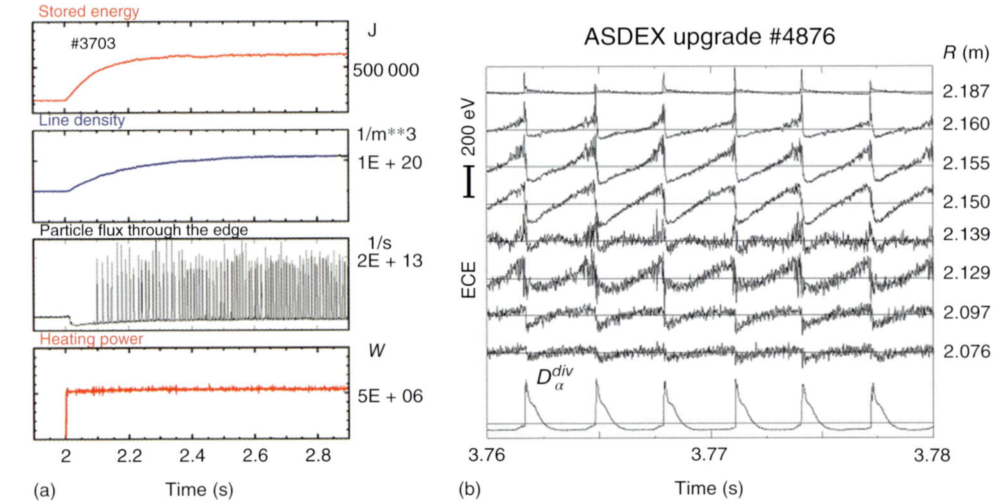

Figure 6.2 Temporal evolution of plasma parametres in an H-mode discharge in ASDEX Upgrade. While the integrated density and stored energy are quasi-stationary on the timescale of the global energy confinement time (a), the edge temperature is modulated by the occurrence of ELMs (b, the outer plasma edge is at $R = 2.17$ m). ELMs can also be seen in (a) as sharp spikes of D_α light, indicating the particle outflux during each ELM event. (This figure also appears on page 85.)

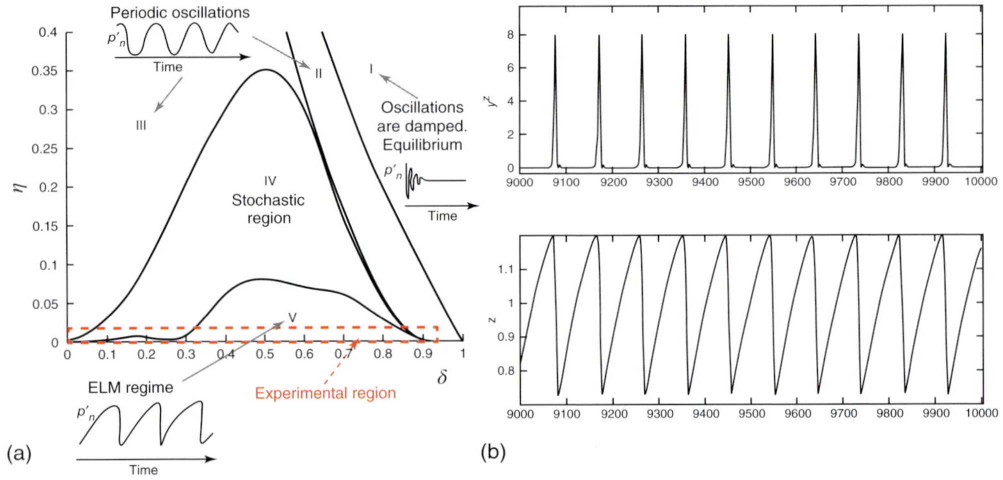

Figure 6.5 Occurrence of different dynamics in parametre space (a) for a simple non-linear model of coupled differential equations describing the interaction between MHD modes and a driving force that is restored on a much slower timescale. In the relevant regime of timescales ($\eta \ll 1$) non-linear cycles are observed that resemble the dynamics of ELMs (time traces of ξ^2 and p', b). *Source*: Adapted from D. Constantinescu *et al.* 2011 [16] with permission from AIP. (This figure also appears on page 92.)

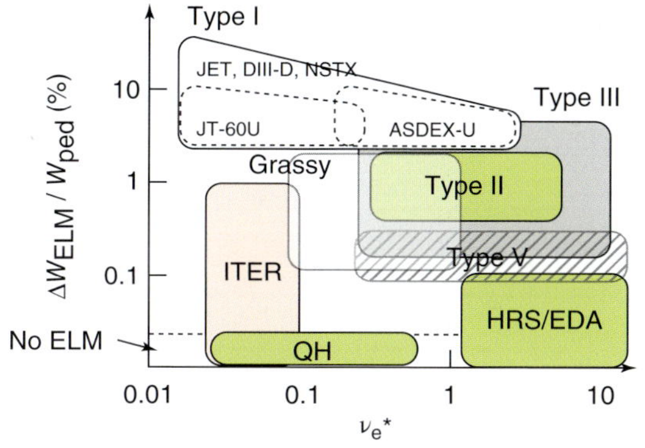

Figure 6.8 Overview of the occurrence of different ELM regimes, ordered by relative energy loss $\delta W_{\mathrm{ELM}}/W_{\mathrm{plasma}}$, as function of collisionality ν^*. *Source:* P.T. Lang *et al.* 2013 [26], reproduced with permission of the IAEA. (This figure also appears on page 97.)

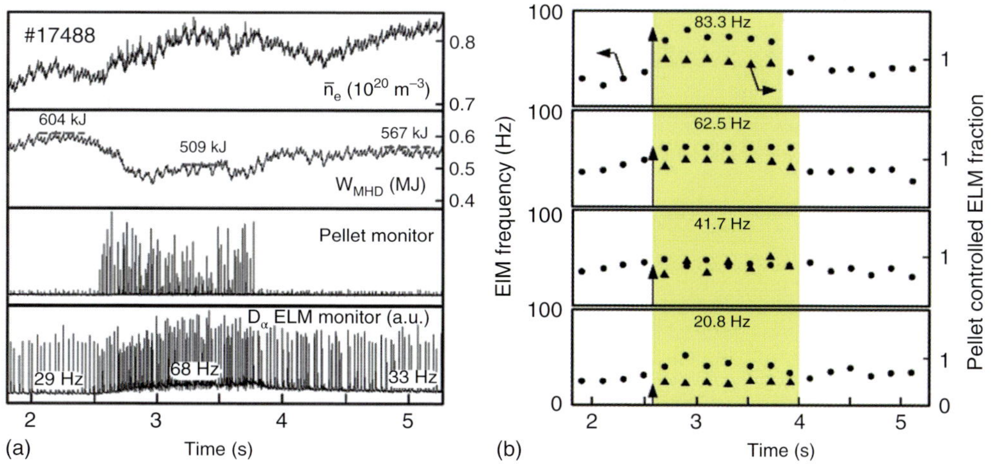

Figure 6.9 Pellet pacing of ELMs: the ELM repetition frequency locks to the pellet frequency, leading to smaller ELMs. The fraction of pellet triggered ELMs, indicated in (b), approaches 1 for sufficiently high $\nu_{\mathrm{pell}}/\nu_{\mathrm{ELM,natural}}$. *Source:* P.T. Lang *et al.* [27], reproduced with permission from the IAEA. (This figure also appears on page 99.)

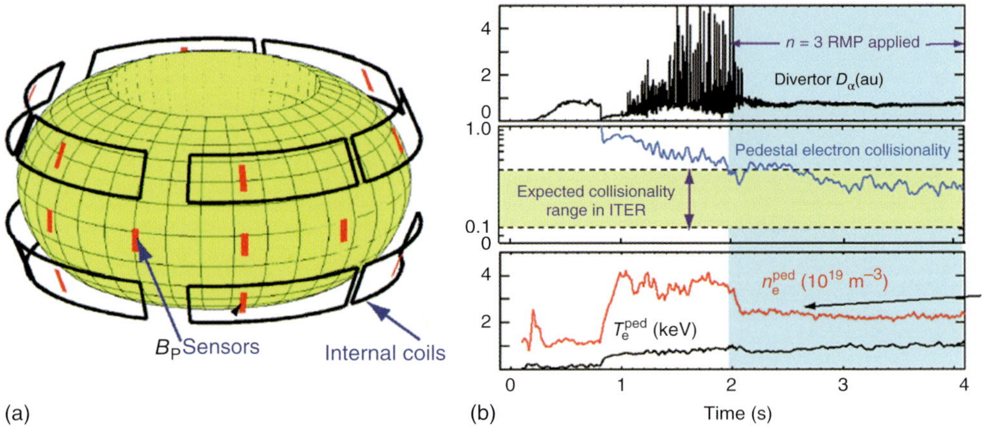

Figure 6.10 Suppression of type I ELMs by application of RMPs in DIII-D. (a) A schematic drawing of the coil set used to produce the RMP; (b) Time traces giving evidence of the complete ELM suppression. *Source*: Adapted from T. Evans 2008 [21], reproduced with permission from the IAEA. (This figure also appears on page 100.)

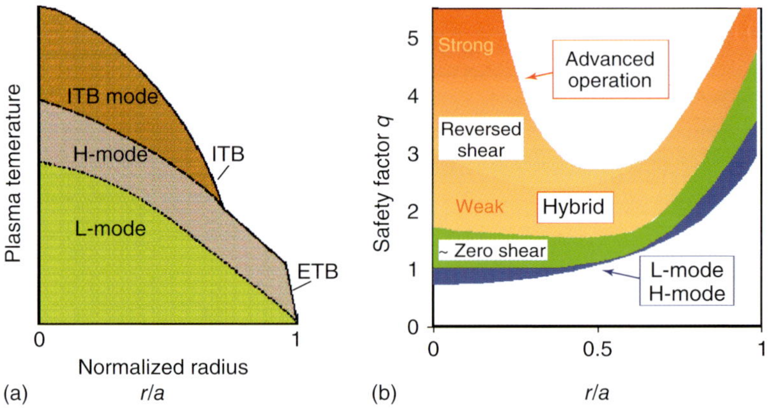

Figure 7.1 Typical pressure (a) and safety factor (b) profiles of different tokamak operational scenarios. Adapted from ITER Physics Basis Update 2007 [29], reproduced with permission of the IAEA. (This figure also appears on page 104.)

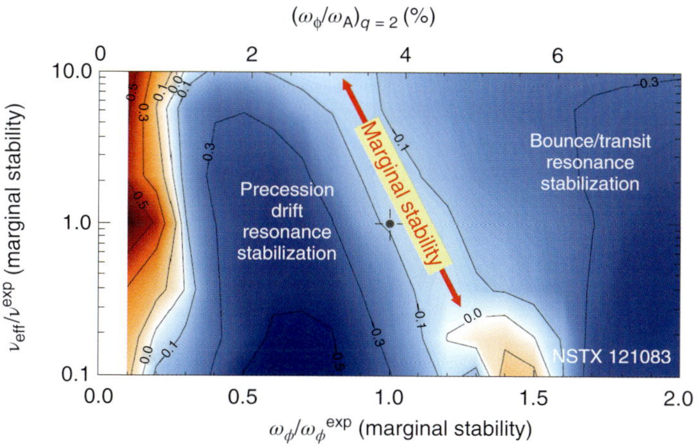

Figure 7.8 Analysis of kinetic stabilization effects on the RWM onset for a discharge preformed on the NSTX tokamak. Two regions of increased stability due to kinetic effects are obtained, modifying significantly the dependence of the growth rate on the rotation of the RWM. *Source:* J. W. Berkery *et al.* 2010 [35], reproduced with permission of AIP. (This figure also appears on page 116.)

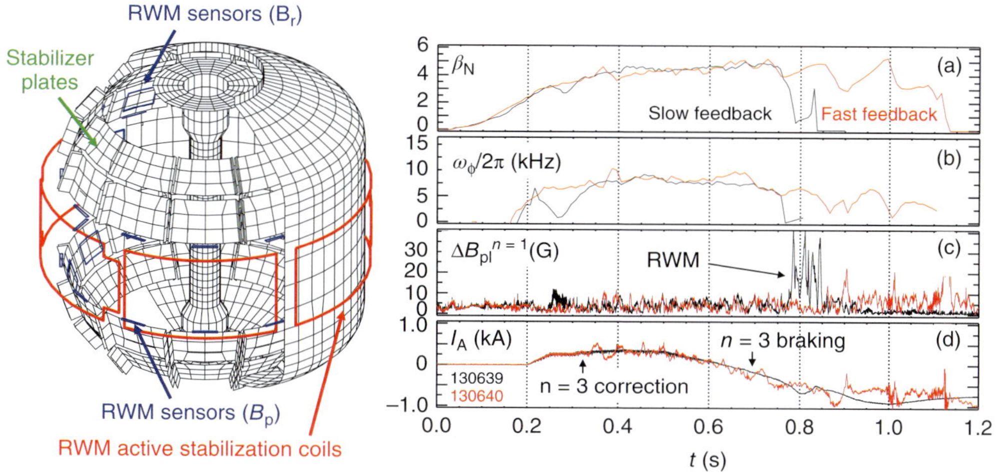

Figure 7.9 (A) Active stabilization of an RWM at low rotation using a set of control coils located on the outside of the vacuum vessel of the NSTX tokamak shown. (B) With additional direct feedback on the $n = 1$ mode (fast feedback), the amplitude of the RWM is controlled while with error field correction only (slow feedback), the mode grows (the mode amplitude is shown in trace (c) and causes a disruption. Trace (d) shows the current in one of the control coils. *Source:* S. A. Sabbagh *et al.* 2010 [36], reproduced with permission of the IAEA. (This figure also appears on page 117.)

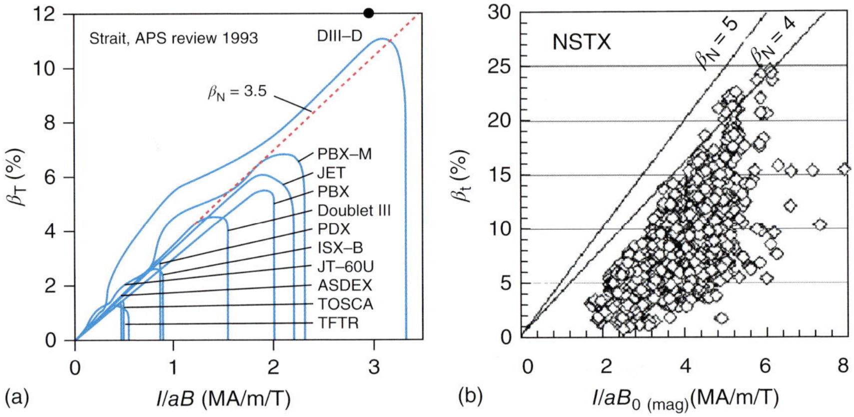

Figure 7.10 (a) Envelope of data points from various tokamaks indicating the validity of the Troyon limit. (b) Including a small aspect ratio tokamak, here NSTX, increases the region in which the scaling is validated significantly. *Source (a):* E. J. Strait 1994 [38] and (right) S. A. Sabbagh *et al.* 2002 [39], both reproduced with permission of the AIP. (This figure also appears on page 119.)

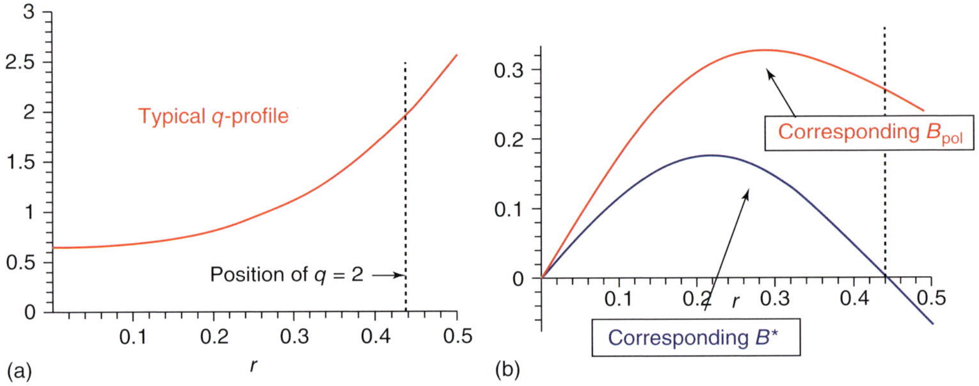

Figure 8.3 (a and b) Visualization of the equilibrium poloidal field B_{pol} and the corresponding $q = 2$ helical field $B^*_{0\mu}$ for a typical q-profile shown. The helical field changes sign at the $q = 2$ surface. (This figure also appears on page 129.)

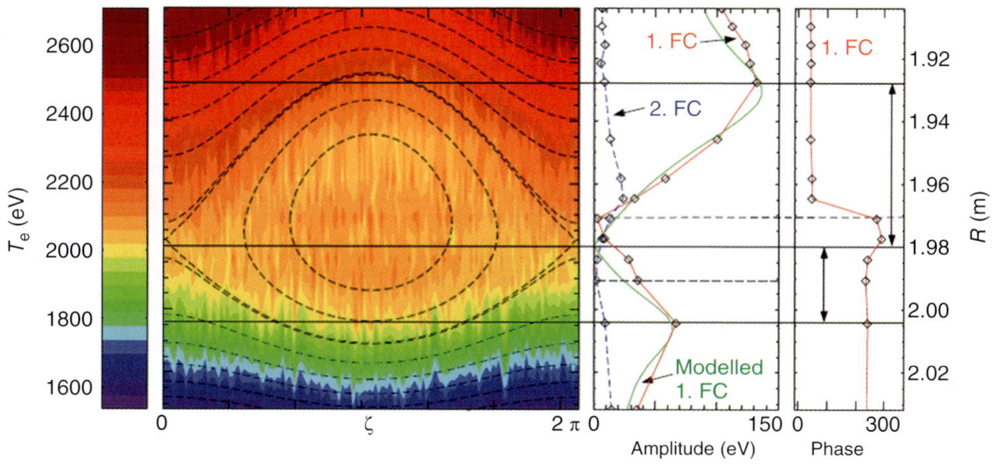

Figure 9.1 Measured temperature distribution across a large island. The contour plot shows the temperature with the modelled asymmetric island overlaid. The panels on the right show how the island structure can be determined from the profiles of the perturbed temperature in the first and second Fourier component with the location of r_s given by the characteristic phase jump. *Source:* Meskat 2001 [46]. Reproduced with permission of IOP. (This figure also appears on page 142.)

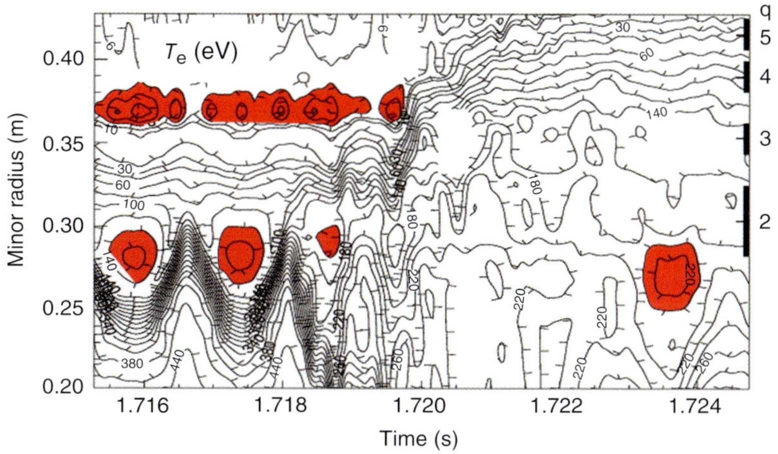

Figure 9.3 Experimentally observed rapid energy loss due to the interaction of two phased locked island chains on ASDEX Upgrade, leading to stochastization. Here, rapid edge cooling when approaching the density limit (see Section 10.1.1) has led to the occurrence of (2,1) and (3,1) current gradient driven tearing modes. Isothermals as function of time and radius are shown. The modes rotate in the laboratory frame. *Source:* Suttrop *et al.* 1997 [50], reproduced with permission of the IAEA. (This figure also appears on page 144.)

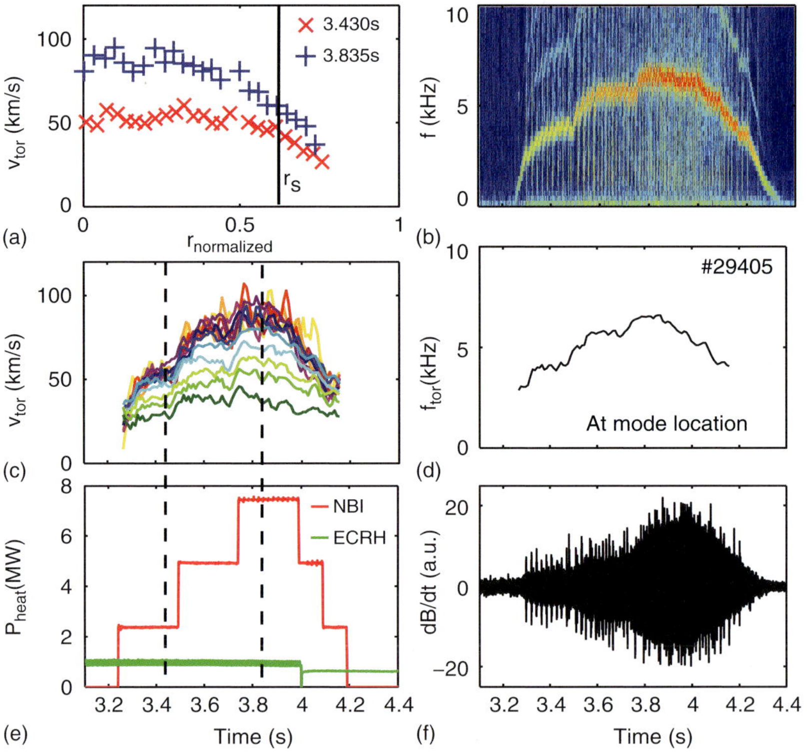

Figure 9.5 Time traces of the evolution frequency of a (2,1) tearing mode in ASDEX Upgrade (spectrogram of magnetic perturbation shown in (b), with the raw signal shown in (f)). (d) The temporal evolution matches the evolution of the fluid rotation at the $q = 2$ surface. (e) The variation of the fluid rotation is achieved by changing the momentum input due to heating by neutral beam injection (NBI). (a) Rotation profiles for different time points are shown together with the temporal evolution at different radii (c). (This figure also appears on page 147.)

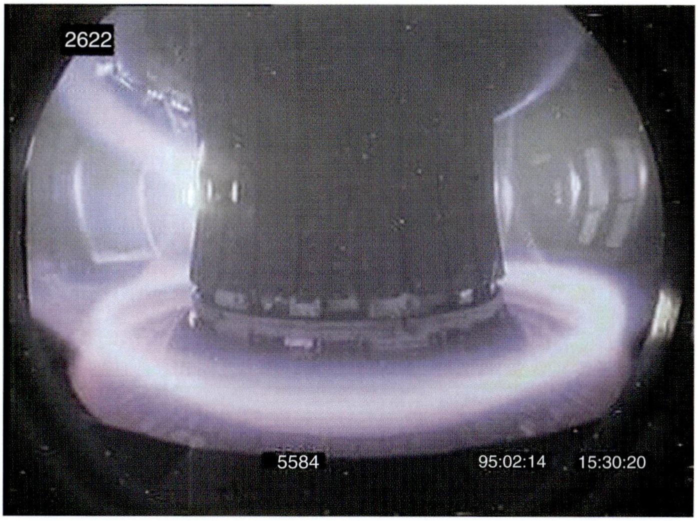

Figure 10.3 Visible light camera image of a plasma discharge in ASDEX Upgrade exhibiting a MARFE that can clearly be recognized by the toroidally symmetric radiation zone in the lower part of the vacuum vessel. (This figure also appears on page 163.)

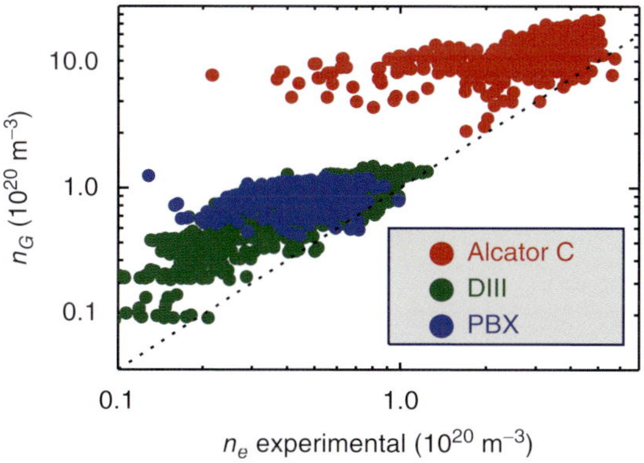

Figure 10.4 Operational points from the Alcator C, DIII, and PDX tokamaks plotted against the empirical density limit n_{GW} described by Eq. (10.1). The data points support the empirical scaling for different cross sections. *Source:* Replotted from Greenwald 1988 [49], reproduced with permission of the IAEA. (This figure also appears on page 164.)

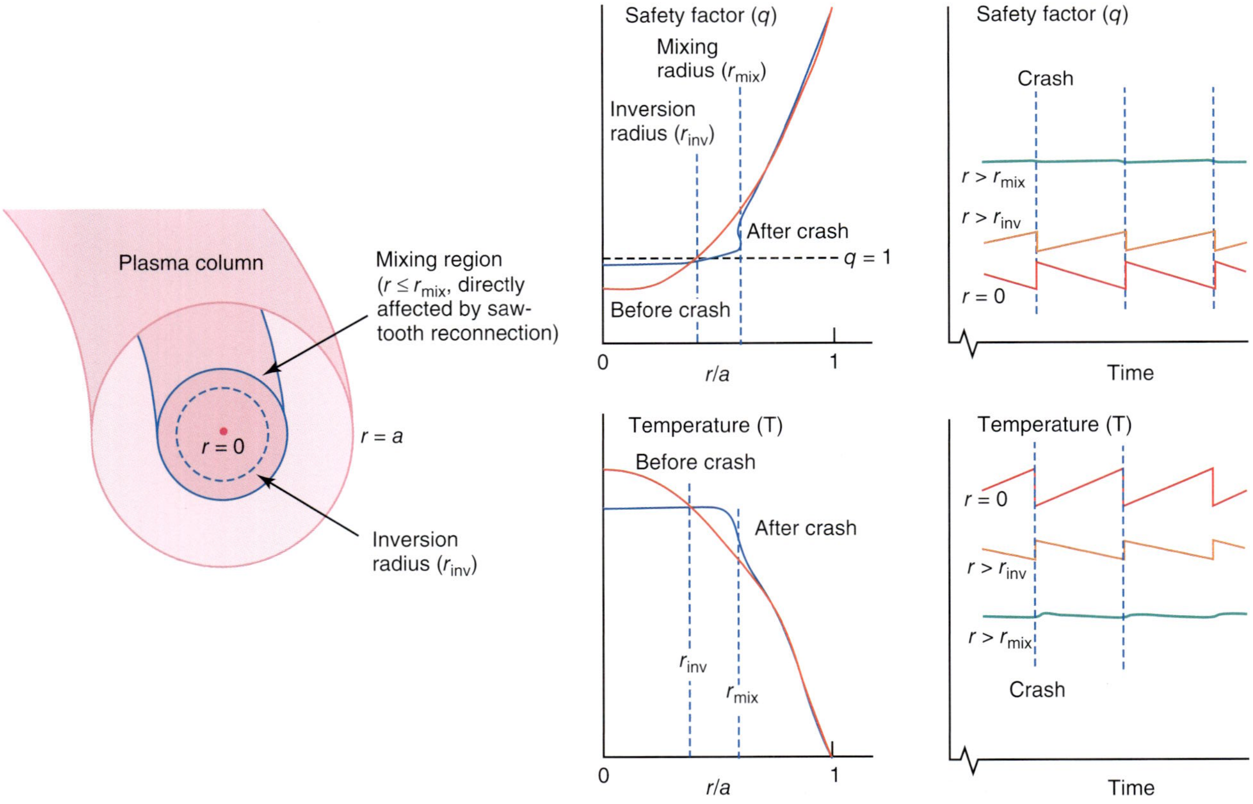

Figure 11.1 Schematic representation of typical time traces and profile evolution during a sawtooth cyle. *Source:* ITER Physics Basis, Chapter 3 [58], reproduced with permission of the IAEA. (This figure also appears on page 177.)

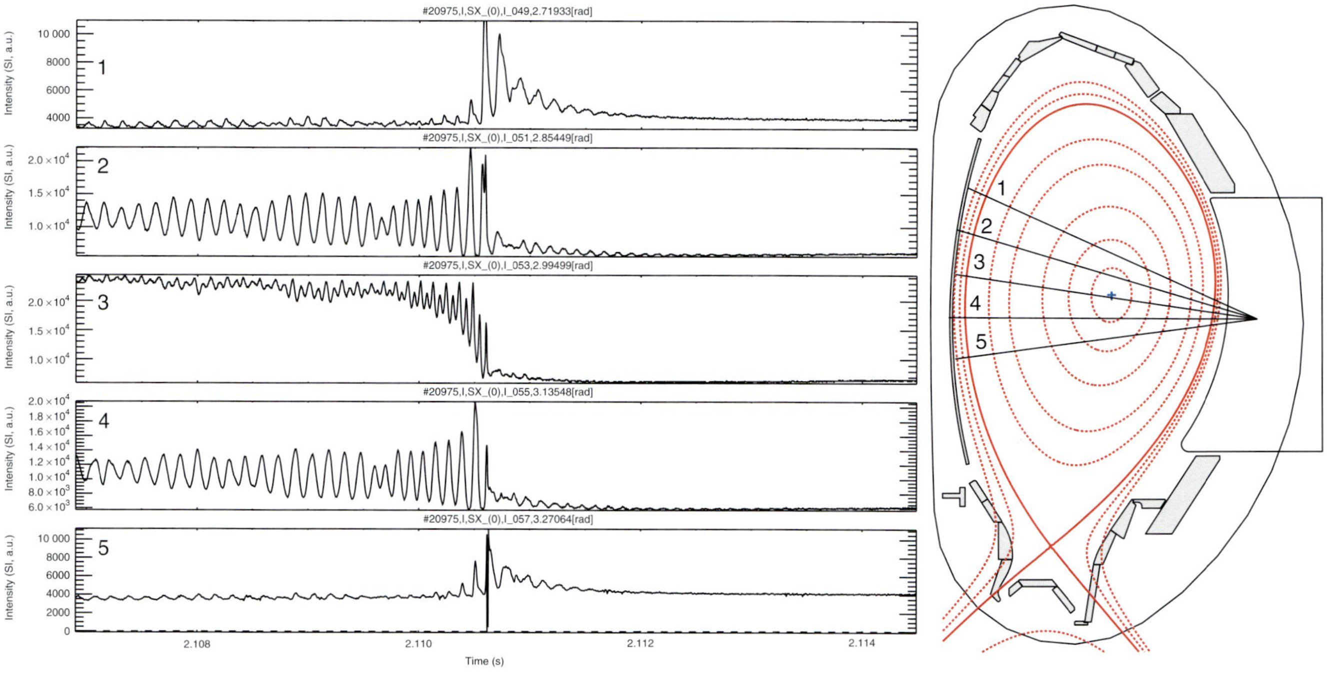

Figure 11.5 Observation of a (1,1) 'postcursor' oscillation following a sawtooth crash on the ASDEX Upgrade tokamak, indicating incomplete reconnection in contrast to the simple Kadomtsev model. Line-integrated signals of the soft-X-ray diagnostic are shown, with the viewing geometry indicated on the right. Here, lines 1 and 5 are outside r_{inv}, lines 2 and 4 roughly coincide with r_{inv} and line 3 views through the centre. Note the suppressed zero of the signals. *Source:* Courtesy of V. Igochine, IPP. (This figure also appears on page 185.)

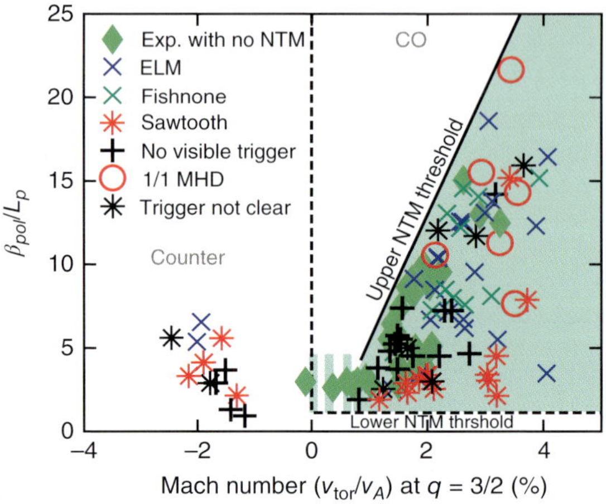

Figure 12.5 Onset β for (3,2) NTMs as function of the normalized rotation of the resonant surface in ASDEX Upgrade. The maximum achievable β has a minimum around zero rotation. The points are labelled according to the triggering event. *Source:* Fietz *et al.* 2013 [74], reproduced with permission of IOP. (This figure also appears on page 197.)

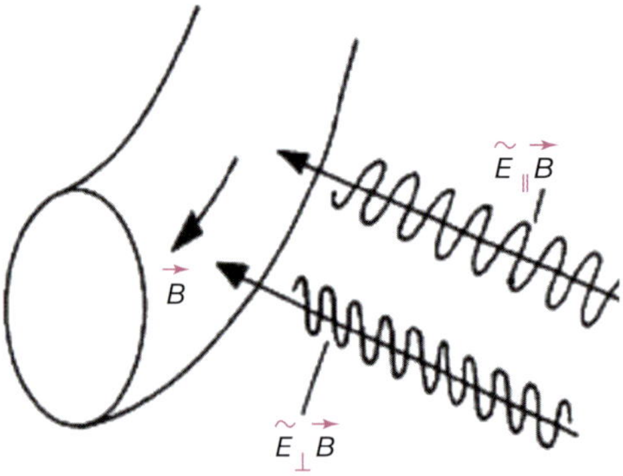

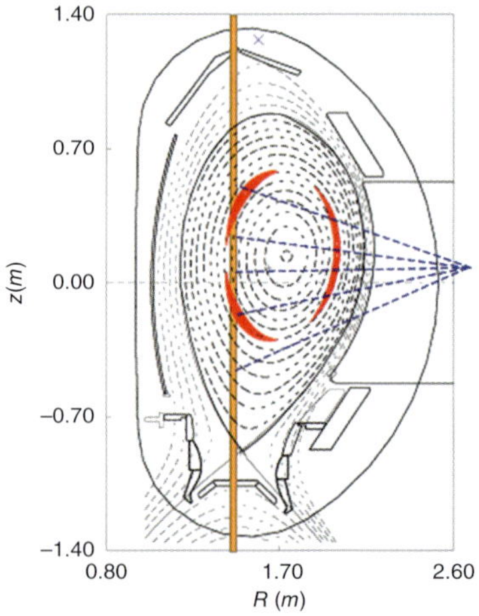

Figure 13.1 Sketch of generating localized current at a location controlled by the injection geometry of a microwave beam at the ECRH frequency. The different beam trajectories correspond to a motion of the injection mirror. (This figure also appears on page 203.)

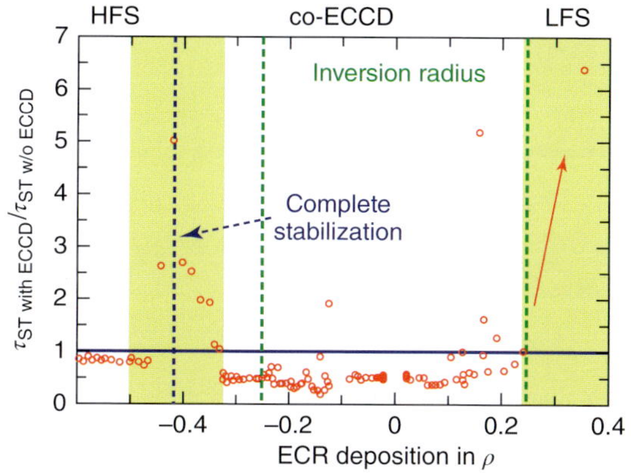

Figure 13.4 Effect of localized ECCD deposited at ρ_{dep} on the sawtooth period τ_{st} in ASDEX Upgrade, normalized to a similar case without ECCD. Negative values of ρ refer to deposition of the high field side, the shaded region is an estimate of the location of the $q = 1$ surface. *Source:* Mück 2004 [79], reproduced with permission of IOP. (This figure also appears on page 209.)

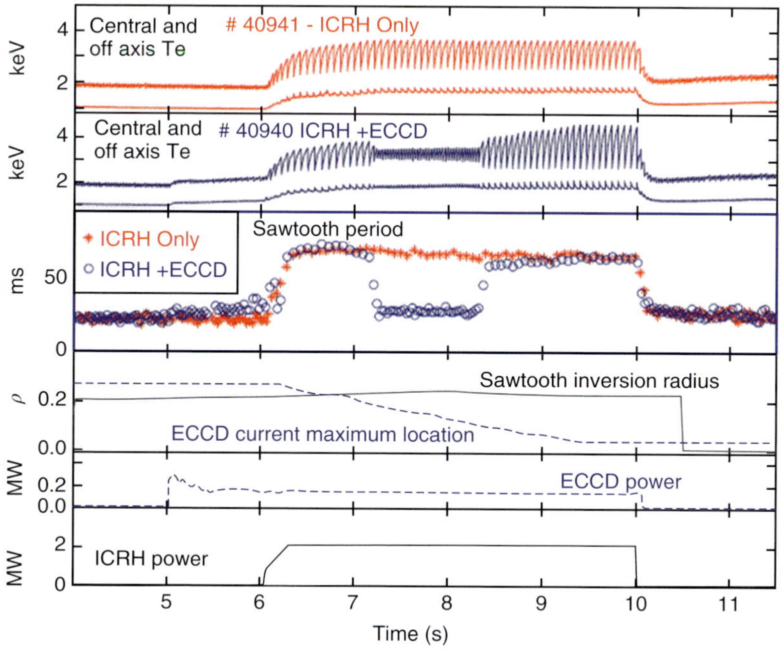

Figure 13.5 Demonstration of change of sawtooth period by application of ECCD on the Tore Supra tokamak in a discharge where fast particles created by ICRH initially lead to large sawteeth with long period τ_{st}. *Source:* Lennholm *et al.* 2009 [81], reproduced with permission of the APS. (This figure also appears on page 210.)

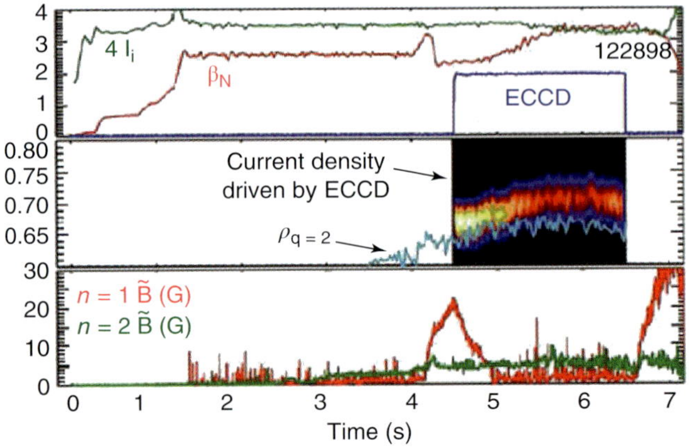

Figure 13.8 Suppression of a pre-existing (2,1) NTM by ECCD. Following complete suppression, β_N (shown in the upper panel) can be increased above the original onset level and the mode, shown by the $n = 1$ amplitude in the lower panel, does not reappear as long as ECCD is on. The deposition is feedback controlled to track the radial motion of the $q = 2$ surface (see middle panel) due to the increasing β. *Source:* Adapted from Prater *et al.* 2007 [85], reproduced with permission of the IAEA. (This figure also appears on page 213.)